RURAL DEVELOPMENT AND POPULATION: INSTITUTIONS AND POLICY

RURAL DEVELOPMENT AND POPULATION: INSTITUTIONS AND POLICY

Geoffrey McNicoll
Mead Cain
Editors

Based on the Expert Consultation on Population and Agricultural and Rural Development convened by the Food and Agriculture Organization, Rome

POPULATION AND DEVELOPMENT REVIEW
A Supplement to Volume 15, 1989

THE POPULATION COUNCIL
New York

New York Oxford
OXFORD UNIVERSITY PRESS
1990

Oxford University Press
Oxford New York Toronto
Delhi Bombay Calcutta Madras Karachi
Petaling Jaya Singapore Hong Kong Tokyo
Nairobi Dar es Salaam Cape Town
Melbourne Auckland
and associated companies in
Berlin Ibadan

Published in the United States
by Oxford University Press, Inc. and The Population Council, Inc.

FIRST PRINTING 1990

The designations employed and the presentation of material in this publication do not imply the expression of any opinion whatsoever on the part of the Food and Agriculture Organization of the United Nations concerning the legal status of any country, territory, city, or area, or of its authorities, or concerning the delimitation of its frontiers or boundaries.

The designations "developed" and "developing" economies are intended for statistical convenience and do not necessarily express a judgment about the stage reached by a particular country or area in the development process.

ISBN 0-19-506847-5 clothbound
ISBN 0-19-506849-1 paperback

Library of Congress Cataloging-in-Publication Data

Rural development and population : institutions and policy/edited by Geoffrey McNicoll, Mead Cain; based on the Expert Consultation on Population and Agricultural and Rural Development convened by the Food and Agriculture Organization, Rome.
p. cm.
"Population and development review, a supplement to volume 15, 1989."
Includes bibliographic references.
ISBN 0-19-506847-5. — ISBN 0-19-506849-1 (pbk.)
1. Rural development—Developing countries—Congresses.
2. Developing countries—Population, Rural—Congresses.
3. Developing countries—Rural conditions—Congresses.
4. Developing countries—Social conditions—Congresses.
I. McNicoll, Geoffrey. II. Cain, Mead. III. Expert Consultation on Population and Agricultural and Rural Development (1987)
IV. Food and Agriculture Organization of the United Nations.
V. Population and development review. Vol. 15 (Supplement)
HN981.C8R86 1990
307.1'412'091724—dc20 90-43047
CIP

PRINTED IN THE UNITED STATES OF AMERICA

CONTENTS

PREFACE

IN THE FOUR DECADES SINCE 1950 the rural population of the Third World has increased in number from 1.4 billion to 2.7 billion, a virtual doubling. This demographic fact has had profound effects on rural development efforts and achievements. In some cases population growth has provided needed human energies to complement land and other material resources in economic production; in many more cases population growth has impeded the development process and slowed the pace of welfare improvement. While the peak rates of rural population increase have passed, in a number of regions—sub-Saharan Africa most notably—growth will continue for many more decades before being halted by the twin effects of lower birth rates and continuing rural-to-urban migration.

Understanding how rural development has been and is being influenced by demographic change is of great importance for wise choice of development strategy —for policies directly affecting the welfare of half of humanity. Development processes in turn modify the components of demographic change: levels of mortality and fertility and rates of migration. What can be said about these relationships at a very general level, however, is of limited policy value. The effectiveness of measures promoting rural development and demographic transition depends on their being tailored to a country's specific patterns of social and economic organization: not only agrarian economic institutions such as tenure systems and labor markets, but also elements of social organization such as family systems and local community and government structures.

In 1987 the Food and Agriculture Organization of the United Nations, in collaboration with the Population Council and with the financial assistance of the United Nations Population Fund and of the Government of the Netherlands, convened a meeting of social scientists to address these issues. The purpose was to have the major gaps of knowledge in the field of population and rural development discussed by leading independent experts, so as to draw broad analytical and policy implications. The exercise was intended to provide guidance for FAO's own population activities, which aim at the integration of population dimensions into rural development policies and programs through technical assistance and policy-oriented research. The present volume includes a wide selection of papers presented and discussed at the meeting, extensively revised in the light of debate, and three new papers by participants at the meeting.

Rural development, with its promise of contributing to overall economic growth, alleviating rural poverty, and limiting environmental degradation, will long remain a subject of policy importance in the Third World. Its demographic dimensions, as these papers attest, warrant close attention. FAO's population program is committed to continuing its support to research and policy analysis in this area, with the aim of expanding the knowledge base upon which sound development strategy can be built.

JOOP ALBERTS
FAO Population Programme Coordinator

INTRODUCTION AND FRAMEWORK

Institutional Effects on Rural Economic and Demographic Change

GEOFFREY MCNICOLL
MEAD CAIN

RURAL DEVELOPMENT IS THE PROCESS of sustained growth of the rural economy and improvement of well-being of the rural population. In conventional wisdom population growth is variously implicated in this process. It is assigned a progressive role in inducing technological advance and allowing scale economies in infrastructure, a retrogressive one in hindering capital-deepening and harming the environment. In turn, rural economic growth supposedly promotes the decline of mortality and fertility along the familiar path of demographic transition, and is accorded some part in generating the massive rural-urban migration of labor that has created the megacities of the Third World.

The working premise of the present volume is that much of this conventional wisdom is inadequate: not necessarily wrong, but seriously underdetermined. What is too often missing or seriously incomplete in studies of how rural development—or its absence, rural stagnation or impoverishment—is interrelated with demographic change is an appreciation of institutional contingency. Population–rural development relationships are modulated by society-specific patterns of social organization and by the rules and routines of economic and political behavior—in short, by the society's institutional structure. To ignore this structure is to assume that institutional patterns are uniform and constant across societies or change in determinate, predictable ways in the course of development—or (a position once associated with neoclassical economists) that they are no more than a veil, obscuring and to a degree interfering with the operations of free markets. These assumptions are unwarranted: they lead to constricted analysis and sometimes dubious policy advice.

Admitting institutional effects in production relations—property rights, factor and product markets, landlord-tenant relationships—is a considerable advance, but by no means fully covers the territory. Yet this is where the frontier of research in rural development now lies. In effect, an "economic" domain of rural life is partitioned off, implicitly assumed to contain all that is needed to understand agrarian outcomes. Certainly there are major insights to be got here, but not without careful attention to ceteris paribus conditions. And the examination of those conditions leads us to other institutional domains, not perhaps intrinsically less accessible to economists but less frequented by them. How to take institutional dimensions into account in analyzing rural development–population interrelations, and the implications of doing so, are the subjects of concern here.

Not all parts of the field of rural development have significant ties to demographic processes and thus may be omitted from consideration. For example, changing factor use and technological innovation in agriculture are clearly pertinent to the subject, since both may be influenced by demographic pressure; most of the details of agricultural practices and farm economy, however, are not. This is not to say that diligent scholarship could not trace out lines of influence from demographic factors to, say, the intricacies of cropping patterns, or from soil fertility to human fertility—it can, and does. But this very readiness to pursue the finest threads of connection has produced a picture in which the main lines are obscured by trivial (though perhaps interesting) issues. Research is thus sidetracked or governed more by methodological interests and "data sets" than by an informed sense of substantive priorities. Policy conclusions turn out to call for measures that would in any event be wholly defensible as contributing to higher productivity and improved levels of welfare, with adjustments in detail and weight in order to garner some additional demographic benefits.

The more important linkages between population processes and rural development lie in the institutional arrangements that govern each. Consider the case of rapid population growth. The modern era of population growth, exogenously generated, in which doubling times have been reduced from the order of 50–100 years to as little as 20–30 years, places unprecedented stresses on family systems and on the rules and conventions in the broader society that regulate social relations, access to the economy, and environmental stability. Traditional patterns of social and economic behavior that evolved in the former demographic regime are manifestly incompatible with the new conditions. Yet the adaptations that emerge, for reasons that are usually not hard to fathom, often do not represent a shift toward sustainability—through faster economic or slower demographic growth.

An analogous argument applies to *economic* change from sources exogenous to the rural sector, whether modern agricultural technologies and inputs or the administrative technologies of government planning and programming. While increased agricultural productivity is clearly a happier condition to deal with than faster population growth, the degree to which the rural sector translates this to routinized welfare improvement and socially appropriate demographic behavior depends on the rural institutional arrangements that dictate the incentives individuals perceive.

In brief, the simple analysis of economic-demographic relationships in an institution-free world, drawing on a technical production function on one hand and an equally technical "fertility function" (the outcome of regressing the birth rate on income, education, occupation, and similar variables) on the other, produces a picture of economic development and demographic transition that omits nearly all its problematic content. The policies it leads to are similarly straightforward and conflict-free: those that speed economic growth have favorable effects on mortality and fertility; those that promote demographic transition have incidental advantages for per capita material welfare. The subject of population and rural development, thus construed, has little in it beyond its two separate components.

No serious student of economic development sees the development problem wholly in terms of technical relationships. Even in those sub-areas where such relationships might have seemed most appropriate, institutional dimensions increasingly are recognized as crucial to a full understanding. Two significant examples are the study of famine and of environmental degradation. The conventional view of famine is a straightforward deficiency in food supply or distribution, reflecting some combination of crop failure, hoarding, and inadequate storage and transport facilities, and, in the longer run, population growth "outpacing" the rate of increase in agricultural productivity. A large literature on population and food, ranging from Club of Rome–style global modeling to detailed studies of population carrying capacity, concentrates on these physical magnitudes and technological (and biological) relationships. Contrast this approach, however, with the analysis of famine by Amartya Sen (1981). Sen has shown, through a careful analysis of modern instances, that aggregate measures of per capita food availability are singularly poor predictors of famine. Major famines—Bengal in 1943, Bangladesh in 1974, Ethiopia in 1973—have occurred, if not amid plenty, at least in the absence of unusual food shortages. The cause of those famines is to be found not in a simple decline in aggregate food availability but in a failure of what Sen calls trade entitlements. Sen's approach focuses on the diverse ways in which particular segments of the population establish entitlement to food and the equally diverse ways in which entitlements may collapse. The crises in both 1943 Bengal and 1974 Bangladesh can be attributed primarily to the entitlement failure of rural wage laborers, through a combination of depressed labor demand and food-grain price inflation. In 1973 in Ethiopia, the famine fell most heavily on pastoralists, whose trade entitlement collapsed with the death of their livestock. Understanding the structure of food entitlements in a particular rural society calls for study of the nature and distribution of property rights, the structure of labor and product markets, and analogous elements of the socioeconomic institutions. (See Hill in this volume.)

A similar shift from the technical/material to the institutional is seen in the analysis of environmental degradation. The unreflecting assumption that excessive population growth leads to ecological deterioration through processes such as deforestation, erosion, siltation, salination, and so on, has given way to careful examination of the breakdown of the societal arrangements that may formerly have acted (and elsewhere continue to act) to maintain land resources, or to finding reasons for the failure of such arrangements to emerge in the face of more intensive resource use. The central analytical task becomes that of delineating and explaining the effects of demographic change on rural institutions.

While such instances of shifts in perspective and expanded problem definition can be pointed to, the institutional dimensions of interactions between population change and rural development receive much less attention than they warrant. Redressing this situation, we believe, is the most promising way forward for both research and policy thinking in the area. That endeavor will not lessen controversy—choice of the weights accorded particular institutional domains in the analysis is fertile ground for dissension—but the arguments will have become more germane to real development issues.

This essay aims to develop and illustrate these arguments and to contrast the resulting analytical approach with some of the prevailing assumptions and methods. As background, we begin by sketching the demographic outlines of the rural demographic situation in the Third World and covariant trends in agricultural resources and technology. The essay then lays out the rationale for explicit treatment of institutional contingency at various levels of economic and social organization. The institutional configurations of chief interest here, in addition to property and labor relations, are family patterns, community organization, and government administration. We draw on both topical and country illustrations of rural development–population change relationships. Following from that discussion, the final section deals with the policy issues thereby raised—variants, for the most part, of the dictum "getting institutions right."

Rural futures: Visions and constraints

The demographic setting

The rural population of the Third World has grown from 1.4 billion in 1950 to an estimated 2.7 billion in 1990, almost a doubling. The 1988 UN medium-variant projections show a continuing but much slower increase in the future: to 3.0 billion in 2000 and a peak of 3.15 in 2015, as the twin forces of fertility decline and outmigration take greater effect. The absolute average yearly increases in Third World rural population over successive five-year periods (estimated and projected, in millions) are as follows:

1950–55	23	1975–80	35	2000–05	19
1955–60	21	1980–85	36	2005–10	8
1960–65	32	1985–90	36	2010–15	1
1965–70	40	1990–95	35	2015–20	−6
1970–75	39	1995–00	28	2020–25	−12

By contrast, the yearly increments in the Third World urban population increase throughout this period: 14 million per year in 1950–55, a projected 45 million per year in 1985–90, and a projected 90 million per year in 2020–25 (United Nations, 1989a).

Tables 1–4 present simple regional analyses of these figures. As a first-level breakdown by broad region, the present (1985–90) rural population of the Third World is divided into East Asia and South Asia, each with some 900 million people and together making up two-thirds of the total; Southeast Asia and sub-Saharan Africa, each with some 300 million; and Latin America and Middle East/North Africa, each with some 100 million. (The present rural population of the more developed countries, in the UN estimates, comes to 330 million.) The principal information conveyed by these tables can be summarized as follows:

— The major predicted change in relative size of regions will result from the continued surging growth of sub-Saharan Africa, which takes the rural population of that region to equality with India's by the end of this projection period. Even if, as a result of the AIDS epidemic, resurgent malaria, or other causes, the UN's mortality assumptions for the region have to be adjusted upward (they currently show a nearly 30 percent reduction in the death rate over 1985–2000 and life expectancy reaching 65 years by 2025), rural population growth there is likely still to be the highest.

— Aside from sub-Saharan Africa, peak rates of rural population growth are now well in the past. On the other hand, no "emptying out" of the countryside is foreseen. Absolute declines in rural population are predicted to have begun in all regions except sub-Saharan Africa by late in the period, but the numbers of people in rural areas in 2025 would be barely below today's levels—and in South Asia and Middle East/North Africa, substantially greater.

TABLE 1 Estimated and projected rural population of major less developed regions, 1950–2025 (millions)

Region	1950	1975	1990	2000	2025
Latin America	97	125	124	123	116
East Asia	517	767	913	982	855
Southeast Asia	155	252	313	338	320
South Asia	402	668	869	995	1031
Middle East and North Africa	63	90	115	128	131
Sub-Saharan Africa[a]	161	265	364	446	604
All less developed regions[b]	1398	2169	2702	3017	3064

[a]Africa excluding Egypt, Libya, Tunisia, Algeria, Morocco
[b]Total also includes developing countries in Oceania (rural population 5 million in 1990)
SOURCE: Data from United Nations (1989a: Table A-5)

TABLE 2 Distribution of rural population of major less developed regions, 1950, 1990, and 2025 (percent)

Region	1950	1990	2025
Latin America	7	5	4
East Asia	37	34	28
Southeast Asia	11	12	10
South Asia	29	32	34
Middle East and North Africa	5	4	4
Sub-Saharan Africa	12	13	20
Total[a]	100	100	100

[a]Includes developing countries in Oceania
SOURCE: Table 1

TABLE 3 Absolute and average rates of rural population growth in major less developed regions, 1950–75 and 1975–2000

	Absolute increase in population (millions)		Average rate of growth (percent per year)	
Region	1950–75	1975–2000	1950–75	1975–2000
Latin America	28	−2	1.0	−0.1
East Asia	250	215	1.6	1.0
Southeast Asia	97	86	1.9	1.2
South Asia	266	327	2.0	1.6
Middle East and North Africa	27	38	1.4	1.4
Sub-Saharan Africa	104	181	2.0	2.1
All less developed regions[a]	771	848	1.8	1.3

[a]Includes developing countries in Oceania
SOURCE: Table 1

TABLE 4 Rural population as percentage of total population in major regions, estimates and projections 1950–2025

Region	1950	1975	1990	2000	2025
Latin America	58	39	28	23	15
East Asia	88	78	75	72	51
Southeast Asia	85	78	71	64	46
South Asia	84	79	72	66	47
Middle East and North Africa	74	55	46	40	26
Sub-Saharan Africa	89	79	69	62	44
Europe	44	31	27	24	18
North America	36	26	25	25	22
Japan	50	24	23	22	20
USSR	61	40	32	29	26
World	71	62	57	53	40

SOURCE: United Nations (1989a: Tables A-5, A-7)

— The present share of total population living in rural areas is about 70 percent in each major Third World region except Latin America and Middle East/ North Africa. At projected growth and migration rates, this fraction will fall to about one-half by 2025. (It is striking that sub-Saharan Africa shows this pattern despite the rapid rural population growth of this region.) Latin America's time-path in proportion rural shows an exceptionally rapid drop over 1950–75 and thereafter conforms with the more developed regions of the world. Middle East/ North Africa presents an intermediate case. For the Third World as a whole, the estimated and projected time trend in percentage rural is as follows:

1950	83
1975	73
1990	66
2000	60
2025	43

No similarly straightforward data can be assembled for rural product or other indexes of sector-wide economic performance. Agriculture, as the dominant rural industry and occupation, is usually made to serve as proxy. Both cross-sectionally among countries and over time for individual countries the share of agriculture in gross domestic product (GDP) and the share of the labor force in agriculture bear fairly predictable relationships to overall economic development measured by per capita GDP. For groups of countries assembled into World Bank categories, we have the following averages for share of agriculture (data for the mid-1980s—World Bank, 1988):

	Percent share of agriculture in	
	GDP	Labor force
Low-income countries	32	72
Middle-income countries	15	43
Industrial market economies	3	7

In many of today's low-income countries it is hard to imagine a future in which agriculture would have fallen to 3 percent of GDP and 7 percent of the labor force. This may be simply a reflection of the scale of their present difficulties: well-informed observers around 1950 were just as skeptical of Japan's industrialization prospects. But it is equally possible that some of these countries will remain predominantly agrarian in employment and substantially agrarian in product into the indefinite future.

What are the uncertainties in the demographic futures sketched in the above tables? Demographers claim to make population projections rather than forecasts, a claim belied by their eagerness to take credit when they turn out to be correct. The UN's medium-variant projection is the UN's best (surprise-free) guess of what the future holds: for the Third World as a whole, the medium-variant projection for 2025 shows a total population of 7.11 billion, of which 43 percent (3.06 billion) is rural. The high and low variants for total populations are 7.97 and 6.33 billion (United Nations, 1989b). If the same urban fraction were applicable in each case (not strictly a defensible assumption, but good enough) then the high- and low-variant projected rural populations in 2025 would be 3.43 and 2.72 billion, a range of 700 million.

Beyond 2025, the already considerable uncertainties reflected in this high–low range mount rapidly. The World Bank, which, less cautious than the UN Population Division, issues country-level projections to 2100, downplays these uncertainties by focusing on a single set of assumptions for each. For Third World mortality, it is assumed that the recent pace of decline will continue more or less steadily until life expectancy reaches the levels of the present-day industrial countries by late next century—a proposition until recently taken for granted but now, with the emergence of new pathogens and drug-resistant strains of some major old ones, seemingly much less well-founded. For fertility, Paul Demeny (1984:115–116) has characterized the assumption as follows:

> In countries where current fertility is low, the index of total fertility is expected to settle permanently at replacement level by the turn of the century. . . . A partial exception is China: there, reflecting current official policy objectives, the projections assume that fertility will remain substantially below replacement level during a 15-year period beginning with 1985. In other formerly high fertility countries—such as Brazil, India, and Indonesia—the projections envisage that the recent rapid decline of fertility will continue unabated, resulting in replacement level total fertility rates around 2015, a level firmly to be maintained thereafter. Countries where fertility is still very high are expected to embark on a course of precipitous fertility decline beginning with 1990 (as in Pakistan and Bangladesh) or with 1995 (as in Nigeria). These countries are assumed to reach replacement levels of fertility in the 2030s.

Commenting on the basis for these assumptions Demeny remarks (1984:120) that:

> [T]he logic of the proposition that fertility trends in all countries will have converged to replacement level well within the first half of the twenty-first century is hardly compelling. What mechanism within a short few decades will induce, at least in terms of average reproductive performance, the same fertility behavior in tropical Africa and in Western Europe? Implicit in the projections for countries with still high fertility is either a more sanguine view than is warranted by history about the efficacy of future socioeconomic development in inducing rapid fertility decline, or the assumption that policy interventions aimed at rapidly reducing fertility will become far more pervasive and effective than has been the case thus far.

In one important respect, the picture of slackening population growth implied for the next several decades in the projections in Table 1 is misleading. The major fertility declines that give rise to this picture do not start to influence the labor force for 15 or 20 years. Hence the labor force growth rate is lagged behind the population growth rate. For those Third World countries where urban fractions are still low and where significant fertility declines began only in the 1970s, rural labor absorption problems traceable to population growth will diminish substantially only after the turn of the century.

The projected rural population size is also sensitive to assumptions about rural-urban migration. The UN projections are based on continuation of observed differentials between urban and rural growth rates and the slow convergence of this differential to a "world norm" level by the end of the projection period. Clearly, no such smooth processes of urbanization and obedience to international cross-sectional experience need apply in any particular country. Many observers detect signs of new patterns of urban development emerging in which there are large roles for secondary cities and rural industry, and burgeoning "circular migration" and rural commuting to industrial jobs. Expectations of vast urban agglomerations, typified by the widely publicized 1980 UN projection of Mexico City reaching 31 million by 2000 (United Nations, 1980; more recent projections have lowered this figure by 5 million or so), are called into question by the immense difficulties of maintaining minimally effective infrastructure and public services even at present sizes. New industrial technologies, moreover, may modify the scale and locational economies that currently favor urban concentration.

Land resources and agricultural technologies

The FAO project on "Land Resources for Populations of the Future" (Higgins et al., 1982) is the main data source for the worldwide status of existing and potential agricultural land. (A few countries are excluded, most notably China.) The importance of the study lies in its combining estimates of opportunities for agricultural expansion at both the extensive and intensive margins, and its relating of these to hypothesized levels of technology with subnational agro-ecological zones classified by soil and climate. The World Bank (1982:59) cites estimates for the developing countries of unused but potentially cultivable land from 0.5 to 1.4 billion hectares, over and above the existing 0.8 billion hectares under cultivation, most of which is currently forested land in tropical Africa and Latin America. The total cultivable area assumed in the FAO study is well to the optimistic end of this range.

Constraints on land expansion include the high capital costs of clearing land and the continuing cost of preservation of often fragile ecological balances (or the cost in foregone output of land degradation); the distances of much of the new land from existing markets, given the transport inadequacies that typically exist; and, in broad swathes of Africa, endemic diseases to be overcome. A new set of constraints, backed by an increasing mobilization of public opinion, derives from concerns about effects of deforestation on climate, on the well-being of indigenous peoples, and on species extinctions. Paradoxically, the value of unexploited land endowments in a poor country may be slight. Only with progress in economic development are the financial resources and infrastructure likely to be found to turn potential into productive assets.

High agricultural inputs as defined in the FAO carrying capacity study consist of "complete mechanization, full use of optimum genetic material, necessary farm chemicals and soil conservation measures, and cultivation of only the most calorie (protein) productive crops" (Higgins et al., 1982:viii). Resulting yields in grain equivalent are put at around 5 metric tons per hectare, in comparison to the present Third World average of 1.5 tons.

A wide range of views exists on how plausible such high technology futures are. A glowing report on Third World agricultural possibilities, framed in terms of resulting problems for food-exporting countries like the United States, is that of Dennis Avery (1985). He writes (p. 408):

> The wheat and rice varieties of the Green Revolution are legend; genetics has gone on to produce the world's first hybrid wheat, cotton, rice, and rapeseed. Triticale, a hybrid of wheat and rye, outyields other cereals by 250% under certain unfavorable conditions. There are new sorghums for Africa that may have Green Revolution potential. . .

and much more in this vein. The existing network of international agricultural research stations and increasingly competent national research establishments would be important contributors to that course of technological advance, complementing public expenditure on rural infrastructure and agricultural extension services. The relatively subdued tones of the 1982 *World Development Report* (focused on the agricultural sector) stand in some contrast. The Report points out, for example, that little research has been done on pulses, rootcrops, and tubers, although, in sub-Saharan Africa, these make up more than a quarter of agricultural output (World Bank, 1982:68). Additional cause for worry comes from the risk of large-scale ecological instabilities generated by human activity: feedbacks as yet poorly understood that may prove difficult or virtually impossible to counter. Chief among these, albeit with a high associated level of uncertainty, is the greenhouse effect of changing atmospheric composition on ambient temperatures, rainfall patterns, and sea level. At a smaller scale, adverse environmental consequences stemming from heavy applications of fertilizer, pesticide, and weedicide also mount up. Looking just at the technical issues, sustaining production increases of 2–3 percent annually—keeping up with or slightly exceeding population growth—may seem relatively straightforward based on recent experience with crop improvement and modern inputs. There are doubts enough, however, at least to seriously qualify that expectation.

Whatever the technological possibilities, the practical problems of achieving such rates of output growth under the institutional and political conditions of many contemporary Third World countries are formidable. How far actual performance lags behind the technically feasible depends in part on the wisdom and effectiveness of government agricultural policies, particularly pricing policies. In the past these have frequently been highly misguided—for a familiar array of reasons. The now routine attention accorded to understanding the price incentives and disincentives facing farmers and the relating of tales of policy competence and incompetence—stock-in-trade in the advice-giving of international development agencies—give hope for policy improvement in the future. Redressing past errors on this score can produce quick gains in productivity through a move toward the technological frontier. The more intractable problems, in that governments have limited skill and purchase in dealing with them, are in getting *institutions* right.

Rural social and economic organization

Construction of scenarios of the global future a generation or more hence has been a popular activity at least since the 1960s. Well-known landmarks are Herman Kahn and Anthony J. Wiener's *The Year 2000* (1967), the first Club of Rome report (*The Limits to Growth*—Meadows et al., 1972), and the US government–sponsored *Global 2000 Report* (1980). Analogous efforts for particular countries or regions are similarly quite common. The series of "Second India" studies prepared in the 1970s—exploring how India should respond to the challenge of the forecast doubling of its population from 1971 to about 2010—is one example (Ezekiel, 1978). The 1979 Monrovia Symposium on the Future Development Prospects of Africa towards the Year 2000, organized by the OAU, is another (Organization of African Unity, 1979).

Perhaps because successful development is typically associated with emergence of an urban-industrial economy, such scenarios seldom have much to say about rural social systems. We get from them no clear picture of the rural socioeconomic or cultural conditions that go with the envisioned futures. Their interest in rural matters, such as it is, centers on food production and its technology and factor inputs. Little else counts.

For a particular developing country, let us assume that, by whatever means and with whatever mode of organization, the agricultural sector approaches the productivity levels found in the rich countries today. Accompanying such an advance, even with efforts to limit any labor-saving bias in technological change and to tolerate (and finance) surplus production, would be a fall in agricultural employment to "modern" levels, say 15 percent or less of the labor force. (In Japan now it is 12 percent; in the United States, 2 percent.) What happens to the rest of the rural work force? Essentially, there are three things people can do: find work in rural-based industries (including jobs in agricultural processing and marketing), commute to urban jobs, or move to the city. Societies differ in the combination of options selected. Moreover, depending partly on its scale and on details of relevant tax policies, farming may be either a specialized activity, as it largely is in the United States and most of Europe, or part-time, as in much of Japan. Of course, present-day Third World farming is also largely part-time, in the sense that a sufficiently detailed enquiry on the time-use of an average farm family would elicit the virtual gamut of occupations. Specialization, if it comes, sets in later.

As shown by the histories of the developed countries, and as forecast in Table 1 for the Third World, even substantial rates of rural-urban migration (and increasing fractions of population urban) are not generally accompanied by significant depopulation of the countryside. In the United States, for example, the rural population has been roughly constant in absolute size (in the range of 50–60 million) since around 1920, a time at which the urban fraction was one-half. It is the *farm* population that has fallen drastically.

High-productivity agriculture sustains a substantial rural support structure to supply inputs and marketing services to the farm sector. Transport developments allow many rural residents access to urban labor markets. Rural-urban boundaries at the fringes of suburban settlement and at the lower bound of town size become fairly arbitrary, with rural society little disadvantaged in access to consumption goods, communications media, and public services, and converging also to urban patterns in health and fertility. Dualistic aspects of the national economy (modern versus "traditional" contrasts in technology and labor productivity), if they still linger, are less and less coincident with either rural-urban or agriculture-nonagriculture sectoral divisions.

While the number of countries that will achieve this sort of rural prosperity over the next several decades can only be guessed, it is fairly safe to say that the majority of the 3 billion or so people who will make up the rural Third World will encounter less favorable outcomes. In their details, country experiences will of course be culture-specific, dependent on particular inheritances of family system, ethnic divisions, political culture, and other such elements that in any society display notable staying power in the face of economic change. Diverse geographical and ecological circumstances similarly leave persisting imprints that hinder generalization. In broad economic terms, however, there may be greater commonality. In deploring simplistic doom-laden scenarios of the consequences of population growth in India, Robert Cassen (1978:331) writes:

> The gradualness [of population growth] leaves room for innumerable processes of adjustment, and so does the spatial dimension. All these extra millions arrive not in huge crowds springing up overnight in city centers, but in small numbers each year in half a million villages and on the edges of towns and suburbs. To a very considerable extent one can answer the question, what does the future hold for India, with the observation that the future has already arrived.

Modest progress, with distinct economic successes appreciably but not wholly offset by losses elsewhere, and attained at ecological and esthetic costs largely uncounted in conventional income measures, is a plausible picture not just of the next few decades in rural India, but of rural trends in a broad middle range of countries in the rest of South Asia, perhaps China, the Middle East, poorer Latin America, and richer Africa.

For completeness, we should also consider rural futures in which productivity levels are either uniformly low or low for most of the labor force. Lengthy periods of stagnation resulting from bad policies or dramatic reversals of performance tied to political turmoil and civil disorder can equally achieve this outcome. That there will be "failed" economies 50 or 100 years hence is likely enough. How they might differ from those that may be so identified today, given a vastly changed international context (with far more relatively rich countries, the easy spread of a range of modern technologies and of information on differential economic circumstances, and another global population doubling), is impossible to predict.

The "standard" institutional content of rural development studies

Studies of the interrelations between rural development and population change usually emphasize agricultural production technology and various aspects of labor market structure and land tenure, with population characterized simply by size, density, or vital rates. Population growth is a source of pressure for technological advance and sometimes for adaptations in tenure systems or employment contracts, and in turn demographic processes may themselves respond to improved economic conditions by progressing along a smooth path of transition from high to low mortality and fertility. Different agrarian systems have different implications for the pace of productivity change, for labor absorption, and for income distribution, hence allowing a ranking of outcomes according to these criteria and suggesting directions for policy intervention that might favor a preferred outcome.

Institutional analysis in population and rural development thus tends to be identified with investigation of land tenure systems and labor relations, their response to demographic change, and their putative demographic effects. This work draws on a growing body of economic theorizing that seeks to explain agrarian institutional structures in terms of assumptions about individual interests and behavior. Pranab Bardhan (1989a) identifies three strands of theory on agrarian economic institutions. One derives from Marx's analysis of property relations; a second from recognition of significant transaction costs in much of economic exchange; and a third from the analysis of information asymmetries and incomplete markets.

Marxian analysis of class interests and their institutional manifestations can contribute much to explanation of agrarian system change (or stasis). Practitioners, however, may tend to see all nonmarket social relations as aspects of class relations, characterized by "unequal exchange"—exploitation or coercion. Thus, in rural economic-demographic change, class-based institutions are the salient mediating structures in the relationship. In consequence, possible roles for complementary mediating institutions that are not class-based are not seriously considered. Robert Brenner on agrarian development in preindustrial Europe is a notable case in point, propounding the thesis that "it is the structure of class relations, of class power, which will determine the manner and degree to which particular demographic and commercial changes will affect long-term trends in the distribution of income and economic growth" (Brenner, 1985:11; for critiques see Aston and Philpin, 1985). Jan Breman, in this volume, presents a study of class-cum-caste mediation of economic-demographic change in an Indian setting (demography entering the picture chiefly through geographic mobility), in which blithe expectations of the decay of feudal labor relations under market pressures are belied by emergence of neotraditional patterns of bonding, backed by "extra-economic" force.

Bardhan's second strand of institutionalist theory is neoclassical institutional economics. Here the premise is that institutions evolve to minimize transaction costs (that is, the costs of entering into and enforcing contracts among economic agents). Changes in technology, changes in scale, and development of markets (inter alia) alter those costs, eventually to the extent that the benefits promised by a different institutional configuration outweigh the costs of change—whether by negotiation or force. The main field of application of these ideas has been agrarian systems. As a few among many examples in a now large literature, Douglass C. North and Robert Paul Thomas (1971, 1973) interpreted the transition from feudalism to landlordism in Europe in those terms; with more modest ambition, S. N. S. Cheung (1969) uncovered the rationale supporting various forms of sharecropping, once seen as generating economic inefficiency through a perverse incentive structure; and Yujiro Hayami and Masao Kikuchi (1982) applied transaction cost arguments to account for shifting forms of labor contract in Southeast Asian agriculture. Population growth has an appreciable role in this appraoch—by altering factor proportions, increasing market size, inducing technological progress, and so on, working through effects on transaction costs to set up conditions for institutional change.

The third theoretical strand of institutional analysis comes from exploring the implications of incomplete information on markets and economic behavior. In this view, institutions emerge "to substitute for missing credit, insurance, and futures markets" (Bardhan, 1989a:7). In the context of rural development, the elaboration of this theoretical approach—highly amenable to formal treatment, as the studies assembled in Bardhan (1989b) attest—is termed by Joseph Stiglitz (1986) the "new development economics." The contrast with the transaction cost approach is more one of perspective than any substantial difference in understanding the nature of institutions: imperfect and costly information and asymmetry in its availability are sources of high transaction costs. By transposing the basis of the problem to the realm of information, however, the analysis is less grounded in the tangible processes of development—a weakness for our purpose since demographic change is very much a part of those processes.

Demographic influences on agrarian economic performance working through property and labor institutions are fairly well accepted (see, for example, the discussions by Lipton, this volume, and Pingali, this volume). The demographic effects of such production arrangements, however, are less demonstrable. A defensible view is that tenure systems (to take one instance) are, in and of themselves, of comparatively minor significance for population–rural development interrelations except insofar as different tenurial arrangements affect the course of rural development and pace of income growth (see Cain and McNicoll, 1988). Whatever the merits of land reform as an instrument or precondition of rural development, the evidence suggests that no clear fertility effect follows from, say, a shift in tenurial status from tenant to owner. (As will be noted below, the *security* of both civil and property rights is an issue of some demographic significance; however, this is more the province of the legal and administrative systems.) Effects on outmigration are more plausible, although here there are alternatives of accommodation of population increase within the rural economy through "static expansion" or through some form of institutional adaptation—the choice among which has as much to do with family patterns as with land tenure institutions (see, for instance, the chapters by Greenhalgh and Mabogunje, this volume).

Any adequate treatment of population and rural development, especially one that aspires to account for intersocietal differences in outcomes, must of course pay due attention to agrarian economic institutions. Two sections of the present volume have this focus. However, agrarian outcomes are conditioned also by other institutional factors—indeed we would maintain that characteristics of family and supra-family social structures may contribute as much to the operation and achievement of the rural economy as do more narrowly "economic" institutions, and perhaps more to the course of demographic change. Yet such factors are rarely brought into consideration. In effect, certain parts of the setting in which economic-demographic change is being investigated are, for unexamined reasons, taken to be structurally elastic and other parts inelastic, not by evidence but by assumption or, possibly as often, by inadvertence. Sometimes this may not matter, sometimes it may.

An expanded institutional configuration

Often-neglected institutional forms that arguably play major roles in agrarian economic and demographic outcomes are: family and gender systems, village and community structures, government administrative arrangements and legal systems, and the international institutional regime. We give a brief gloss on each of these "clusters" and set out the rationale for a greater relative weighting of their significance.

Family systems

Family systems differ in how they deal with intergenerational property transfer, their control over the establishment of new households, the role differences between men and women they perpetuate, and their marital fertility responsiveness to the changing economics of children. Gender systems—patterns of organization of sex roles and the "mode of reproduction"— can for our purposes be subsumed within family systems. (See Boserup, this volume.) Family systems also are notably resilient in the face of economic, demographic, and cultural change in societies. The essential forms persist relatively intact over many generations. Any comparative discussion of population and rural development issues across societies characterized by different family systems that does not take account of the implications of those systems is prima facie suspect.

Rules of property transfer and household formation have important potential implication for the integrity of agricultural holdings over time. In Northwest Europe, rules of household formation and inheritance were such as to prevent land fragmentation or the proliferation of households. Systems of primogeniture were in effect over much of this area, but even where partible inheritance was the norm, the division of land was avoided. Furthermore, in the Northwest European family system the timing of marriage and thus the formation of new households were closely tied to economic conditions. This responsiveness created a rough homeostasis, such that periods of economic contraction induced later and fewer marriages, and as a result lower fertility and more moderate rates of population growth. Economic prosperity had the opposite effect. The Japanese family system was similarly "successful," exerting strong controls to prevent land fragmentation and the proliferation of households. A demographic-economic equilibrating mechanism comparable to Europe's can be inferred for rural Japan.

Family systems in most developing countries of today exhibit neither of these characteristics. Many of these societies are governed by "joint" household formation systems (as defined by Hajnal, 1982) and by norms of partible inheritance, which, in recent decades, have contributed to the rapid fragmentation of agricultural holdings. In contrast to the historical systems of Europe and Japan, the joint family system is relatively inflexible with respect to the timing of marriages (the system encourages marriage at a young age for males and at an even younger age for females) and the formation of new households; it is relatively unresponsive to changing economic conditions. How the European and Japanese systems would have coped with an exogenously induced mortality decline of proportions similar to that experienced by Third World countries since the 1940s is a matter of speculation. However, given the premium that these systems placed on the preservation of agricultural holdings and the subservience of marriage and inheritance practice to this goal, it seems likely that a better accommodation would have been witnessed than is evident in the joint-family societies of the present day.

This likelihood appears stronger when one considers a third important difference between Northwest European and joint family systems: in the former, there seems to have been little connection between the number of surviving children and the welfare of parents in old age, while in the latter the connection is strong. In Northwest Europe the welfare of the indigent elderly was the province of public institutions rather than the family even in preindustrial times. (Persons with property typically negotiated formal retirement contracts.) In contrast, in joint family societies old-age support is very much a family concern, with sons most often bearing the responsibility for the care of the elderly parent. In the event of substantial mortality decline, the economic calculus underlying any fertility response would be quite different in the two regimes: in the European case, substantial security benefits of children would not be a factor in the fertility response of parents, while in joint family societies they would be an important consideration.

Sub-Saharan African family systems stand apart from the joint family systems of the rest of the developing world. (See Goody, this volume, however, on the hazards of any such generalization for Africa.) One way of characterizing this distinctiveness is by reference to the boundary of the corporate group: in joint family systems the boundary, for important practical purposes, is that of the nuclear or the vertically extended family, while in sub-Saharan Africa the boundary tends to encompass a larger kin or lineage group. In the joint family system the conjugal bond is strong; in African family systems this bond is weak relative to ties with one's natal family and lineage group. In Africa the latter group is the locus of control over the demographic and economic behavior of individual members to a much greater extent than, for example, in Asia. (This reflects other important institutional and economic realities in sub-Saharan Africa, including systems of communal property rights and the traditional importance of livestock as a form of wealth.) With respect to rural development and demographic transition, African family systems appear to be the least facilitating of the three general types. The persistence of corporate lineage groups permits the diffusion of reproductive costs and sustains the separation of fertility interests of husband and wife, impeding fertility decline. Lineage interests also constitute a source of resistance to the introduction of individual private property rights, which many would argue was a necessary condition for agricultural development.

Community organization

Community forms, often tied to settlement patterns and kinship affiliations, are, like family systems, long-lasting. Institutions rooted in local community structures have a potentially significant bearing on rural economic and demographic outcomes. In some cases such institutions in effect make up an agrarian system—as in corporate villages, for example—but more often the two are conceptually separate and impose distinct kinds of constraints on the direction of change. Where there is a sizable nonagricultural rural economy overlapping (in location and labor force) with the agricultural economy, the role of community-based institutions in both demographic change and rural development may be larger than that of specifically agrarian institutions.

The stereotype of traditional rural society posits solidary hamlets or natural villages as the next important social unit after the family. While the "village republic" is no longer taken seriously as a depiction of historical reality, in most rural societies local supra-family groupings did exert considerable influence over their members' behavior in certain domains of life, including matters bearing on the family economy and fertility. That influence has weakened and its domains have narrowed, but in many societies an appreciable residue remains today.

The important distinguishing characteristics of local organizational systems for the present discussion are their degree of corporateness and territoriality (see Cain and McNicoll, 1988). Corporateness governs the capacity of the group, or of an elite within it, to influence the behavior of members to suit group interests, however these may be defined. Territoriality affects the likelihood that demographic behavior will be included in the kinds of behavior subjected to group pressure. Territoriality also facilitates orderly governance: land stays where it is.

Kinship is the chief competitor to territory in defining community systems in traditional rural societies. Natural villages may have kin ties linking many of their members, but their principal identification is as a territorial unit with more or less fixed boundaries. There are instances, however, where kinship takes precedence over territory, where clans or other kinds of corporate kin-groups dominate the local-level social landscape. Also common are cases where there are several distinct bases of affiliation, no one of them dominant, each generating a system of local groupings with a particular range of interests.

Preindustrial Japan and Switzerland can be taken as illustrations of strong corporate-territorial community systems. Villages in both could and did exact a high degree of conformity from members and could presumably cope readily with free-rider problems. Modern examples, different from each other in many respects but not in these, are villages in China and South Korea—mobilized in support of government policies and programs but still far from being mere instruments of government authority. These cases contrast strikingly with, say, villages in contemporary Nepal, which appear to be nearly helpless to enforce sound agricultural and forestry practices under rapid population growth, or in Bangladesh, where the prominent local-level organizational roles are played by kin-groups and factions rather than territorial communities. In sub-Saharan Africa, similarly, village organization has traditionally been quite weak in comparison to tribe and lineage—there reflecting the history of pastoral rather than agricultural economies and longstanding migratory patterns. (See Hill, this volume, on the Sahelian case.)

The clear presumptive conclusion is that corporateness and territoriality are community characteristics that facilitate demographic restraint and may do so even under an "adverse" family system. They appear to be valuable in safeguarding environmental stability (see McNicoll, this volume). Whether they also help to promote vigorous economic performance is more doubtful: indeed, a strong case can be made that they as often as not impede it.

Local government administration

The lower rungs of the government's apparatus of civil administration (together with the parallel local apparatus existing in some polities that is associated with military control or dominant political party organization) are also relevant to rural economic and demographic change. These structures often date from colonial or even precolonial periods; they may have been designed for purposes wholly unlike those they have today and for conditions similarly far different. Even where new local government systems are established supposedly de novo, this prehistory, reflected tangibly in interest groups and individual and community expectations, may push the system into the earlier procedural patterns, de facto if not de jure. Local administrative realities may be significant not only in constraining the freedom of movement of the rural economy but also in influencing the effectiveness of central government programmatic efforts to modify economic and demographic behavior.

An important determinant of the effectiveness of local administration—with respect both to regulation and control and to the provision of services—is the societal level down to which it operates. It is arguably impossible for government fully to coopt natural villages as administrative units, since the face-to-face contact that characterizes them generates local pressures and loyalties strong enough to capture any officialdom at this level. On the other hand, there are problems in having the lowest-level administrative unit too distant from the natural village. Historical evidence suggests that the administrative vacuum set up by such a gap will be quickly filled by informal political entrepreneurs, ready to broker relationships between government and governed and in the process forging a new, extra-administrative organizational system designed to maximize brokerage—a familiar "rent-seeking" outcome. Such systems grew up in the colonial period in South Asia as a result of attempts by the colonial authorities to extract revenues with minimal effort. The beneficiaries of the system comprise a powerful coalition that can block change. This administrative legacy has been a major impediment to rural development and achievement of lower fertility in both Bangladesh and parts of northern India.

The forms of local administrative organization, although at first sight something changeable at will or whim by government, in fact tend to persist. Successive governments may differ greatly in political orientation but be quite alike in their interests at the grass-roots level (in civil order, revenue raising, and mobilization for economic growth). Administrative designs tend to gravitate to those that have worked or been in place before. Promoting this structural inertia at supra-village levels is the spatial logic of markets and administration. In the former case this logic dictates an economic structure of marketing areas, typified by the hexagons of Christaller-Lösch central-place theory (see Whyte, 1982). Local administrative systems, not surprisingly, often make such areas into administrative units, adding selected government functions to the economic functions of market towns, seeking thus to transfer the stability of the one to the other. The commune in China was an administrative division roughly superimposed on the marketing area (see Skinner, 1965). In the African case, discussed in this volume by Goran Hyden, traditional kin- or clan-based affiliative ties permeate the administrative structure and strongly influence its performance.

Two particular arenas of government activity at the local level warrant explicit mention: public finance and the legal system. The development process, seemingly by necessity, entails a proliferation of line ministries and secretariats with increasingly specialized functions, each of which, once established, guards its administrative territory and actively seeks to expand it. What might once have been a modest and relatively unified civil administrative structure, with responsibility for police, a court system, revenue collection, and other basic public functions, multiples into many parallel, often competing lines of authority. Public finance, on the expenditure side, becomes similarly specialized. Moreover, because power is husbanded at the center, any devolution of expenditure authority to lower administrative levels away from the center tends to be resisted. Even when an appreciable share of revenue is directly raised from the local population (the larger revenue sources are typically the more-easily administered royalties and foreign trade duties), the rural taxpayer is unable to discern how his tax money is spent. In particular, he sees no evident connection between local revenue contributions and the quality or quantity of public services.

The absence of such a connection deprives government of a potentially powerful policy instrument. Just as a poorly designed fiscal system can be seriously detrimental to agricultural productivity, so too it can destroy (or prevent the emergence of) local initiatives in public services or infrastructure development that may be no less consequential. Moreover, a possible role in demographic outcomes may be envisaged. Demeny (1982:5–6) takes the case of education as an illustration: "[T]he rules of central resource allocation ensure that if a local community somehow managed to have half as many children as their equally numerous neighbors, they would end up with half as many school rooms, teachers, and school books, all of a quality no better than their neighbors' children have access to." Centralized public finance systems thus tend to discourage quality-quantity trade-offs at the local (and individual) level. The reverse would potentially be true of a decentralized system in which a significant measure of public financial accounting is confined to the local level. (See the later discussion of this policy direction.)

The legal system is another source of social-administrative incentives bearing on population change and the pattern and pace of rural development. Clearly, in its procedural manifestation, it may largely coincide with local community organization, in the case of much customary law, or with local government administration, for statutory law. Separate treatment of this array of pressures and sanctions, however, is analytically helpful. In content, the areas of law that are most relevant for the issues of present interest are those concerned with property rights (including inheritance), contracts, and the family (especially marriage and divorce). As important as the formal content of legal dictates are the de facto processes of application: differential perceptions and realities of legal sanctions and accessibility to redress, extent of extra-legal undermining of judicial processes or of their intended results, and the interplay of competing systems of law (customary, religious, and modern civil codes).

Even a minimalist view of the appropriate role of government in rural development would give high priority to its functions as guarantor of civil and property rights through the establishment and maintenance of an effective system of law and law enforcement. Much of what is thought to be economically inefficient in traditional agrarian systems is, at least in part, an institutional response to insecurity in property and civil rights. Demographic behavior is also at issue: both with respect to the effective uses of laws bearing directly on such behavior, and indirectly because the demand for children is partly security-based.

The international system

Finally, we should note for completeness that rural development and population change in the contemporary world are not only linked through the domestic economy and society but also through the international system. International institutions—the trade regime, capital markets, international law (supplemented by numerous covenants and treaties on matters ranging from the environment to human rights), and the organizations through which these are sustained or put into effect, play an independent mediating role in agrarian economic and demographic outcomes at the national level. A dense web of nongovernment organizations pursues an overlapping set of agendas in the international arena. Together with these tangible organizational realities, this external setting has important informational and cultural dimensions, conveying "lessons" (many perhaps misleadingly drawn) of policy successes and failures elsewhere, images of rich-country lifestyles, and suggestive routes of political development that increasingly bypass the filtering efforts of national governments.

International influence is transmitted most directly through the prices set in world markets, which countries are pressed to acknowledge domestically. The scope of policy action in various fields that can be contemplated by governments may also be constricted by international practice or by conventional wisdom—and a political or economic cost exacted for exceeding those limits. And consumption preferences and normative expectations of what economic growth should bring to a large extent reflect behavioral patterns established elsewhere. Paul Demeny (this volume) discusses the various routes and patterns of international influence on Third World agrarian progress and on the direction of domestic economic and demographic policy. The lost opportunities have been manifold; the need for greater acuity in distilling lessons from international development experience is compelling.

Patterns of institutional mediation

The topics dealt with above we believe encompass the main institutional variables that should be weighed in analyses of interrelations between population change and rural development. No suggestion is intended that there are not other such variables that play a significant and even sometimes a salient role in the relationship. Identification of the institutional forms that *are* pertinent for a particular setting, time frame, and comparative purpose should be a critical part of the analyst's task.

With the need to take account of institutional contingencies in broad additional domains such as these, it might be thought that we risk having too many degrees of freedom for useful theorizing. A rural setting is now to be characterized not only by the standard economic measures—of factor endowments, technology, the organization of production, and intersectoral economic relationships—but also by family system, community structure, the design of local government, legal codes, even the international institutional environment. What once may have seemed a tractable, if complicated, analytical problem—amenable, e.g., to simulation modeling—now eludes any such formal treatment. There are, however, enough commonly observed links among these various institutional domains to delineate some "typical" patterns of rural development and population change, and, perhaps more important, some typical ways in which development and demographic transition are blocked.

The analytical task is that of dealing with institutionally contingent structures of incentives. For economic productivity, the simple assumption of the individual rational actor remains the same, but the actor is now set in an intricate and shifting environment. For demographic behavior, individual economic incentives also matter. Children are expensive, and, as they enter their working life, become economic assets. How these costs and returns are typically distributed within the family and the broader population, and the level of risk and uncertainty seen as attaching to them, influence how parents, together or separately, evaluate (ex ante) their interests in particular family sizes and paces of childbearing. Different social systems impose different distributions, the unraveling of which by careful institutional analysis can shed much light on fertility outcomes. Analogous statements apply to migration.

This is not to claim that behavior in either domain is fully captured within an economic calculus. Even when such a calculus is conceptually stretched to encompass what to many might appear utterly noneconomic variables, it cannot in itself provide satisfactory explanations without additional ceteris paribus encumbrances. These are cultural in nature: the bases of preference systems, the categories through which the world is perceived, the unconscious culturally imposed limits of individual agency. (Amartya Sen, in this volume, examines such matters as they shape gender relations within families.) It is uncontroversial to hold that fertility is influenced by factors enmeshed in the cultural substructure that carries a society's values and beliefs and is reproduced in each new generation. It is more controversial, because resisted by economists, to admit some residual cultural influence in more directly "economic" behaviors such as migration and activities directed at raising productivity. Cultural change of course does take place, but not necessarily in directions that can be predicted from a knowledge of changes in conventionally observed social and economic conditions. The once common notion that rural development entails transformation of hidebound, subsistence-minded traditional peasants into profit-maximizing, price-taking small farmers, or of semi-feudal clients into market-oriented wage laborers, has dissolved almost totally under the scrutiny of research over the last several decades. What has taken its place is an acknowledgment both of the broad "rationality" of decisions across this development continuum and of the context-specific "boundedness" of that rationality in any cultural system.

Some illustrations can help to show how attention to institutional contingency at various levels of social organization fills out an explanatory picture of population–rural development interactions. Given the preoccupation with land tenure and labor market in the literature on those interactions, we will focus instead on family, community, and government. Three country experiences are drawn on: Indonesia, Bangladesh, and Kenya. In the very brief compass available these are no more than outline sketches, highly simplified renderings of obviously complex patterns of social structure and change, albeit derived from a larger research base. While any such exercise risks imposing retrospective coherence on a process in which fortuity plays an appreciable role, we would maintain that the broad institutional "logic" portrayed in these cases properly reflects the empirical realities.

Local government in "New Order" Indonesia

Indonesia's economic recovery and fertility decline since the late 1960s can be traced in considerable part to the political and administrative reforms undertaken at that time. (These paragraphs summarize several chapters in McNicoll and Singarimbun, 1983.) In the early 1960s severe economic mismanagement in Indonesia was reflected in a declining volume of exports, a balance-of-payments crisis, chronic food deficits and stagnant agricultural productivity, a visible deterioration of capital stock, and very high inflation—all making for an apparently bleak economic future. Over 1966–67, following rising political turbulence and (in 1965–66) extreme anti-Communist violence, a new political order was established. Under this "New Order" regime the economy flourished, Green Revolution technologies took hold and food surpluses appeared, and a rapid fertility decline set in.

Contributing factors to this dramatic turnabout were renewed large-scale foreign assistance and the (coincidental) surge of oil export revenues from the OPEC price increases. But particularly important for the rural situation were the changing character and roles of local government. The lowest formal administrative level in Indonesia's rural areas is the administrative village (*kelurahan*), usually comprising a group of hamlets (6–8 on average) with a total population in the range of 300–800 households. Its head, the *lurah*, is an elected official, although his term is indefinite and in some regions the post is virtually hereditary. The higher administrative levels—subdistrict, regency, and province—are headed by appointed officials (*camat, bupati*, and provincial governor). Roughly paralleling this civil administration is a military hierarchy extending downward from the commander of each military district, the counterpart in the larger provinces to the governor.

Two significant changes took place in this system after the mid-1960s: first, a considerable strengthening of its role and increase in its effectiveness qua administrative control, and second, devolution of responsibility for many aspects of development program performance to its lower reaches—regency, subdistrict, and kelurahan.

The first of these changes can be seen as a series of responses to the political events of the mid-1960s. A thorough shake-up in the civil bureaucracy, especially in provinces such as East Java and Bali where the Communist Party had been strongest, led to a greatly increased proportion of military men, both active and retired, in provincial and local government positions. The security concerns of the military bureaucracy at all levels were intensified—toughening the civil bureaucracy, as one observer (Emmerson, 1978:83) put it, "with an exoskeleton of military command." Both civil and military administrations were enlisted in support of a new state party designed to establish a broad political base for the New Order government. The activities of other political parties were sharply curtailed. The combined result was an administrative system much more able than before to translate government policy into action.

The second change, increasing local program responsibility, was made feasible by this enhanced efficacy and necessary by the mounting demands of the development effort. Indonesia traditionally had a large array of line-ministry programs extending down to the kelurahan, each attempting to work independently. One field study (Sinaga et al., 1976) identified no fewer than 60 programs under 12 separate government departments that in theory operated down to this level. While no single kelurahan could have had all or close to all of these programs in operation, the scope for confusion and the work overload on local officials are evident. When the policy process no longer stopped (as it once did) at the stage of planning and target-setting but was concerned too with resulting performance, there was little alternative to devolution of program authority and local program coordination. Increased power of the administrative hierarchy at the expense of functional departments was thus to be expected. The programs that were the main success stories of the New Order government were those that were able to work through or count on the active support of this hierarchy: rice intensification, a public works/employment creation scheme financed by per capita revenue sharing grants, and the family planning program.

Mobilization of local government and local community groups to bring informal pressures on eligible couples to use contraception was the chief distinctive feature of the Indonesian family planning program. The more routine logistical and reporting systems of the program also depended markedly on the newly coherent local administration. It is difficult to assess the degree to which fertility might have declined in the absence of such efforts. To some extent the economic and political disruptions of the 1960s may have impeded a decline that would otherwise have begun in that decade. The intense competition for secondary education that emerged with the 1970s expansion of primary schooling (and thus devaluation of primary school completion as a labor market credential) would plausibly have generated a strong trade-off of quantity for "quality" of children irrespective of family planning program activity. At least in the early stages, however, few would deny a prominent role for local government in speeding the adoption of modern contraception in Indonesia. Fertility dropped from about 5.5 children per woman around 1970 to below 4 in the mid-1980s. Over the same period, agricultural production roughly doubled, again with significant assistance from an extension program that drew on government administrative resources.

This style of "top down" development has certain costs. At the kelurahan level there was a diminution of consensual village politics, as the lurah and other village officials took on more administrative (and shed representational) functions. The lurah had always held high status vis-à-vis his constituents; since the 1970s he tended in addition to become (de facto and in self-identification) a full-time government official. This trend toward cooptation of local community leaders into the national administrative system moved Indonesia away from concepts of village autonomy and self-reliant development made familiar from the experience of rural development in, say, Meiji Japan or (less certainly) post–

1950 Taiwan. Local leaders were increasingly conditioned to look to outside, mainly government, assistance to stimulate "development"—in the form of funds, organization, and ideas.

Insecurity and underadministration in Bangladesh

Bangladesh's rural population was expected to reach 100 million in 1990, having doubled over the preceding 30 years. It is growing at a rate that would double it again in the next 30. In contrast to this demographic dynamism is an agricultural economy that has performed disappointingly for many decades. In significant measure explanations for both trends can be found in the structure of rural social organization. (This description is drawn from Adnan, 1976–77; Arthur and McNicoll, 1978; Cain et al., 1979; and Cain and Lieberman, 1982.)

In any agrarian society, ties of kinship, patronage, and neighborhood cut across economic status divisions. They strengthen or weaken a person's claim on social product and modify his sense of risk or security. In Bangladesh, the nuclear or patrilineally extended family is the basic social and economic. Beyond his immediate household, a person has well-defined duties to his *bari* (homestead) and somewhat weaker ties to larger, kin-based groups (*paribar* and *gusthi*). For example, the bari often operates as a corporate entity with land held in the name of its head, who exercises patriarchal control over members. Kin-groups may dominate in such matters as selection of spouse and negotiation of dowry or bride-price, the upbringing of children, and disposal of assets.

These extended family and lineage groups have more significance among the bigger landowners. Families with no land or only a house compound, for whom there is little economic rationale for emphasizing relations with close but equally impoverished kin, are more likely to seek alliances with leading surplus farmers in patronage groupings (*shamaj*, or *reyai*). In exchange for allegiance the small landholder or landless worker may receive preferential employment, some support in bad years, seasonal credit, and other benefits. The leader in turn obtains a following to support him in local elections and disputes, as well as a sense of importance. The nucleus of a patronage group consists of the leader and his close relatives; wider membership typically comes from adjacent households within the village.

"Natural" villages or hamlets, the basis of rural settlement and local allegiance in much of Asia, are weak in Bangladesh. The natural village (*gram*) is socially defined, and its inhabitants have a clear perception of its territorial boundaries; but these units appear to have no corporate features, little cohesive identity, and only a residual degree of solidarity. An anthropologist (Islam, 1974:75) observes: "A man's duties are, in order, to his own family (bari), then toward his paribar, then to his gusthi and then to his village." Geographically determined settlement patterns are partly responsible: in land subject to extensive seasonal flooding, small groups of homesteads are clustered on raised ground built up from surrounding flood plains; hence dispersed or linear settlements rather than nucleated villages are the rule. But the relative lack of function of villages in the society can also be traced to the colonial failure to provide effective local administration. In consequence, village life is segmented. A man may reside in one village, attend a mosque in another, patronize a market in a third, and cultivate plots of land in any or all of them. For adjudication or minor disputes he may call on the head of his gusthi or on the leader of the shamaj to which he belongs; for assistance in ploughing or harvesting he may turn to other members of his paribar or to wage labor from distant villages.

Territoriality is not completely missing from Bangladesh social organization. The subvillage neighborhood cluster, the *para*, is sometimes a cohesive social unit, although it has little explicit role. And as already noted, patronage groupings also have a kind of territorial basis, albeit a fluid one. But for the most part, functions usually ascribed to the village community in descriptions of other peasant societies are filled in rural Bangladesh by a variety of nonresidential and overlapping groups with more or less specialized concerns. Kinship and patronage ties stand out as the most powerful organizing forces of rural society. They act as a simple, nonmarket distributive mechanism, channeling access and security downward. By binding together people at various social levels, they have tended to diffuse any strong manifestation of class. They have given the system some measure of stability.

It is not hard to see why fertility might be resistant to change in such an environment. First, there is or has been an apparent economic rationale for large family size. For affluent landowners, children represent opportunities for the family's occupational diversification and hence for expansion or consolidation of its local power. Lower down, among middle and poor peasants, the evidence has suggested that children become net producers at a young age, while the consumption costs of early childhood tend to be sheltered within a patriarchal family; in addition, sons who have reached maturity by the time their father dies are an important source of security for the widow and indeed for the family's assets.

Reinforcing these economic considerations, the pattern of social organization in rural Bangladesh, sketched above, militates against the emergence of social pressures at the local level (or administrative pressures from higher levels) able to oppose high fertility. High fertility is no direct threat to the economic or political interests of kin and patronage groups—interests which in essence are those of the dominant families within them. The numbers and rights of the fringe membership of such groups can adjust to permit maintenance or further accumulation of per capita resources at the core. Families at or beyond the margins of the patronage system bear the major part of the short-run costs of continued high fertility in the society, although costs are shared more widely through the high levels of economic and mortality risk and through the uniformly disadvantaged position of women. In the longer run, the society has in a sense been transferring demographic costs into the future, mortgaging its own coming generations.

For a transition to low fertility to occur in Bangladesh, if this analysis is correct, the course of social and economic change would have to be such as to lessen either the opportunity for shedding demographic costs in this manner or the perceived advantage in doing so. It is far from certain that the natural pressures on the system generated by continued population growth will effect such a shift. More hope would attach to the prospect that productivity gains (from new crop varieties and better irrigation and flood control) will increasingly offer families routes of economic security and perhaps advancement not tied to the traditional order and not facilitated by high fertility. Cultural change toward the values associated with consumer societies and improved status and opportunities for women, interacting with economic growth, would likely also contribute.

Obstacles to present proprietorship in Kenya

Kenya is demographically renowned for its rate of natural increase in the 1980s of 4 percent per year. Its rural population is estimated to have grown at over 3 percent per year in this decade and to have increased threefold, from 5.5 million to 17.2 million, in the period 1950–85.

Beginning in colonial times, Kenyan governments have encouraged and established programs to promote the orderly conversion of traditional systems of land rights into freehold titles. These efforts were accelerated as independence approached, with the departure of expatriate farmers. In effect, a new land frontier was opened. Large settlement schemes were set up to create smallholdings on much of this land; other programs financed the purchase by Kenyans of large farms from departing settlers. By the 1970s most of the country's agricultural land and a significant fraction of its pastoral land were registered as private freehold.

With population growth at its current rate (total fertility of around 7 children per woman, 5–6 of them surviving to adulthood) any agrarian system has obvious built-in instabilities. Two questions that immediately require answers in understanding the situation are: how is this growth being accommodated? and why does it persist? (Sources for the following are Livingstone, 1981; Dow and Werner, 1983; and Frank and McNicoll, 1987.)

On accommodation, both urban migration and rural labor absorption have been important, the latter more so. Urban migration typically took the form of individual, usually male, family members living in towns while their families remained on rural smallholdings. Leys (1984:181) wrote of this practice:

> [For] most of the urban work-force the relations of production of their smallholdings . . . still predominated over those of their urban jobs. This is not to deny the reality of the urban culture of Nairobi, Mombasa or even other towns. But it was a

> transient culture; even highly-placed civil servants commonly looked on some rural plot as "home"; four out of five wage workers wanted to retire to the countryside. . . .

Leys saw this as a transitional stage in the evolution of a fully commercialized wage labor economy. However, later studies such as Livingstone's (1981) show the phenomenon continuing. The massive rural-urban migration that many observers had forecast, based on the pace of rural natural increase, has not (yet) come about.

In explaining how the rural sector has thus far managed to absorb and support its demographic expansion, Livingstone (1981:ch.10, and this volume) describes what he calls a "sponge effect," operating chiefly through the subdivision of farms into smaller and smaller plots. Remarkably large gains in output per hectare are attainable by more intensive application of labor—even on half-hectare holdings. Other contributors to rural labor absorption are yield increases from improved crop varieties (notably, hybrid corn), exploitation of nonfarm activities such as wage-labor on estates, and migration to poorer agricultural or pastoral lands in semi-arid regions. The limits to these outlets appear close to being reached, however: at some point not far hence there could be a rapid expansion in landless poor and a much larger rural exodus.

Under such circumstances, why is there no drastic downward adjustment in fertility? The reasons presumably must lie in the family system and the specific fertility incentives that are refracted through it. Kenya's traditional family system, like those of most other countries of sub-Saharan Africa (discussed above), is organized by lineage rather than by conjugal ties. Marriage was an alliance of lineages, the husband obtaining the labor and childbearing services of his wife (or wives) in return for bridewealth paid to her family or lineage. The household did not have a "pooled" economy. The wife's responsibilities were concerned with providing subsistence; the husband's, with cash crops or wage labor (often entailing physical absence). As privatization of land progressed, the subsistence farm economy might have been expected to generate the incentives for fertility control that are seen in systems of peasant proprietorship in other parts of the world as child labor benefits diminish and child costs (particularly costs of education) rise. But in this case the subsistence farmer, the wife, was not the head of household. Her husband's preferences, sheltered from those immediate childraising burdens, continued to favor a large family.

Families in Kenya must of course greatly reduce their fertility before long if the agrarian economy and the social system are to survive. (Historically, Kenyan fertility must have been well below the 1980s level.) The family context in which they will do so remains somewhat speculative. Two directions of change can be suggested, both of which entail erosion of most remaining lineage influence and the reordering of the incentive structure away from high fertility. The first is the gradual nuclearization of families, with greater emphasis on conjugal ties and the merging of formerly separate gender interests. This is the trend forecast by John

Caldwell (1982) in his West African studies, tentatively expected by Jack Goody and by Akin Mabogunje (this volume), and explored by Thomas Dow and Linda Werner (1983) in the Kenyan case. Alternatively, it is possible that future family patterns in Kenya will most closely resemble those found in the (relatively) low-fertility Caribbean countries today, where female-headed households predominate and visiting unions are common (see Frank and McNicoll, 1987).

To restate the point made above, these highly compressed historical anecdotes are presented in order to suggest the plausible significance of an array of institutional variables that too rarely enter the analysis of rural development–population relationships. The essays collected in this volume, by being far more circumscribed in topic, take that task considerably further.

Rural population and development policy

We turn, finally, to policy issues. What policy directions and emphasis follow from this perspective on rural development and population change and how do they differ from or add to the familiar armamentarium of measures?

Policy goals are of course not affected. The objective of rural development qua *process* is rural development qua *state*, defined by technological, structural, and welfare criteria. The objective of population policy is to modify demographic behavior and outcomes in ways that contribute to individual welfare and facilitate the development process (not just rural development). This would usually translate into efforts to lower fertility and promote health and longevity. In countries where there are no near-term resource constraints on rural population size, the case for moderated levels of fertility may still be strong, both to reduce pressures of rural-urban migration and to facilitate a more ecologically stable mode of resource use in rural areas. This is not, however, to rule out occasional instances where there is little reason for fertility limitation to be ranked high on the development agenda.

Mainstream rural development and population strategies

Consider next rural development strategies. "Mainstream" policies in rural development are principally concerned with issues of agricultural-sector scale of production, ownership, provision of technology, and "integration." In population, the mainstream comprises clinic-based programs for family planning and primary health care, and associated information and extension activities.

Scale and ownership In Bruce Johnston and William Clark's (1982) important treatise on rural development, the chief classification of policies is according to the distribution of size-classes of holdings ("unimodal" versus "bimodal" or dualistic) and the type of ownership (private versus collective). The "East Asian" rural development model, derived from the experience of Japan, South Korea, and Taiwan, is built around the creation and support of a thriving small-farm sector. (This is the strategy advocated at length by Johnston and Kilby, 1975.) A structural precondition is a moderate degree of equality in the distribution of landholdings—a situation that, perhaps fortuitously, emerged from Japanese feudalism and that was created by the post–World War II land reforms in Korea and Taiwan. Interestingly, the drastic early-1950s land reforms in China, combined with the early-1980s restoration of limited private tenure, to a considerable degree make China another, if belated, instance of the East Asia model (on this, see Susan Greenhalgh, this volume).

In each of these cases, the political circumstances that made effective land reform feasible were highly exceptional. Land reforms elsewhere have typically been much less thoroughgoing—aiming, for example, at a more uniform distribution of operational units rather than at equity goals—and in practice more subject to circumvention. Moreover, where rural densities are very high and there is a substantial fraction of landless families in the agricultural sector, land reform may accomplish little for equity and less for productivity.

Peasant proprietors cannot, of course, be created by fiat. Neither the attitudes and values on the part of a farm family that make for the long-run stability and growth of a smallholding enterprise nor the system of support services required to sustain it in material terms is easily generated. Numerous instances of failures of frontier settlement schemes could be drawn on to illustrate this point.

"Bimodal" land distribution, in which a large-farm sector (occupying perhaps most of the cultivated area) coexists with a smallholder sector, is the norm in Latin America and is seen in parts of South Asia (in Pakistan, for example, and to a lesser degree in the Indian Punjab). It has been a goal of the Kenyan resettlement programs. While "success" cases are decidedly rarer, especially if success is defined in terms of poverty alleviation, complementary strategies to promote aggregate rural productivity growth and welfare improvement are clearly feasible under this land distribution pattern.

Technology A second broad area of rural development policy is the provision of technology. Choices among available technologies, influenced by government factor pricing decisions and programs of agricultural extension, and, in the longer run, by government support of agricultural research, have implications for rural employment and rural-urban migration. Clearly, effective policy in this domain must mesh with the existing agrarian system—at least closely enough to seek to move it in directions better able to meet development objectives. Since much agricultural technology is also location-specific, an analogous meshing with ecological conditions is required. Implicit incentives to use inappropriate technology, research biased to favor the demands of large commercial enterprises at the expense of smallholders, or extension programs directed at men in agricultural systems where much farming is done by women, are examples of the kinds of misconceived policies that are frequently seen.

Integrated rural development programs Finally, also in the mainstream of existing rural development strategy, we should take note of integrated rural development programs. An integrated rural development strategy is "a coordinated approach . . . to improve infrastructural facilities and to provide credits, inputs, extension and managerial services to small farmers. . . . In a 'best case' scenario, this strategy leads to productivity and income increases for small holders, agricultural improvements and diversification, and to the emergence of new community-based institutions which take the place of the many services and controls introduced by state agencies" (Lieberman, 1980:4–5). Emerging from the disappointing experience of "community development" and failed efforts to establish broad-purpose producer cooperatives in the 1950s and 1960s, integrated rural development in variant forms has been widely promoted in South Asia and Africa (see Food and Agriculture Organization, 1977). Less all-encompassing programs—food-for-work and guaranteed employment schemes, for example—can be designed along similar lines: the programmatic focus or span of concern is in many ways less significant than the managerial design and organizational content of the particular effort. The strategy is administratively complex and prone to ineffectiveness as a result of too-heavy demands put on all-purpose field operatives. While such programs remain in place in a number of countries, the early enthusiasm for them has waned.

Family planning and health Extension programs in some respects analogous to those in agricultural technology and practice operate also in the population and health field. The modern technology of contraception is purveyed through elaborate distribution networks of family planning clinics and supply depots; that of preventive and curative medicine through health clinics. Family planning "motivators" and nurses carry information and services to individual households. Nutrition campaigns disseminate information and sometimes food supplements to eliminate specific deficiencies. To a varying degree, in part depending on the inclinations and ability of governments and the medical profession to monopolize services, the private sector operates a parallel distribution network through local practitioners and pharmacies. In addition, a competing array of traditional treatments for illness and pregnancy, with corresponding traditional healers and midwives, may exist.

In the case of nutrition and health, the strength of underlying demand can be taken more or less for granted. (Medical anthropologists and social scientifically attuned public health experts, however, are now probing and in some cases questioning that strength—see, for example, Mosley, 1985.) That assumption is less readily justified in the case of modern methods of contraception, despite the improved convenience, effectiveness, and safety they offer—enough perhaps to induce a switch to them by users of traditional methods. Childbearing is obviously a highly consequential activity that is responsive both to economic and social structural changes affecting the costs and benefits of children to parents and to cultural changes affecting views of individual autonomy, efficacy, and aspirations and instilled images of the family. The extent of additional program-specific effects is hard to gauge—although that does not prevent routinely expansive claims by program operatives.

Strategies to deliver public-sector family planning services can entail separate, vertically organized administrative structures or be integrated with health or other programs at the local level. The arguments for and against integration mirror those in agricultural development: whether extension workers are overloaded, whether a clearer line of authority and responsibility is needed, and so on. The range of actual program designs reflects more the inconclusiveness of this debate than purposeful tailoring of programs to different local conditions.

Mainstream population policy embraces more than these supply-oriented program activities, although its remaining content is often neglected. Rules of marriage can affect fertility levels, and to a limited extent may be modifiable through government action. Expansion of educational opportunities for women has many justifications, and hoped-for demographic benefits (lower child mortality and lower fertility) are among them. Immigration is responsive to policy given a minimal degree of government will and capacity. Regulation of internal migration, calling for more of both, is less often attempted.

Policy reinforcement To what degree do these rural development and population policies reinforce each other? Many of the effects of policy success in rural development are associated with lower fertility: more efficient markets that facilitate transactions outside family and kin, the shift to nonagricultural activity where child labor is less helpful, increased incomes and resulting greater use of modern consumer goods, and so on. (Not all are. The belief that a more equal income distribution—say, through land reform—leads to lower fertility is now called into question: see, for example, Boulier, 1982 and Cain, 1985.) With some exceptions (for instance, the continued high fertility seen in the Soviet Central Asian republics despite, by Third World standards, relatively advanced agriculture, low mortality, and high incomes), success seems to breed success: with rural development comes falling fertility.

The style of rural development also affects rates of rural outmigration. Agrarian reforms typically influence capital intensities of production, wage levels, and the extent of use of family labor, all bearing on the level of agricultural employment. Latin America offers some striking illustrations. Under different agrarian policies, the massive rural exodus seen in much of that region in recent decades could perhaps have been lessened and hence also the high social costs that were (and are still being) incurred. Attention to potential labor displacement is now a routine part of mainstream agrarian policy debate.

Less can be said on the economic side-effects of population policy, chiefly because of the major uncertainties as to what family planning programs accomplish in demographic terms. The fact that fertility declines by means of greater use of contraception supplied through such a program says nothing about the determinative role of the program in the decline. The same fertility decline, through the same or through other means, might have occurred without it. Moreover, the presence and operational effectiveness of a program in any given community may in part be a response to local demand—in turn reflecting shifting economic conditions and cultural patterns wholly independent of population policy. Most efforts to disentangle family planning program effects from socioeconomic effects through statistical means, popular as they are, cannot be taken seriously, given these large possibilities for misspecification (see Demeny, 1979).

At this juncture we might conclude that current mainstream strategies of rural development, focused on rising productivity with side-attention to poverty alleviation and economy-wide labor absorption, have favorable demographic effects in helping to bring down fertility; and that direct antinatalist intervention through family planning programs may well contribute appreciably to this end. Such a conclusion, as we observed at the outset, would leave little content to the subject of "population and rural development": policy would be free of both conflict and controversy. It is not in fact the conclusion we reach below. The policy considerations that emerge from the analytical approach taken in this essay suggest a more complex relationship, and one where policy conflict as well as policy reinforcement may be present.

"Getting institutions right"

We provide here only generalities whereas policy problems are intrinsically location-specific, rooted in a particular environment, culture, and historical experience. Yet generalities may still help in guiding policy thinking: poorly founded generalities (rife, we believe, in this subject) are a cardinal source of policy weakness.

Family systems: the limits of policy Family structure, we have argued, plays a critical role in rural economic outcomes and in demographic transition. The family systems found in Northwest Europe and Japan were peculiarly suited to preservation of peasant agriculture under the fluctuating mortality of preindustrial times and to a fertility response to the economic threat represented by secular mortality decline. The different forms of adaptation found in the family systems of much of the contemporary Third World delayed fertility declines, although ultimately the economic and cultural changes associated with urban industrialization, impinging on even quite remote rural populations, have brought about lower fertility within those systems. The distinctive family systems of sub-Saharan Africa, in which conjugal ties are weak and lineage ties traditionally strong, are still less amenable to fertility decline in response to changing socioeconomic circumstances. Continuing average family sizes in excess of six children among small landholders are the norm, even in land-scarce economies such as Kenya. As economic pressures intensify, adaptations in African family arrangements conducive to or at least accommodating low fertility are likely to occur, but, as we noted in the Kenyan case, it is by no means clear what form those adaptations will take.

Even if we knew more about incipient changes in family structure, the scope for facilitating demographic transition through policy action directed at that level seems extremely narrow. One thing we do know about family systems is their resilience and persistence. The same is true of gender systems. The kinds of interventions that are available to governments and can pass muster on political grounds appear to have little purchase: typically, unenforced and unenforceable provisions of a civil code enacting "modern" family law. Another arena of action consists in efforts to alleviate women's disadvantaged status vis-à-vis men in pursuing their vision of their own and their family's economic and demographic interests—efforts that are defensible on many other grounds but that in the main have had indifferent success.

Community capacity and mobilization The lessons of experience in rural productivity growth point to the strong advantage of individual, private incentives over group incentives. Even at the height of China's enthusiasm for collectivization, production brigades in that country converted the collective targets they were expected to meet into an intricate internal economy monetized with "work-points." But if collective effort is a poor competitor to an individualized reward structure, it very clearly is not so in one other function: that of expressing demands on a bureaucratic or politico-economic system. Where conformity of behavior is socially advantageous, as often may be the case with environmental protection and sometimes perhaps with low fertility, the social pressures that a solidary community can exert on its members are also not to be dismissed.

For effective community response, the existence of a social interest must be apparent, not simply asserted. There would be no question of such an interest, for example, for day-to-day law and order: hence the harshness of the instant punishment often meted out by villagers to apprehended offenders. The case of fertility is much less simple. The existence of a community interest in control of population size depends on the extent to which the community can "mine" the rest of the economy. At one extreme is the community with a fixed territorial boundary and limited migratory options for its labor force—the best case might be English parishes in the seventeenth and eighteenth centuries, compelled by government to take responsibility for their own indigent. Social control of demographic behavior, notably marriage but sometimes also fertility, is then likely to be tight. At the other extreme—Bangladesh and much of sub-Saharan Africa might approach it—is the nonterritorial community, perhaps with a fluid fringe membership, concerned to increase the per capita assets of its core members; for it, fertility is unlikely to be an object of social control.

How do community forms therefore fit into rural development and population policy? That policies can effect changes in those forms is apparent: local government procedures and the permitted degree of local community autonomy and freedom of political activity are policy choices. Combining a free-market, individualized economy with a mobilized community demand on government services—a major factor, it seems, in attaining a decent quality of performance for services such as schools and health clinics—is a difficult balance. Adding a community role in population policy may destroy that balance. China, it can be argued, drew on community pressures in achieving its radical fertility decline in the 1970s, but found that the economic costs of mediocre agricultural performance necessitated a retrenchment in community roles and restoration of individual enterprise. Fertility policy implementation became more a purely government concern in the 1980s.

As a policy area, the functions of local (typically village) communities and their interaction with local government are potentially of great importance for both rural development and demographic change. The inclination of many governments to leave such matters of social system design—one of the few parts of social structure where some degree of deliberate design might indeed be possible—to interior ministries or their equivalents, whose interests may extend little beyond political stability and security, is to forgo action in a broad field of policy development. (See the essays by Jodha and McNicoll in this volume on the management of common property resources at this level of social organization.)

Government functions and competence A third possible field for rural population and development policy suggested by this analysis is that of rural administration. Constituent policy elements might encompass issues of devolution of program authority and accountability to local government, improvements in the design of local fiscal systems, the extent of administrative interference in factor and product markets and more generally in the local economy, and the effectiveness of the legal system as it bears on economic and demographic incentives.

Devolution can only be contemplated if there exists an appropriate structure of local administrative units to begin with. Variations in cultural patterns, in dictates of climate and topography, in colonial administrative systems, and in the nature and degree of politicization that has accompanied emergent nationalism all make for a wide divergence of administrative structures among nations and often among regions within nations. Examples of relatively successful administrative devolution that can be drawn from Japan, Taiwan, South Korea, and in some respects China, and of the comparative failure of that strategy in Bangladesh, suggest the importance of this inheritance. But much may be achievable in administrative policy virtually de novo. As we noted earlier, few observers of Indonesia in the early 1960s would have predicted the emergence of strong and relatively effective local government that subsequently took place. Those who see no basis for a significant local governmental contribution to the development effort in contemporary African societies might be similarly undervaluing new organizational possibilities.

A policy prerequisite that may be missing in this whole field of administrative reform is a simple recognition of its importance, and hence a marshaling of efforts toward institutional design. Raising the salience of these issues in policy debate, deliberate searching of the empirical record of experience elsewhere for relevant insights, and readiness to experiment, would be critical steps forward.

Culture, institutions, and policy

The intent of rural policy is to establish institutional settings conducive to socially desired patterns of economic and demographic change; the realities of the policy problem are far less cut and dried. Perhaps because in one of its meanings the word "institution" can refer to a public building, it is common to exaggerate the tangibility of institutions in the more general, abstract sense of the term that is used here and throughout this volume. This tendency in turn breeds too sharp a conceptual division between institutions and culture. Culture comes to be seen as something ethereal, pervading but only marginally constraining the individual-level transactions and interests that give rise to institutional patterns—and largely irrelevant to the hard-edged social engineering that supposedly is the real stuff of policy (McNicoll, 1978). While important insights into institutional forms can be gained by analyzing their constituent transactions—such is the program of the so-called new institutional economists and of methodological individualists in sociology—the imputation of a thoroughgoing present-time rationality to those existing forms is unwarranted. To do so is to trivialize the role of historical accretion in forming a social structure and to ignore the variety of cultural forces sustaining behavioral patterns. It is also to understate the potential sources of resistance to efforts at institutional reform.

The problem is not resolved by proclaiming culture, still institution-free, to be the real source of social change and hence the proper if elusive object of policy action. In this view, as extreme as the institutional determinist's, major behavioral change is seen as a manifestation of cognitive shifts in apprehending the physical and social world. Institutions are subordinate, and dependent phenomena and the efforts to make sense of them an empty endeavor. Policy must therefore win hearts and minds, not cater to material interests within a logic of collective action.

If we reject such all-or-none treatments of cultural influence, and recognize the historical layerings of institutional forms and their economic and cultural supports that make up the organization of any complex society, what follows for policy? Perhaps the main implication is that policy designs will not be readily transferable among settings. In rural development as in many other fields, a welter of policy advice and instruction is showered upon Third World governments, occasionally sought, more often proffered gratuitously or with implied sanction. The lessons, transmitted with casual certitude, are selective distillations of policy experience elsewhere. Yet what has "worked" in facilitating economic growth and demographic transition in any particular circumstances may often have been inadvertent—the outcome attributable to features of the local setting that had no part in any policy deliberation or weighing of alternatives. What should be of more benefit to genuine policy debate in such countries are the materials from which to draw relatively low-level insights into policy accomplishments and failures. The essays in this volume, grouped by the institutional sphere with which they are primarily concerned, can be seen as contributing to these materials.

References

Adnan, Shapan. 1976-77. [Dacca University Village Study Group Working Papers]. Dacca.

Arthur, W. Brian, and Geoffrey McNicoll. 1978. "An analytical survey of population and development in Bangladesh," *Population and Development Review* 4: 23–80.

Aston, T. H., and C. H. E. Philpin, (eds.). 1985. *The Brenner Debate: Agrarian Class Structure and Economic Development in Pre-Industrial Europe*. Cambridge: Cambridge University Press.

Avery, Dennis. 1985. "US farm dilemma: The global bad news is wrong," *Science* 230: 408–412.

Bardhan, Pranab. 1989a. "Alternative approaches to the theory of institutions in economic development, in Bardhan (1989b).

———— (ed.). 1989b. *The Economic Theory of Agrarian Institutions*. Oxford: Clarendon Press.

Boulier, Bryan L. 1982. "Income redistribution and fertility decline: A skeptical view," in *Income Distribution and the Family*, ed. Yoram Ben-Porath, Supplement to *Population and Development Review* 8: 159–173.

Brenner, Robert. 1985. "Agrarian class structure and economic development in pre-industrial Europe," in Aston and Philpin (1985).

Cain, Mead. 1985. "On the relationship between landholding and fertility," *Population Studies* 39: 5–15.

————, Syeda Rokeya Khanam, and Shamsun Nahar. 1979. "Class, patriarchy, and women's work in Bangladesh," *Population and Development Review* 5: 405–438.

————, and Samuel S. Lieberman. 1982. "Development policy and the prospects for fertility decline in Bangladesh." Center for Policy Studies, the Population Council, New York, Working Paper No. 91.

————, and Geoffrey McNicoll. 1988. "Population growth and agrarian outcomes," in *Population, Food, and Rural Development*, ed. Ronald D. Lee et al. Oxford: Clarendon Press.

Caldwell, John C. 1982. *Theory of Fertility Decline*. New York: Academic Press.

Cassen, Robert H. 1978. *India: Population, Economy, Society*. New York: Holmes and Meier.

Cheung, S. N. S. 1969. *The Theory of Share Tenancy*. Chicago: University of Chicago Press.

Demeny, Paul. 1979. "On the end of the population explosion," *Population and Development Review* 5: 141–162.

———. 1982. "Population in Nepal's Sixth Development Plan: A comment," CPS Notes No. 46, Center for Policy Studies, the Population Council, New York.

———. 1984. "A perspective on long-term population growth," *Population and Development Review* 10: 103–126.

Dow, Thomas E., Jr., and Linda H. Werner. 1983. "Prospects for fertility decline in rural Kenya," *Population and Development Review* 9: 77–97.

Emmerson, D. K. 1978. "The bureaucracy in political context: Weakness in strength," in *Political Power and Communications in Indonesia*, ed. K. D. Jackson and L. W. Pye. Berkeley: University of California Press.

Ezekiel, Hannan. 1978. *Second India Studies: Overview*. Delhi: Macmillan.

Food and Agriculture Organization. 1977. *Report on the FAO/SIDA Expert Consultation on Policies and Institutions for Integrated Rural Development*. Rome.

Frank, Odile, and Geoffrey McNicoll. 1987. "An interpretation of fertility and population policy in Kenya," *Population and Development Review* 13: 209–243.

The Global 2000 Report to the President: Entering the Twenty-First Century [1980]; a report prepared by the Council on Environmental Quality and the Department of State. Washington, DC: US Government Printing Office.

Hajnal, John. 1982. "Two kinds of preindustrial household formation system," *Population and Development Review* 8: 449–494.

Hayami, Yujiro, and Masao Kikuchi. 1982. *Asian Village Economy at the Crossroads: An Economic Approach to Institutional Change*. Tokyo and Baltimore: University of Tokyo Press and Johns Hopkins University Press.

Higgins, G. M., et al. 1982. *Potential Population Supporting Capacities of Lands in the Developing World*. Rome: Food and Agriculture Organization.

Islam, A. K. M. Aminul. 1974. *A Bangladesh Village, Conflict and Cohesion: An Anthropological Study of Politics*. Cambridge, Mass.: Schenkman.

Johnston, B. F., and P. Kilby. 1975. *Agriculture and Structural Transformation: Economic Strategies in Late-Developing Countries*. New York: Oxford University Press.

———, and William C. Clark. 1982. *Redesigning Rural Development; A Strategic Perspective*. Baltimore: The Johns Hopkins University Press.

Kahn, Herman, and Anthony J. Wiener. 1967. *The Year 2000: A Framework for Speculation and on the Next Thirty-Three Years*. New York: Macmillan.

Leys, Colin. 1984. *Underdevelopment in Kenya: The Political Economy of Neo-Colonialism 1964–1971*. Berkeley: University of California Press.

Lieberman, Samuel S. 1980. "Alternative rural development strategies in South Asia." CPS Notes No. 21, Center for Policy Studies, the Population Council, New York.

Livingstone, Ian. 1981. *Rural Development, Employment and Incomes in Kenya*. Addis Ababa: International Labour Office.

McNicoll, Geoffrey. 1978. "Population and development: Outlines for a structuralist approach," *Journal of Development Studies* 14: 79–99.

———, and Masri Singarimbun. 1983. *Fertility Decline in Indonesia: Analysis and Interpretation*. Washington, DC: National Academy Press.

Meadows, Donella, H., et al. 1972. *The Limits to Growth*. New York: Universe Books.

Mosley, W. Henry. 1985. "Will primary health care reduce infant and child mortality? A critique of some current strategies, with special reference to Africa and Asia," in *Health Policy, Social Policy and Mortality Prospects*, ed. Jacques Vallin, Alan D. Lopez, and Hugo Behm. Liège: Ordina.

North, Douglass C., and Robert Paul Thomas. 1971. "The Rise and fall of the manorial system: A theoretical model," *Journal of Economic History* 3: 777–803.

———, and Robert Paul Thomas. 1973. *The Rise of the Western World: A New Economic History*. Cambridge: Cambridge University Press.

Organisation of African Unity. 1979. *What Kind of Africa by the Year 2000?* Final report of the Monrovia Symposium on the Future Development Prospects of Africa towards the Year 2000, February 1979. Addis Ababa.

Sen, Amartya K. 1981. *Poverty and Famines: An Essay on Entitlement and Deprivation.* Oxford: Clarendon Press.

Sinaga, R. S., et al. 1976. "Indonesia case study: Rural institutions serving small farmers and labourers in the village of Sukagalik, Garut Regency, West Java," paper prepared for the ESCAP Expert Group Meeting on Rural Institutions Serving Small Farmers, Bangkok.

Skinner, G. William. 1965. "Marketing and social structure in rural China—Part III," *Journal of Asian Studies* 24: 363–399.

Stiglitz, Joseph. 1986. "The new development economics," *World Development* 14: 257–265.

United Nations. 1980. *Patterns of Urban and Rural Population Growth.* New York.

———. 1989a. *Prospects of World Urbanization, 1988.* New York. (Population Studies, no. 112).

———. 1989b. *World Population Prospects 1988.* New York. (Population Studies, no. 106).

Whyte, R. O. 1982. *The Spatial Geography of Rural Economies.* Delhi: Oxford University Press.

World Bank. 1982. *World Development Report 1982.* New York. Oxford University Press.

———. 1984. *Toward Sustained Development in Sub-Saharan Africa: A Joint Program of Action.* Washington, DC.

———. 1988. *World Development Report 1988.* New York: Oxford University Press.

FAMILY AND GENDER SYSTEMS

Population, the Status of Women, and Rural Development

Ester Boserup

Major increases in population modify the man–land ratio, facilitate specialization and communication, and provide other economies of scale. Whether the impact of these changes on rural development is positive or negative, and the likely feedback effects on demographic trends, depend in the first instance on macroeconomic factors, notably including government policy. But, microeconomic factors enter the picture as well. This essay deals with one of these microeconomic factors, namely the subordinate status of rural women as compared with men of the same age and social group. I consider how women's status varies under three types of family organization and land tenure arrangements prevalent in rural areas, and argue that the response of rural populations to economic and demographic change is more or less flexible depending on the type of family organization.

The status of women varies between rural communities and sometimes between families in the same village. The roles of women are also strikingly different. Christine Oppong distinguishes between seven roles that compete for women's time and energy, but here I shall focus only on those that are most relevant to my subject.

Of course, it is impossible to discuss the multitude of local patterns in a brief review. Nevertheless, family organization and the status of women are related to the agricultural system, which in turn is related to population density and technological levels. Therefore, it is possible to simplify the analysis by distinguishing a few major patterns of interrelationships between population, status of women, and rural development that together describe most rural communities in the Third World.

The conflict between motherhood and work

Rural communities at low levels of economic development usually have high levels of fertility. Young children can substitute for adults in many forms of agricultural and domestic work, including child care, animal husbandry, and gathering and transport of food and fuel wood. Therefore, children are likely to contribute more to family production of goods and services than they consume, at least if they are not sent to school.

Moreover, child mortality continues to be high in rural areas of most Third World countries in spite of the spread of modern medicine and improvements in sanitation. In the recent past, mortality was often so high that many families had few surviving children in spite of high fertility, and many women died before the end of their childbearing years. In regions with low levels of economic development, a sufficient reproduction of the population was a condition of survival both of the parents and of the community. The young generation had to support their elders, and in communities where the maintenance of law and order was left largely to members of the family or the village, adult sons had to defend family and village against enemies.

Until recent times, the need for defense against enemies motivated not only parents, but also governments, local and national, to be pronatalist. The social pressure on parents to maximize family size was bolstered by granting a high status to fathers and mothers of large families, and a low status to the unmarried and infertile, especially women. Under these circumstances, family organization was based on the institution of early and universal marriage for girls, with men tending to divorce or abandon infertile and subfecund women.

Families were organized as autocratic age-sex hierarchies. Younger family members owed obedience to older persons, and the oldest man in the family was the decisionmaker. In daily life men and women could dispose of the labor power of younger members of their own sex according to the customary distribution of labor, but when a change in the distribution of labor became necessary—because of changes in the method of production or in the availability of resources, for example—the family head decided on the new distribution of labor. By assigning as much work as possible to women and younger family members, he could increase family income while reducing his own work burden.

Either custom or the marriage contract stipulated that the wife must do the work that was needed to serve male family members and children. This obligation included not only child care and domestic duties, but also carrying water, gathering fuel, and often caring for domestic animals and processing crops. In most rural communities, women were obliged to work in the fields, and in some communities women and children did nearly all the agricultural work.

Frequent pregnancy and prolonged breastfeeding tax a woman's health. If a woman married early and spent the entire period between puberty and menopause being either pregnant or breastfeeding, and if she also worked hard at multiple tasks in the household and in agriculture, the strain on her health was heavy. Maternal and infant morbidity and mortality and the frequency of involuntary abortion were attributable to or aggravated by such patterns of childbearing and work. Thus, a conflict existed between the family's interest in a large number of surviving children and the family head's interest in obtaining as much work as possible from the adult women.

Rural societies have dealt with this problem in a variety of ways. One is spacing of births, another is polygyny, and a third is customary strictures on women's work in agriculture.

In communities where modern contraception is not practiced, a family can prolong the period in which the wife can recover strength between the end of breastfeeding and the onset of a new pregnancy by abstinence or other traditional means of fertility restriction. Traditionally most African societies had enforced periods of abstinence after each birth, and in many parts of India women moved back to their parents' household after child birth.

The custom of marital abstinence after child birth provides an inducement to polygyny. Moreover, the conflict between motherhood and work is attenuated when a number of wives share the burden of serving the husband and performing tasks related to domestic duties, child care, and agricultural labor. In sparsely populated regions with free access to cultivation of common land, polygyny is highly advantageous to the family head, because he can combine large family size with a negligible work burden, all the work being done by his wives and children. Where market access is available for surplus products, he can expand the area under cultivation in step with the increase in the family labor force and become rich by means of unpaid family labor. In other words, polygyny is a means to create family wealth in regions without land shortage and without the availability of labor for hire. A likely outcome for the society as a whole is increased income inequality.

The sexual division of labor is different in rural areas where increasing population density has led to the replacement of long-fallow production on common land with more intensive systems of agriculture using animal draught power. The use of animals for land preparation and transport is always a male monopoly. Men perform the operations for which animals are employed, while women and children perform some or all of the tasks for which only human muscle power is used. In this system of agriculture, women and children contribute a smaller share of total agricultural work and animals and men a larger one. Moreover, since additional land can only be obtained by purchase, if at all, women's agricultural work is less of an economic asset than under common land tenure. There is less economic motivation for polygyny, which at most is practiced only by a small number of rich men.

In many rural communities, men exercise control over women to such an extent that they prevent them from working outside the home. Women who do no work in the fields can devote all their time to child care, domestic work, and other activities that can be performed in or close by the home. With this type of family organization, as well, only a small minority of rich men can afford to be polygynous.

The interrelationship between demographic and economic change differs in these various rural settings. The sections that follow describe the relationship between population change, female status, and rural development under each of the three basic patterns of rural organization: first, men and women farming common land; second, men farming with women assisting on private land; and third, men farming while women remain secluded. The last section deals with the interrelationship between the status of women and economic and social services available in rural areas.

Men and women farming common land

In regions where land is held in common, a large increase in the local population results in a major expansion in the cultivated area. If market access is available for cash crops, a share of the increase in the rural labor force will be absorbed in expanding production of cash crops for urban use or for export. Since men have the right of decision, they leave the production of food crops for family consumption to the women and devote their own labor to cash-crop production. When women are obliged to feed men and children, men can earn money by cash cropping or wage labor. Thus, the first pattern is characterized by a division of labor, with men as producers of cash crops and women as producers of subsistence crops.

Although cash crops are produced mainly by men, who in turn reap the income from sales, men in most cases require their female family members and children to do some of the manual work in production and processing. Many cash crops are produced with the help of large amounts of family labor provided by women and children. Even in cases where women do all the work on a cash crop, men often sell the crop and pocket the income or oblige the women to hand over the money to them.

This organization of family labor, in which male farmers require women and children to do a large share of the work without pay, provides a strong inducement to the expansion of commercial agriculture, but it is a handicap to the production of food crops. When the population in a village increases, some of the surrounding pasture and other uncultivated land are brought under cultivation. The men appropriate the best land for their cash crops, while the women's food crops must be grown on poorer land or land at a greater distance from the village. As the cultivated area is expanded, fuel, wild food, and fodder must be gathered further away from the village. All the additional work is usually the responsibility of women and children. In these circumstances women may be unable to produce enough food for the family; either the diet deteriorates or imported food is purchased.

Men are unlikely to fill the gap in food production, because this low-status activity is women's work and depends on primitive techniques, which the men avoid using whenever possible. Thus, the unequal burden of family labor contributes to the shift from subsistence production to commercialized agriculture and to increased dependency on imported food. Although the main cause of this shift to cash-crop production for export is the distortion of world market prices, due to the farm support policies in the industrialized countries, the subordinate status of women in the exporting countries also contributes to the change in the production pattern.

As population growth and increases in cash-crop production make good land scarce, some family heads migrate to find wage labor on plantations or large farms or in urban areas, leaving the women behind to support the rest of the family. In many villages such deserted wives account for a large share of the agricultural population. A low-status group without male support, female heads of households are usually discriminated against in access to land, agricultural inputs, and credit. The female-headed families become a village proletariat with too few resources to produce sufficient food.

Growing population density and increasing cash-crop production cause a gradual change from common land tenure to private ownership of land. Men who plant trees or intensive cash crops on common land are likely to register this land as their property as the amount of uncultivated land in the area is reduced. When the reduction in uncultivated land shortens fallow periods, a family cultivating the same piece of land with only short interruptions comes to consider the land their own, even when no formal registration of property is undertaken. With few exceptions, privatization of land leads to a deterioration in the status of rural women. Under the system of common tenure, both male and female community members had the right to use the land for cultivation either by simply farming it or by having it assigned to them by the village chief. But only in exceptional cases is land registered as the property of female cultivators. The right of ownership becomes vested in the husband, as head of the family.

As mentioned earlier, common land tenure encourages large family size, because women and children represent free labor; the more of them a man has, the more land his family can cultivate. This advantage disappears, of course, when women lose their cultivation rights through privatization. A larger family no longer confers a right to additional land. Moreover, a private landowner can mortgage or sell land to weather emergencies and support his old age, while cultivators of common land have no such recourse, but must rely on help from adult children. Thus, private tenure provides less inducement to large family size than does common tenure.

However, if privatization makes men less dependent on help from adult children, it renders women more dependent on both adult children and spouses. This has wide implications, especially for women who live under the marriage systems that are traditional in Africa. Most African men have no legal obligation to support a wife and children, even while the marriage lasts, and they are free to terminate the marriage at will. The age difference between spouses is large—around eight years in first marriages and much greater when a married man takes additional wives. In a family system based on the principle of an autocratic age-sex hierarchy, a large age difference implies a low status for the wife, especially the youngest wife. A wife's lack of adult status makes her extremely vulnerable, particularly if the husband exercises his right to abandon her.

Because men are the decisionmakers, African societies tend to make progress in modernization through routes such as privatization of land, where men win and women lose, but they avoid forms of modernization that would be to the advantage of women—as demonstrated by the nearly total failure to modernize family legislation. Thus, rural African women are more than ever dependent upon adult children for help in old age and in case of abandonment by the husband. For such women, voluntary restriction of fertility is a risky affair, given the persistence of high child mortality and widespread subfecundity in many regions (the latter due to a variety of sexually transmitted diseases).

Male farmers with female labor on private land

The second pattern of rural organization is one of private property in land and use of animal draught power. The male family head and other male workers use the animals, while female family members perform some or all of the operations for which only human muscle power is used. In communities of this type the main means of accommodating population increase is by intensification of agriculture. Continued population increase in an area with no unutilized land leads to reduction in farm size with subdivision between heirs, and the smaller farm size leads to intensification of land use. High-yielding crops are introduced, and short fallow agriculture is replaced by annual cropping. When this reduces the area available for pasture, fodder crops may be introduced in the rotation. In the most densely populated areas, annual cropping may be replaced by multicropping.

When extensive systems of agriculture are applied and animals gather their own feed, most of the work is concentrated in the sowing and harvesting seasons with long periods in between without any work in the fields. Women's participation is limited to the relatively short peak seasons. But, when population growth results in intensification of agriculture, many traditional operations require more labor, and additional operations that were unnecessary with more extensive production must be introduced in order to obtain high yields or to avoid declines in yields and damage to the land.

When both cultivation and animal husbandry become intensive, many operations must be performed in the dry season, and the growing season may be prolonged by means of irrigation and by cultivation of crops with overlapping growing seasons. Much of the additional work involves hand operations of the type usually done by women, so that women perform agricultural work for a much longer part of the year. Many of the female operations, like weeding and transplanting rice to permit multicropping, are onerous, highly labor-intensive tasks. When intensification results in high crop yields, the work involved in harvesting and processing of crops increases. Usually, women are obliged to do all the additional work.

When infant and child mortality declines, more children survive to an age at which they can produce more than they consume. If children do not attend school, the male head of household can freely deploy the labor force of several family members and reap considerable benefit from labor-intensive agricultural production with relatively little additional work on his part. The extensive use of women and children for the additional operations required by labor-intensive production provides a greater inducement to intensification than would be the case if the adult male labor force had to do all the additional work or to hire agricultural workers.

If the availability of female family labor promotes labor intensification of agriculture, it also retards industrial inputs. In some cases, the potential for increasing yields and speeding up seasonal operations that such inputs offer makes them economical, notwithstanding the availability of unpaid female family labor. But in other cases, such changes are prevented or delayed because wives and children work without pay.

Use of mechanized equipment is a status symbol reserved for males. When tools and other techniques are upgraded, female labor is systematically replaced by male labor. The application of chemical inputs is considered high-status work reserved for men. This helps to explain why chemicals are used mainly on cash crops, although such inputs would also raise the yields of subsistence crops, thus saving both land and female labor.

The green revolution in Asia combined labor-intensive and modern inputs. Chemical inputs raised crop yields, and tractors speeded up the operations to allow multicropping; but once green revolution techniques were introduced in rice farming areas, women had to do much more handweeding and transplanting. The availability of female family labor may help to explain why herbicides were little used while handweeding was extensively employed in many regions. In their choice of techniques, the decisionmakers, be they male farmers or male agricultural experts and advisers, take little account of the disutility of backbreaking work for the women in the family, and the women themselves have no say in the choice of technique.

Use of tractors, pumps, and other machinery is spreading rapidly in many developing countries. As soon as an operation becomes mechanized, men take over, while women continue to use only hand tools. Examples of this abound in the development literature. In Korea, even transplanting, once exclusively performed by females, became a male operation with the introduction of a rice planting tractor. A man drives the tractor, while women follow after and handplant seedlings. In other agricultural tasks the same trend is seen: with the introduction of hand-tractors and threshing machines, women became assistants to the male operators. Their work burden was not necessarily reduced by this type of mechanization; it even increased in some cases, because the machine could cover a much larger area than could the old techniques.

In villages where land is held privately, those with small holdings and the landless work for large landholders in the peak seasons. Female family members of these laborers also often work for wages in the peak seasons, still performing traditional female jobs. Because of their inferior status, traditional female jobs command much lower wages than traditional male jobs, although the former are often the most unpleasant.

As mentioned above, when population increase leads to intensification of agriculture, the demand for additional female labor is large. The same is true for diversification of the cropping pattern because most secondary crops rely on hand labor, while the basic crops use animal-drawn or mechanized equipment, at least for some operations. Low female wages help to make labor-intensive crops profitable even on larger farms, but since the small farms have access to unpaid female family labor, while large farms must pay for their marginal labor, small farms are usually more intensively cultivated and have a more diversified cropping pattern. The specialization in labor-intensive crops helps the small farms to survive the competition from larger farms. Female family members on small farms are usually prevented by male heads from accepting employment as wage laborers on large farms.

When farms are subdivided among heirs, some small farmers survive on reduced landholdings by intensification of land use, while others change from full-time to part-time farming supplemented by wage labor on larger farms. Increasing pressure on agricultural land can lead to land concentration. Larger farmers or persons in nonagricultural occupations buy up small farms, and the former owners become full-time wage laborers. These alternative responses to the problem of increasing man–land ratios have different effects upon demographic trends, because the attitudes of landowning farmers and full-time agricultural workers toward family size are likely to differ.

As mentioned in the previous section, privatization of land weakens motivations for large family size because land ownership provides security against destitution in old age and emergencies, thus making the landowner less dependent on help from adult children. Moreover, if a family has too many male heirs, it risks losing status through subdivision or sale of the farm. Therefore, in communities where land ownership is long established, family-size attitudes are likely to differ from those in communities that have—or until recently had—common land tenure. Among landowning farmers the parents of a large family are held in high esteem only if the landholding is large enough to support the family. There is evidence in many societies of customary family limitation among landowning farmers, the means ranging from late marriage, celibacy, and prohibition of remarriage of widows to induced abortion and infanticide. Voluntary family planning programs in rural areas are most successful in such societies.

The increasing incomes and monetization in rural areas during the spread of green revolution technologies in recent decades permitted a growing share of the population to count on property income and access to credit as security, instead of relying on help from adult children. In Japan, the decline of family size accompanied very high rates of private saving. Where private landownership is the norm, marriages typically are more stable and the wife's right to support from her husband is better secured by legislation than is the case where land is commonly held. In these circumstances fathers assume greater responsibility for their families, and both men and women are more inclined to practice contraception.

Agricultural workers have less motivation for fertility limitation than landowners in the same community: the workers experience higher child mortality; having no property and less access to credit, they are much more dependent upon adult children as economic support; and they command more help from young children, who rarely go to school. Children of agricultural workers help their parents both by engaging in wage labor for landowning farmers and by assisting in subsistence farming. If population growth results in increasing landlessness, this may retard the fertility decline, unless the harder labor performed by wives substantially increases the number of involuntary abortions.

The relative numbers of landowning farmers and landless workers vary widely among developing countries. In most of Latin America, the share of landless or nearly landless workers in the rural population is high; in most Asian countries it is much lower. This may help to explain why rural fertility is lower in Asia than in Latin America. Another feature that may help to explain the slow pace of fertility decline in much of rural Latin America is the marriage system. Most lower class Latin American women, like African women, live in unstable marriages, and many of them cannot count on much, if any, support from the fathers of their children. Therefore, they must rely on their young children for income support and on adult children for old-age support. Men who feel no obligation to support either their children or the mothers of these children lack motivation to practice contraception, and most rural women in Latin America have recourse to no better means of fertility control than illegally induced abortion.

Men as farmers, women secluded

The main purpose of the seclusion of women, or purdah, is to ensure the fidelity of wives and the chastity of daughters and other unmarried women. But this institution is also a radical means of giving priority to motherhood over field work by women. Women who avoid outdoor activities devote their time to child care, domestic work, processing of crops, care of stabled animals, and home crafts. When women are secluded, small farms forgo free labor in the fields—a serious impediment to intensification of agriculture and diversification of the cropping pattern. Agricultural production is usually less intensive in regions without a supply of female labor than in regions with similar conditions in which women do participate in field work. Moreover, where female seclusion is practiced, farmers plant mainly crops that use animal draught power, avoiding crops that require the manual labor performed only by women.

When population in a densely settled rural region increases without concomitant intensification of agriculture, the population becomes poorer in the absence of employment opportunities for men on large farms, in rural industries, or through migration. Bangladesh is one such region with severe restrictions on female labor force participation. In spite of the very high population density, agriculture is less intensive in Bangladesh than in many other parts of Asia. There is more broadcasting and less transplanting of rice, and the cropping pattern is little diversified, being concentrated on rice and jute. Rural poverty is extreme, and the country is heavily dependent upon food imports. India also has regions where the cultivator's wife and other female relatives do not participate in outdoor work. But where these are multiethnic communities, the negative effects of lack of female family labor are less severe. Women from landless families belonging to an ethnic group in which both men and women do field work find employment as day laborers for the cultivator families with secluded women.

When women only work indoors, there is little motivation to introduce changes to raise their labor productivity. Food preparation and crop processing are done with primitive equipment and highly labor-intensive methods. Women continue to make clothing and other household items even after the region has reached a stage of development at which such activities are usually replaced by purchases. In some cases, women in seclusion add to family incomes by engaging in domestic production of, for instance, textile or tobacco products, which are sold in the market by male family members or children. Or such women perform services like handweaving or sewing for customers. But, if seclusion is widely prevalent in a region, there will be an oversupply of such products and services. Extremely low prices will yield derisory hourly pay, and women's work will contribute little to family income, even though they work exceedingly long hours.

A more profitable solution, much used in rural communities where women are secluded, is for men to migrate to seek work. If the husband migrates and leaves a secluded wife and children behind, he appoints another male family member as guardian for the wife and sends remittances to support the family. Under seclusion, family organization is highly autocratic and the age hierarchy is pronounced. The status of the wife usually deteriorates and she is subjected to even stricter supervision when she is left under the charge of other family members instead of her own husband.

Owners of larger farms can in theory compensate for the lack of female family labor by recruiting male wage labor to do the work done by women in other regions. But in fact, men are often unwilling to do these jobs. Men not only support their higher status by monopolizing the use of mechanized equipment; they also defend their status by refusing to do jobs that are usually performed by women. Here again, multiethnic societies may provide a solution. In some regions of India, male wage workers belonging to ethnic minority groups perform the handweeding that neither the men nor the secluded women from the landowning majority groups will do.

Men who perform work that is also done by women receive substantially higher wages than the women. The large differences between male and female wages arise from the higher status of men rather than from lower female productivity: it is generally agreed that for many operations, such as transplanting and cotton plucking, female labor is more efficient than male. In communities where seclusion is practiced, a competitive labor market between the sexes, which equalizes male and female wages, appears only under particularly acute shortages of labor. Statistics from North India and Egypt reveal that when female labor is crucial for agricultural operations, yet scarce because of widespread seclusion, female wages may be equal to male wages. Because the higher male earnings are status wages and not efficiency wages, female seclusion is a handicap to intensification and diversification of production not only on small farms relying on family labor, but also on large farms relying on wage labor.

When women are secluded, girls are considered an economic burden, because dowries tend to be high and adult daughters cannot provide economic assistance to their parents. Therefore, in communities with restrictive attitudes to use of female labor, preference for sons is strong and in many cases girls are neglected by their parents. Poor nutrition and health care for girls, if not open infanticide, limit family size.

In communities where prejudices concerning women's work are strong, women must rely either on property income or on help from adult sons or other male family members in case of divorce and widowhood. Because of son preference, the risk of divorce is larger, the fewer living sons a woman has. Moreover, the earlier a woman begins childbearing, the more likely she is to have a son old enough to support her in case of widowhood or divorce. So, both parents and daughters have an interest in early marriage. Age differences between spouses are large, thus increasing the risk of early widowhood. School attendance is low for girls in such settings. When women are allowed to work only in the house, education is a luxury that only rich parents may choose for family prestige.

For these reasons, there has usually been little, if any, decline of rural fertility in those Arab and Asian countries in which the secluded wife continues to be the ideal.

Status of women and access to rural services

The extent to which national and local governments provide services to rural areas varies widely between and within countries. Some governments supply mainly health and educational services; others make large investments in rural transport and communications, large-scale water control, and major land improvement schemes. Some governments have established extensive networks of agricultural services, including research, training, and extension facilities. Of course, agricultural development is much more successful where government investment is higher, with the result that private services and private investment in ancillary rural crafts and industries are also larger in these areas.

Rural women benefit indirectly from agricultural investments and services because of improved family incomes, but the direct benefits of the research, training, and extension facilities have virtually all accrued to men. In this field, as in many others, the lower status of women has been the main cause of the uneven distribution of resources. Male extension agents address themselves only to male heads of households; neither female producers nor female family workers benefit from their advice. Research activities focus on the cash crops produced by men; the secondary crops and subsistence food produced by women are neglected. Finally, trainees have virtually all been men.

The passing on of skills from parents to children is an important part of agricultural training in most developing countries, but when women are bypassed by the extension agents and the recruiters to training courses, they have no knowledge of modern methods to pass on to their daughters, while the sons learn readily from their fathers. Everywhere, the male status in agriculture is enhanced by reserving formal training for men and male youth; women are refused access, or only a token number are admitted. In many countries, agricultural work has little prestige among young men, and attendance at agricultural schools and courses is often low. In spite of this, the concern for male status among local personnel and foreign advisers prevents women from obtaining agricultural training, even in countries where they do most of the agricultural work. Where courses for rural women exist, they provide instruction in health, nutrition, family planning, and domestic skills, not in agricultural or other income-earning skills.

The exclusion of women from agricultural training and extension has been discussed and lamented at numerous international conferences and meetings. As a result, some national governments and multinational and bilateral donors have set up rural projects aimed at better integration of women in agricultural modernization. Some of these projects have been directed at "women only"; others have been "general" projects meant to take special account of women. Some projects have been sabotaged by local communities or by governments or donors. In other cases, there has been no open resistance, but women have hardly benefited. Few "women only" projects have been outstandingly successful, and in most of the "general" projects the benefits accrued mainly to men.

It is difficult to avoid the conclusion that these projects failed either because the male decisionmakers were convinced that women were unable to learn, or because they wanted to preserve the gap in status and prestige between men and women. The projects were unable to stem the general trend toward a widening distance between male and female agricultural qualifications, resulting from the sex discrimination in training in modern methods and in access to modern equipment and other inputs. It seems that general improvement in female status by means of political action, legal reform of family organization, and female education are preconditions for substantial changes in women's position in the agricultural sector.

If women have reaped few benefits from agricultural services, the situation is quite different with respect to social services. The most important improvement in the lives of rural women was the decline in maternal and child mortality resulting from the spread of medical and sanitation services. The physical and psychological strain on women is reduced when they produce fewer children for the graveyard. Also access to modern means of birth control in rural areas of many developing countries reduces the strain of frequent motherhood and of induced abortion.

Health improvements and mortality decline for women and girls have been large nearly everywhere, and girls' use of educational services has increased rapidly. In some countries boys still represent a large majority of students at rural schools, but the gap between the rates of school attendance for boys and girls has been narrowing everywhere. In regions where women do much of the agricultural work, however, they pay a high price in terms of labor input forgone for the education of their children, especially girls.

Many African mothers have been willing to forgo help from their school-age daughters and to pay their school fees in order to improve the girls' future earnings. Under the optimism prevailing in Africa in the period immediately following independence, parents did not view a large family and education of children as incompatible goals. Only later did economic crisis radically change the prospects for the growing numbers of educated young people. It remains to be seen whether the shortage of jobs for the educated will influence the attitude of rural parents toward education or toward fertility control.

As a result of the rapid spread of education, age–power relations between women are changing. Illiterate mothers and mothers-in-law have less authority over literate daughters and daughters-in-law than they did over illiterate ones. But it is uncertain to what extent increasing school attendance of girls contributes to greater equality between the sexes. Most of the development literature concerning female education focuses on the quantitative aspects: How many girls go to school, and for how many years? It is tacitly assumed that school attendance helps to improve women's status whatever the ideological content of the curriculum. But indoctrination in national culture is usually an important element in the curriculum of rural schools, and if the principle of female inferiority is an important feature of the national culture, the influence of school attendance on the attitudes of boys and girls to the status of women is not what might be hoped for. Young women may be more inclined to use their educated status to justify withdrawing from agricultural work than to compete with educated young men for more prestigious employment on farms or in rural services. And women who want to improve their position may be more inclined to migrate to towns than to face the rural community's opposition to changes in the status of women.

The occupational choice of female rural school leavers depends not only on the national ideology, but also on the opportunities for female nonagricultural employment in rural areas. Development of rural services and small-scale industries provides possibilities for accommodating a large family on a small landholding. In parts of Asia many girls add to family income by means of nonagricultural work in rural areas. They earn money to pay their own dowry, and this may give them more say in the choice of a marriage partner as well as contribute to a rise in the age at marriage, thus reducing fertility.

Conclusion

Autocratic methods of family organization, with men as decisionmakers and women obliged to obey them, promote adaptation to population growth by labor-intensive methods, and they help small farms to survive the competition with large farms that must pay for their marginal labor supply. The exception is regions in which women are prevented from working in agriculture for status reasons. In these regions, the need to pay male wages for marginal labor provides a disincentive to agricultural intensification. The subsistence crops that are grown by women are likely to be neglected in efforts to improve productivity, with serious effects on nutrition and family welfare in many countries.

Because women in most underdeveloped rural settings are forbidden to use mechanized equipment and other modern inputs, low (or zero) female wages for labor-intensive operations compete with much higher wages for male use of modern inputs, and the resulting distortion of cost calculations is a deterrent to agricultural modernization. Under these conditions, agricultural work has even less attraction for female than for male school leavers, and the former are likely either to migrate to towns, to become housewives without agricultural work, or, in areas where there are such opportunities, to seek local employment in service or processing industries. The incentives to male migration are strongest in areas where women are secluded and where women produce the family food without male help.

The subordinate status of women is conducive to large family size. Women's dependence upon the good will of their husbands and older family members makes them hesitant to practice contraception, and, because their opportunities for self-support in case of divorce and widowhood are limited by labor market restrictions and wage discrimination, they run a great risk of becoming dependent upon help from their children. However, when rural development raises men's interest in family limitation, the inferior status of women may actually serve to reduce fertility, because it is usually the husband who decides on family size.

Bibliography

The stylized treatment of rural settings and their concomitant family and fertility patterns in this essay is not appropriately tied to specific references. The bibliography following provides general supporting material for the themes and arguments adduced.

Abdullah, T. A., and S. A . Zeidenstein. 1982. *Village Women of Bangladesh.* Oxford: Pergamon Press.

Ahmed, I. (ed.). 1985. *Technology and Rural Women.* London: Allen & Unwin.

Boserup, E. 1980. "Food production and the household as related to rural development," in *The Household, Women and Agricultural Development*, ed. C. Presvelou and S. S. Zwart. Wageningen: Veenman and Zonen.

———. 1984. "Technical change and human fertility in rural areas of developing countries," in *Rural Development and Human Fertility*, ed. W. A. Schutjer and C. S. Stokes. New York: Macmillan.

———. 1985. "Economic and demographic interrelationships in sub-Saharan Africa," *Population and Development Review* 11, no. 3.

———. 1986. "Shifts in the determinants of fertility in the developing world," in *The State of Population Theory: Forward from Malthus*, ed. D. Coleman and R. Schofield. Oxford: Basil Blackwell.

———. 1989. *Woman's Role in Economic Development.* London: Earthscan Publications.

Bukh, J. 1979. *The Village Woman in Ghana.* Uppsala: Scandinavian Institute for African Studies.

Bunster, X. B., et al. (eds.). 1977. *Women and National Development.* Chicago: University of Chicago Press.

Buvinic, M., M. A. Lycette, and W. P. McGreevey (eds.). 1983. *Women and Poverty in the Third World.* Baltimore: The Johns Hopkins University Press.

Chipande, G. 1984. "The impact of demographic changes on rural development in Malawi," paper presented at the Seminar on Population, Food, and Rural Development, IUSSP.

Epstein, T. S., and R. A. Watts (eds.). 1981. *The Endless Day.* Oxford: Pergamon Press.

Federici, N. 1985. "The status of women, population and development," *IUSSP Newsletter* 23.

Fortmann, L. 1981. "The plight of the invisible farmer," in *Women and Technological Change in Developing Countries*, ed. R. Dauber and M. L. Cain. Boulder, Colo.: Westview Press.

Huston, P. 1985. *Third World Women Speak Out.* New York: Praeger.

International Rice Research Institute. 1985. *Report of the Project Design Workshop on Women in Rice Production.*

International Service for National Agricultural Research. 1985. *Women and Agricultural Technology.* The Hague.

Jahan, R., and H. Papanek (eds.). 1979. *Women and Development: Perspectives from South and South East Asia.* Dacca: Asiatic Press.

Jejeebhoy, S. J. 1987. "Reproductive motivation: The extent to which women differ from men," paper presented at the Seminar on Development, Status of Women, and Demographic Change, Italian Institute for Population Research.

Loutfi, M. F. 1980. *Rural Women: Unequal Partners in Development.* Geneva: International Labour Office.

Mason, K. O. 1985. *The Status of Women: A Review of Its Relationship to Fertility and Mortality.* New York: The Rockefeller Foundation.

Mickelwait, D. R., M. A. Riegelman, and C. F. Sweet. 1976. *Women in Rural Development.* Boulder, Colo.: Westview Press.

Miller, B. D. 1981. *The Endangered Sex.* Ithaca: Cornell University Press.

Oppong, C. 1980. *A Synopsis of Seven Roles and Status of Women: An Outline of a Conceptual and Methodological Approach.* Geneva: International Labour Office.

Pala Okeyo, A. 1985. *Toward Strengthening Strategies for Women in Food Production.* United Nations: INSTRAW.

Papanek, H., and G. Minault (eds.). 1982. *Separate Worlds: Studies of Purdah in South Asia.* Columbia, Missouri: South Asia Books.

Sen, A. 1985. *Women, Technology and Sexual Divisions.* United Nations: INSTRAW.

Sharma, K., S. Hussein, and A. Saharya. 1984. *Women in Focus.* India: Sangam Books.

"Symposium on law and the status of women," *Columbia Human Rights Law Review* 8, no. 1 (1977).

Tinker, I. 1985. "Women's participation in community development of Korea," in *Toward a New Community Life,* ed. M. G. Lee. Seoul: National University.

Utas, B. (ed.). 1983. *Women in Islamic Societies.* London: Curson Press.

Westergaard, K. 1983. *Pauperization and Rural Women in Bangladesh.* Comilla: Samabaya Press.

Cooperation, Inequality, and the Family

AMARTYA SEN

MCNICOLL AND CAIN ARGUE AGAINST SEEING economic-demographic relationships in an institution-free way, yielding technocratic policy conclusions that are straightforward and conflict-free.[1] Institutions, conflicts, and inequalities are central to understanding rural development. In this essay I shall concentrate particularly on the institution of the family and generally on issues of conflict as well as congruence of interests. The presentation is largely theoretical (even though empirical illustrations will be given), and I shall try to place the role of the family as an institution in a wide theoretical framework.

Social relations between different persons typically involve both conflict and congruence of interest. Economic analysis of social problems cannot go very far without coming to grips with both the combative and the cooperative aspects of interpersonal and intergroup relations. It is, however, possible to emphasize one of these two aspects more than the other, and indeed it is easy to see that various economists have chosen a rather different balance of what to stress and what to neglect.

Adam Smith's focus on the congruent aspects of interests of different people is, of course, well known, with his pointer to the advantages that each gains from the other's pursuit of self-interest.[2] On the other hand, Marx's analysis of class conflicts and exploitation focused particularly on combative aspects of intergroup relations.[3] Of course, Marx too devoted much attention to exploring the congruent elements in social relations (e.g., the widely shared benefits that capitalist development may bring to a feudal society). Indeed, both elements figure in the writings of all the major economists, and the differences lie in the emphasis placed on congruence versus conflict. Walras's investigation of the mutual benefits from the general equilibrium of production and trade, and Keynes's study of how the effective demand of one may create employment for another, primarily focus on elements of congruence. In contrast, Ricardo's analysis of the adverse effects of profitable machinery on workers' employment, and Pigou's investigation of negative externalities imposed by polluting factories on the community, point the finger at conflict. Even though the emphasis varies, in any rich investigation the elements of congruence and conflict both tend to be firmly present.

Social institutions can be seen as dealing with elements of congruence and conflict in particular ways. For example, the market mechanism provides an opportunity for different parties to benefit from each other's complementary activities through exchange, and, at the same time, the relative prices serve as means of dealing with interest conflict, yielding particular distributions of mutual benefits, rejecting others. While both buyer and seller may gain something from the transaction (nontransaction will hurt the interests of both), a higher price can favor the seller and hurt the buyer (different terms of trade yield different distributions of benefits). When exchanges and prices are determined by the market mechanism (e.g., bringing demand and supply into balance), definitive use is made of cooperative opportunities arising from congruent interests, and at the same time some distribution of benefits is selected while others are rejected, thus addressing the issue of conflict.

Family relations, similarly, involve a combination of congruence and conflict. Obvious benefits accrue to all parties as a result of family arrangements, but the nature of the division of work and goods determines specific distributions of advantages and particular patterns of inequality. It is important not to lose sight of either of these functions that families fill, since a model of pure conflict (e.g., a "zero sum" game being played by the man and the woman) or a model of pure congruence (e.g., every member of the family having shared goals and identical interests) would undoubtedly miss something of substance in family relations. It is fair to say that in the standard economic literature, there have not been many models of pure conflict (even though journalistic discussions of conflicts between the sexes may give the impression of unalloyed conflict), whereas models of pure congruence are indeed common.

While Gary Becker has explored the possibility of seeing family relations *as if* they were market relations, dealing with congruence and conflicts in a special—and in this context rather implausible—way, much of his "treatise on the family" turns on the assumption of induced congruence: "In my approach the 'optimal reallocation' results from altruism and voluntary contributions, and the 'group preference function' is identical to that of the altruistic head, even when he does not have sovereign power."[4] The so-called household production models often have this character, as do models of "equivalence scales," even though it is not necessary to develop these models in this particular form.[5] The conflicts are effectively "eliminated" by agreed preferences and goals. The pursuit of efficiency then becomes, in this reduced framework, the central problem, with problems of inequality and injustice getting little recognition or attention.

In analyzing the role of "institutional mediation" in influencing social and economic development and demographic change, it is important to avoid the limitations of frameworks that would emphasize pure congruence (even in an induced form), or pure conflict (even between particular groups, such as men and women). This essay is devoted to exploring the requirements of a more adequate framework and seeking insights that we may get from such a framework in reading the experiences of India's rural economy and society.

Isolation and cooperation

The combination of congruent and conflicting elements in social relations can take many forms. A special class of games has attracted a good deal of attention in the form of the so-called Prisoner's Dilemma.[6] In this two-person game, the pursuit of each party's own goals leads to substantial frustration for both, since each inflicts more harm on the other through this pursuit than the gain that each receives from selfishness.

This type of conflict can be extended and analyzed in terms of the Isolation Paradox.[7] This is an n-person version of the conflict of which the Prisoner's Dilemma is a special case when $n = 2$, and it has the feature of permitting a *plurality* of "collusive solutions" that happen to be superior to the atomistic outcome for both parties.[8] In general, this plurality of collusive solutions introduces an aspect of the problem that is neglected in the classic case of the two-person Prisoner's Dilemma. Indeed, in terms of the analysis presented in the preceding section, it should be obvious that a fuller understanding of the combination of congruent and conflicting interests requires us to consider a richer class of problems than the Prisoner's Dilemma would allow, even though the basic conflict—between individual and collective pursuits of goals—captured by the Prisoner's Dilemma remains and has to be incorporated and integrated in this social analysis.

The Isolation Paradox has an important application in the study of population, in capturing the jointly self-defeating nature of strategies based on the pursuit of individual advantages. It is often believed—not without reason—that a larger number of children may enhance a family's own security, and especially the security of the older members of the family. The private costs of raising additional children may not be very great, and with a larger number of children born, the probability of receiving support from one's children in old age is indeed, ceteris paribus, enhanced. There are also other possible private benefits from larger families—for example, providing employment support even when the parents are not retired and old, and providing greater protection against land usurping and crop robbery.[9] While an expansion of family size may well suit the interests of the respective families or of their heads *taken in isolation*, it is quite possible that the total population expansion as a result of having larger families may exert a downward pull on the living standard of all families involved. The overall impact may be a worsening of the position of *all* families, even though each family benefits from its size given the size of other families.[10] Similar examples of the Isolation Paradox can be found in other inter-family relations, for example, overutilization of land through excessive grazing and the consequent deterioration of land quality and communal assets.[11] Still other examples can be found in problems of taxation, savings, work effort, and incentives; and the general problem indeed has very wide relevance.

Problems of "isolation" provide an argument for "interventionism," as has often been stated. All parties may benefit from the restraint imposed on each against the pursuit of its individual goals, and the state may try to bring about such an improvement through its controlling powers. There are, of course, important problems in determining the nature of control and intervention needed for producing a collusive solution, and also the nature of effective enforcement. But there is a prior and more basic problem, namely, that of selection of one particular collusive solution among many that may be better for everyone than the atomistic outcome. While rules of thumb may be of some use, there are many different rules of thumb, and there is a problem of choice involved here too. Of course, it is possible to impose a simple formula as the basis for finding a "salient" solution, for example, insisting on "no more than two children" in practical population policy (rules of this kind have been tried particularly in China). But there is a substantive issue of fairness and distributional judgment that has to be tackled in taking even the first step toward responding to the Isolation Paradox through public action.

On the other hand, it has been widely noticed that in many real-life situations in which the goals of different members take the form needed for the Isolation Paradox (and the Prisoner's Dilemma) to hold, actual behavior may deviate from pure self-interest and may be more directed toward mutual advantage. Codes of conduct aimed at joint benefits seem to influence individual behavior, even in the absence of formal contracts and enforcement.

Even in experimental games such departure from allegedly "rational" behavior has been systematically observed.[12] It may take the form of acting according to *as if* goals for the sake of mutual benefit, treating joint actions as the relevant instrument of execution, and adopting a less "individualistic" view of the choice of action. There are obvious problems in squaring such conduct with the normal tenets of "rational" behavior. Often this has been done by considering repetition of games over time, with people behaving implicitly cooperatively in search of *future* cooperation from others. There is, however, an argument that this cannot work when the game is repeated a finite number of times.[13] The problem has usually been dealt with in the game-theoretic and related literature through assuming some kind of a "defect" in people's knowledge, understanding, or memory.[14]

This way of solving the problem is rather ad hoc, aside from being quite peculiar for a theory of rationality that insists on informed goal-maximizing logic for other results. If success of achievement depends on failure of knowledge or memory, greater success in knowing or recollecting may bring about failure in results. There is an internal tension in all this *within* the standard model of rationality, aside from doubts as to whether this is why people actually behave the way they do in situations of the Isolation Paradox.[15]

It can be argued that a better approach to understanding social solutions to problems of isolation can be found by making a more radical departure in the characterization of individual behavior and action. I have tried to discuss the nature of the departure elsewhere,[16] but I shall present the main features of the alternative approach briefly. It is helpful to think of the alternative approach as based on three different—but interrelated—leads coming from three thinkers belonging to altogether different traditions, namely, Kant, Smith, and Marx.

Kant's analysis of the categorical imperative is, of course, mainly ethical. The need to "universalize" the basis of individual action is, primarily, a moral demand. The relevance of this moral concept in the present context lies in the fact that a person's action may well be influenced by ethical considerations in addition to prudential ones. The notion of what we "ought" to do is not, of course, the same as what we "will" actually do, but it would be remarkable if the former had no impact on the latter. The problem of isolation and the social failures resulting from it had engaged the attention of classical writers—including Hobbes and Rousseau—a great deal, and Kant's analysis of ethics and the behavior norms derivable from it does address this question.[17]

The second lead comes from Adam Smith's discussion of the role that "rules of conduct" play in saving society from what Smith called "misrepresentation of self-love."[18] Smith argued that by "habitual reflection" we come to accept certain rules of behavior, and do not depart from them even when we can see some immediate advantage from that departure. The approach here is only partly ethical, and the pragmatic issue of the important instrumental role that rules can play is the main point of focus.

The third lead relates to Marx's analysis of the nature of self-interest and perception. In this analysis our own perception of self-interest is in fact a "socially determined" perception.[19] What we see as "our own interest" already includes certain identities with others, and our consciousness of what makes us better off and what we want to achieve incorporates distinctly social elements.

In understanding the empirical observation that problems of "isolation" often get implicitly resolved in actual behavior in many situations, use can be made of Kantian, Smithian, and Marxian departures from the models of "individual rationality" in terms of which much of the discussion of modern economics is conducted. The Prisoner's Dilemma and related problems are just one class of social situations for which these departures are quite crucial. There may be nothing particularly difficult in explaining the committed behavior of individuals in some situations of the Prisoner's Dilemma type if the determinants of individuals' behavior are modified in the ways suggested by Kant, Smith, and Marx. Crucial to this broader view of human behavior is the importance of norms, rules, and perceptions.

Before completing this section, I ought to make two clarificatory statements. First, I do not claim that every situation of the Prisoner's Dilemma type will be implicitly resolved in this "cooperative" way, yielding the right collusive solution, because of the hold of norms, rules, and social perceptions on our behavior. Indeed, that would altogether "over-explain" the nature of the reality under discussion. Many conflict situations of the Prisoner's Dilemma type *do not* get resolved. The role of norms, rules, and perceptions can be important, but there is no guarantee that they will yield the collusive solution in every case. The point, rather, is to note the possibility of the dilemma being resolved in some situations if appropriate norms of behavior, based on rules of conduct or social perceptions, happen to emerge and happen to act powerfully enough. The impasse identified by narrow models of "individual rationality" is disestablished, but there is no guarantee that the difficulty will invariably disappear. Problems of "isolation" can be very deep indeed, and the fact that social values and perceptions try to respond to these challenges does not mean that the problems will, in fact, invariably be solved. For my present purpose, the social responses involved are as important as the nature of the solutions that emerge.

Second, as was discussed earlier, the Prisoner's Dilemma is a very limited type of game, in which there is a unique collusive solution that is superior to the atomistic outcome based on self-interested behavior. Even the *n*-person extension of this game in the form of the "Isolation Paradox" does not entail that feature of uniqueness of the collusive solution. Given the multiplicity of collusive solutions that exist, an important issue is the relation between alternative norms, rules, and perceptions and alternative cooperative solutions that may exist—some more favorable to one and others more favorable to another.

The coexistence of congruent and conflicting elements applies even to the choice between different cooperative solutions. That fact is indeed quite central to the analysis of inequality sustained by value systems, including intrafamily inequalities and sex bias. The value system that leads to implicitly cooperative behavior within a group may well be directed toward a particularly unequal solution in the choice between different cooperative outcomes. Even though the unequal solution may be superior for all parties to the result of fully atomistic and individualistic behaviors, nevertheless one group may systematically receive a lower share of the benefits of cooperation than another. Once the attention is shifted from the resolution of the Prisoner's Dilemma (with a unique cooperative outcome) to the general problems of "isolation" (with multiple collusive possibilities), the question of inequality becomes central.

The general issues raised in this section are pursued in the analysis that follows on the institution of the family and its role in rural development, with particular reference to Indian rural society.

Cooperative conflict and the family

The coexistence of congruence and conflict of interest in family relations makes it tempting to consider the nature of family allocation in terms of the so-called bargaining problem formalized by J. F. Nash.[20] In the classic formulation of this problem, there are two persons with well-defined and clearly perceived interests in the form of two cardinal utility functions. If the two individuals fail to cooperate, the outcome (sometimes called "the status quo position" or "the breakdown position") is one that can be improved for both parties through cooperation. Many cooperative solutions are superior to the breakdown position, but they are not, of course, equally good for both parties. Concentrating only on the "efficient" solutions, if solution A is superior to solution B for person 1, then for person 2, B is superior to A. Even though both parties seek—explicitly or implicitly—a collusive outcome (this is the element of congruence in the relationship), they have strictly divergent interests in the choice among the collusive solutions (this is the element of conflict). Nash proposed a particular solution of the "bargaining problem" as the likely outcome, and other contributors have proposed other solutions.

There have been several attempts at using the framework of the "bargaining problem" to understand the institution of the family.[21] This format has the advantage of tracing the problem of conflict of interest within the family, rather than assuming it away as do models of "household production" and related theories developed by Becker and others. The "bargaining problem" format also has the advantage of combining the conflicting elements with the cooperative ones in a single framework, so that the issues of male–female conflict and related matters do not have to be faced in isolation from the cooperative aspects of family life.

The main difficulty of the "bargaining problem" format as applied to male–female divisions arises from the fact that it gives no role to problems of perception of interest and legitimacy, which can be seen as central to social situations of this type (as discussed in the preceding section). If the notion of self-interest of the individual in a social context is itself "socially determined," that fact has to be incorporated in the formulation of the cooperative conflict itself. With notions of obligation and legitimacy affecting the ethics of behavior (without necessarily taking a fully Kantian form) and influencing actual choices, the perceptions underlying these ideas must be given some room in the formulation of the problem.

It is, in fact, possible to distinguish between some *objective* aspects of a person's interest, and his or her *perception* of self-interest. The latter may incorporate not merely sympathy for others, but also certain senses of "identity" with other members of the group. Insofar as that identity plays an important logistic part in family behavior, it has a status of its own, but perceptions of self-interest based on such an identity may not pay adequate attention to the particular person's own well-being, including such elementary matters as work and leisure, health and morbidity, undernourishment and adequacy of diet.

It is, in fact, reasonable to see a person's well-being in terms of that person's capability of achieving valuable "functionings."[22] A person's "functionings" are his or her "doings" and "being," e.g., eating, sleeping, resting, being entertained, having fun, being happy. The relevant functionings include, inter alia, the most elementary ones, such as being well nourished and avoiding escapable morbidity, but also more sophisticated forms of fulfillment, including psychological and cultural functionings. The *valuation* of these functionings is a matter for reflection and cannot be simply inferred from the prevailing perceptions that pre-reflectively dominate day-to-day actions of individuals. This is, of course, precisely the context in which Marx invoked the notion of "false consciousness" and pointed to an important dichotomy.

It is possible to characterize "cooperative conflict" taking note of this dichotomy. The interpersonal division of "benefits" in the form of functional achievements and the capability to achieve may be influenced by—among other things—the nature of perceptions that prevail in the community about self-interest and legitimacy of behavior. The prevailing atmosphere of cohesive and well-integrated family perspectives—without any perceived combats—may go hand in hand with great inequalities emerging from perception biases. The choice among cooperative solutions may be distinctly unfavorable to a group—women, for example—in terms of objective criteria of functional achievements, without there being any *perceived* sense of "exploitation," given the nature of perceptions of self-interest and conceptions of what is legitimate and what is not.

This is not the occasion to discuss the details of this formal structure of "cooperative conflicts" in a social setting, which I have done elsewhere,[23] but two of the points of departure of this structure from the standard model of the "bargaining problem" may be briefly mentioned here.

Perceived interest response Given other things, if one person's perception of self-interest attaches less value to his or her own well-being, then the collusive solution will be less favorable to that person, in terms of well-being.

Perceived contribution response Given other things, if in the accounting of the respective outcomes, a person is perceived as making a larger contribution to the overall well-being of the group, then the chosen collusive solution will become more favorable to that person.

The inferior position of women in traditional societies may have much to do with perception problems associated with these two elements in the selection among collusive solutions. Consider the issue of self-interest in the particular context of sex inequality in India. The existence of widespread inequality between men and women, and between male and female children, has been observed in terms of various criteria of differential mortality, morbidity, and so on. One of the more stark reflections of that inequality is the simple ratio of females to males (the sex ratio) in the Indian population, which is only 0.93 (i.e., there are only 93 women per 100 men).[24] It has, however, been pointed out that these inequalities are not seen as real inequalities in Indian society, especially in rural Indian society.

Some authors have, in fact, disputed the viability of the notion of "personal welfare" in societies with value systems of the kind that prevail in rural India.[25] It has even been argued that if a typical Indian rural woman were asked about her personal welfare, she would find the question unintelligible, and if she were able to reply, she might answer the question in terms of her understanding of the welfare of the family as a whole. The idea of personal welfare, it has been argued, may itself be unreal in such a context.

This lack of a sense of inequality and of exploitation is undoubtedly an important part of the reality of Indian rural society. On the other hand, it is not at all clear that the right conclusion to draw from this is the nonviability or nonreality of the notion of personal welfare. In fact, this perception bias is a part of the "cooperative" outlook that helps achieve collusive solutions that operate in the family system—collusive solutions that often happen to be also characterized by inequality. Objectively, inequalities often survive by making allies out of the deprived.

This process of "false consciousness" is particularly important in those cases of conflict where the different parties happen to live together, as in the family. The role that nonindividualistic, self-denying perceptions play in the exploitation of Indian women has to be more fully understood in order to appreciate the survival and strength of intrafamily inequalities in rural India. The perception problems are not, of course, immutable. With politicization and educational change, and also better understanding of the objective inequalities between the sexes in India, substantial shifts in perceptions can be expected to come about. To some extent this has been happening already.[26]

In addition to the problem of perception of self-interest, there is also that of perception of individual contributions to family well-being. In the division of labor in which women are often confined to working within the household, while men earn an income from outside, the male productive role may well be perceived as very much stronger than the female contribution. This is, of course, an extremely complex problem, and notions of who is "contributing" how much have many causal antecedents that call for closer scrutiny. However, the "deal" that women get vis-à-vis men is clearly not independent of these perception problems regarding contribution.

In her pioneering analysis of women's role in development, Ester Boserup noted that women appear to fare relatively better in those societies in which they play the major part in acquiring food from outside, for example, in some African regions with shifting cultivation.[27] Outside earning may also play an important role in the contrast within India between the North and the South. It has been noted that in India in the regions where women have little outside earning or "gainful" employment (e.g., Punjab and Haryana), sex disparities are on the whole sharper than in the Southern states, where women have a much greater role in outside activities and earnings. Indeed, even the crude indicator of sex ratio (females per 100 males) is as low as 87 and 88 in Haryana and Punjab respectively, in contrast with the Southern Indian states (96 in Karnataka, 98 in Andhra Pradesh and Tamil Nadu, and 103 in Kerala). There are many factors in the North–South contrast, but outside earnings and activities seem to be an important part of the story.[28]

It can, of course, be argued that activities *within* the household are in no sense less productive than activities elsewhere, since the latter are—in an important way—parasitic on the former. It is only because some people cook and rear children that others can work in the fields or the factories and earn income from outside. Division between "gainful" and other activities is quite arbitrary; there is no way of describing one category of these activities as "unproductive" and the other as "productive," when the latter would be impossible without the former. But the issue is not whether activities within the household are *really* less productive, but whether they are *perceived* as such. And insofar as they are so perceived, that fact itself will have its own influence on the division of benefits within the family.

Various empirical studies at the micro level have indicated that women's status and well-being within the family tend to increase quite sharply when, as a result of economic change, women start having a greater role in outside activities and have outside sources of income.[29]

The importance of outside earnings in incomes in the calculation of "returns" from rearing male and female children has been emphasized in other empirical studies.[30] This consideration of "returns" does relate statistically to differences in child care, especially to the neglect of female children in rural North India.[31] The rewards that elderly members of the family may obtain from surviving children may be particularly influenced by the ability to earn outside income; and the male superiority in that respect, given the market traditions and biases, may thus be influential. Further, social arrangements whereby daughters move away after marriage, and are thus less able to support their own parents, may also be an important factor.

However, the model of "cool, calculating decisions" by parents to neglect female children on grounds of lower rates of "returns" in rearing them has appeared to be unconvincing to many scholars.[32] There is a big issue here concerning the social origins of family behavior. The same statistical observations can also be given other causal explanations, involving social perception of women's contributions and accepted ideas of legitimate behavior. More involvement of men vis-à-vis women in "gainful" activities can affect social perception of what is "due" to men (vis-à-vis women) and to boys (vis-à-vis girls).[33]

Outside earnings can influence intrafamily divisions both directly, by influencing returns to childrearing, and indirectly, through their effects on perceptions of contributions and legitimacy of division of benefits among members of the family. The broad cultural features of different societies reflect traditions of work division and related perceptions. One of the factors that may be relevant in understanding the contrast between Africa and India in the relative survival of women vis-à-vis men is the extent of female participation in so-called gainful activities. India has a much lower sex ratio than Africa, to wit, 93 versus 102. Many factors can be introduced in explaining this difference, including the fact that the reduction of mortality and the expansion of longevity in India have been differentially beneficial to males and females.[34] But the greater role of women in outside economic activities in Africa (especially in sub-Saharan Africa) may also be important in the contrast between Africa and India (and South Asia in general).

There are, however, variations within Africa, and in drawing this contrast between Africa and Asia, regions of Asia other than South Asia would also have to be considered. There are grounds for dividing Africa into Northern and non-Northern regions, and similarly dividing Asia into Southern Asia (including India), Western Asia, and Eastern and Southeastern Asia.[35] In the fivefold division of Asia and Africa,[36] the ranking of life expectancy of females vis-à-vis that of males turns out to be (in descending order): non-Northern Africa, Eastern and Southeastern Asia, Western Asia, Northern Africa, Southern Asia. The ranking of the ratios of outside activity rates of females vis-à-vis males is much the same: Non-Northern Africa, Eastern and Southeastern Asia, Western Asia, Southern Asia, Northern Africa. Only the last pair is reversed in position. The

TABLE 1 Gender differences in life expectancy and female earning activities: Africa and Asia

	Life expectancy ratios (female/male) 1980–85		Activity rate ratios (female/male) 1980	
	Value	Rank	Value	Rank
Non-Northern Africa	1.071	1	0.645	1
Eastern and Southeastern Asia	1.066	2	0.610	2
Western Asia	1.052	3	0.373	3
Northern Africa	1.050	4	0.158	5
Southern Asia	0.989	5	0.336	4

SOURCES: Calculated from United Nations tapes on "Estimates and projections of population" (1985) and from ILO's *Yearbook* (1986) of statistics, and presented in A. Sen, "Africa and India," cited in note 33.

figures for life expectancy ratios, sex ratios, and activity rate ratios are given in Table 1.

It would be a mistake to draw a very clear lesson from this broad interregional contrast of Asia and Africa. However, the importance of female outside activity in the empirical findings is not inconsistent with expectations on the grounds of the theoretical analysis of cooperative conflicts involved in family divisions, especially as mediated through perceptions of contributions and interests.

Concluding remarks

This has been a largely theoretical essay, discussing the problem of cooperative conflicts in the evolution and working of social institutions, and concentrating particularly on the family as an institution. The coexistence of congruence and conflict of interest makes the problem of institutional mediation a particularly complex one in the process of economic development. Benefits to all parties can accrue from the emergence and use of particular institutions, but the division of these benefits calls for systematic investigation. The process of development has to be judged not only in terms of average improvement, but also in terms of inequalities and disparities, and their causal antecedents.

I have argued that institutional models that eschew the problem of conflict within the family (as in some models by Gary Becker and others working in similar traditions) miss something quite central to an understanding of the family as an institution. The absence of *overt* recognition by family members of conflict within the family is, of course, an important fact, but its importance lies in the way this overt sense of "harmony" influences the real *nature* of cooperative conflicts. It is particularly an issue of perception and of perceived legitimacy. Indeed perception bias is itself one of the important parameters in the determination of intrafamily divisions and inequalities.

Models fashioned in the tradition of the "bargaining problem" (pioneered by J. F. Nash) have advantages over the unified objective function approach, making room for congruent as well as conflicting elements in interpersonal relationships. However, they too do not incorporate *perceptions* of well-being and of contribution as explicit variables in the analysis. The approach defended here starts off from the "bargaining problem," but enriches it by bringing in notions of legitimacy related to perceptions of contributions and perceptions of well-being as determining variables.[37] These perceptions and the related norms and rules of conduct are instruments of resolution of action conflicts and are also mechanisms for legitimizing and sustaining unequal resolutions of interest conflicts.

The approach outlined here is a very general one and is particularly concerned with identifying important (and often neglected) variables, rather than specifying particular quantitative hypotheses. I have tried to argue that variables such as perceptions of contributions and legitimacy can have a profound impact on the cooperative conflicts that influence intrafamily divisions and the well-being of family members. The subjective perceptions are among the objective determinants of family behavior and its far-reaching consequences.

Notes

1 McNicoll and Cain, this volume.

2 "It is not from the benevolence of the butcher, the brewer, or the baker, that we expect our dinner, but from their regard to their own interest" (Adam Smith, *An Inquiry into the Nature and Causes of the Wealth of Nations*, 1776; reprinted in R. H. Campbell and A. S. Skinner, eds., Oxford: Clarendon Press, 1976, pp. 26–27). While Smith chose to focus on this aspect of interpersonal relations as far as economic exchanges are concerned, he did not, in the least, ignore conflicts of interest that people have to face in a society and the need to go beyond the pursuit of self-interest to achieve a harmonious and good society. In fact, moving away from the particular issue of economic exchanges, in a more general context Smith argued that while prudence is "of all virtues that which is most helpful to the individual," "humanity, justice, generosity, and public spirit, are the qualities most useful to others" (Adam Smith, *The Theory of Moral Sentiments*, 1790; reprinted in D. D. Raphael and A. L. Macfie, eds., Oxford: Clarendon Press, 1975, p. 189).

3 See particularly Karl Marx, *Capital: A Critical Analysis of Capitalist Production*, 1883; English translation of third edition, London: Sonnenschein, 1887.

4 Gary S. Becker, *A Treatise on the Family* (Cambridge, Mass.: Harvard University Press, 1981), p. 192.

5 For an illuminating account of the various components in these different approaches to family behavior, see A. Deacon and J. Muellbauer, *Economics and Consumer Behavior* (Cambridge: Cambridge University Press, 1980). The underlying issues are clarified further in John Muellbauer's comments on my Tanner Lectures, included in A. Sen, *The Standard of Living*, ed. Geoffrey Hawthorn (Cambridge: Cambridge University Press, 1987).

6 See particularly R. D. Luce and H. Raiffa, *Games and Decisions* (New York: Wiley, 1957).

7 See my "Isolation, assurance and the social rate of discount," *Quarterly Journal of Economics* 81 (1967); reprinted in my *Resources, Values and Development* (Oxford: Blackwell, and Cambridge, Mass.: Harvard University Press, 1984).

8 *Resources, Values and Development*, cited in note 7, pp. 136–138.

9 Some of these issues were taken up in the paper presented by Christopher Bliss at the Eighth World Congress of the International Economic Association in December 1986, and in the discussions that followed its presentation.

10 These issues were discussed and assessed in the *World Development Report 1984*, produced under the leadership of Nancy Birdsall.

11 On this see M. H. Glantz, *The Politics of Natural Disaster: The Case of the Sahel Drought* (New York: Praeger, 1976); G. Dahl and A. Hjort, *Having Herds: Pastoral Herd Growth and Household Economy* (Stockholm: University of Stockholm, 1976). See also my *Poverty and Famines: An Essay on Entitlement and Deprivation* (Oxford: Clarendon Press, 1981), pp. 104–105.

12 See, for example, L. B. Lave, "An empirical approach to the Prisoner's Dilemma game," *Quarterly Journal of Economics* 76 (1962); A. Rapoport and A. M. Chammah, *Prisoner's Dilemma: A Study in Conflict and Cooperation* (Ann Arbor: University of Michigan Press, 1965).

13 The argument is simple enough. Any interest that a person has in behaving "well" must (within the tenets of standard rationality models) arise from the effect that such good

behavior may have on the future behavior of others. Obviously such an argument cannot give a reason for behaving "well" in the final round, since there is then no point in trying to influence the conduct of others (there being no further rounds). But if people are not going to be influenced in their behavior in the last round, there obviously is no argument for behaving "well" in the last-but-one round for the sake of influencing future behavior of others. Thus in the last-but-one round also, no one has any reason to behave well, within the tenets of standard rationality. This argument can be extended backwards all the way to the first round, and no one ends up having any incentive—in this reasoning—for behaving "well" in any round whatsoever.

14 It is sometimes assumed that people do not know how many times the game will be played. Another possible assumption is that each player thinks that the other player may, in fact, enjoy cooperating (even though he or she really does not). Another assumption takes the form of a limited memory of each player in recalling what happened. See, for example, R. Axelrod, *The Evolution of Cooperation* (New York: Academic Press, 1984); D. Kreps, P. Milgrom, J. Roberts, and R. Wilson, "Rational cooperation in finitely repeated Prisoner's Dilemma," *Journal of Economic Theory* 27 (1982); R. Aumann and S. Sorin, "Cooperation and bounded rationality," mimeographed, 1986; Kaushik Basu, "Modeling finitely-repeated games with uncertain termination," *Economic Letters* 23 (1987).

15 On this general issue, see my *On Ethics and Economics* (Oxford: Blackwell, 1987), Lecture 3.

16 *On Ethics and Economics,* cited in note 15, Lecture 3.

17 Immanuel Kant, *Fundamental Principles of Metaphysics of Ethics* (1785); English translation by T. K. Abbot (London: Longman, 1907).

18 Smith, *The Theory of Moral Sentiments,* cited in note 2, p. 160.

19 See particularly K. Marx and F. Engels, *The German Ideology* (1845–46); English translation (New York: International Publishers, 1947).

20 J. F. Nash, "The bargaining problem," *Econometrica* 18 (1950); Luce and Raiffa, *Games and Decisions,* cited in note 6.

21 See, among others, M. Manser and M. Brown, "Marriage and household decision-making: A bargaining analysis," *International Economic Review* 21 (1980); M. B. McElroy and M. J. Horney, "Nash-bargained household decisions: Toward a generalization of the theory of demand," *International Economic Review* 22 (1981); A. K. Sen, "Women, technology and sexual divisions," *Trade and Development* (UNCTAD) 6 (1985).

22 See my "Well-being, agency and freedom: The Dewey Lectures 1984," *Journal of Philosophy* 82 (1985); *Commodities and Capabilities* (Amsterdam: North-Holland, 1985); *The Standard of Living* (Cambridge: Cambridge University Press, 1987).

23 See my "Women, technology and sexual divisions," cited in note 21; and "Gender and cooperative conflicts," in Irene Tinker, ed., *Persistence Inequalities: Women and World Development* (New York: Oxford University Press, 1990).

24 See my *Commodities and Capabilities,* cited in note 22, Appendix B. Among the many other pertinent investigations of women's status are Barbara Miller, *The Endangered Sex: Neglect of Female Children in Rural North India* (Ithaca, N.Y.: Cornell University Press, 1981); Pranab Bardhan, *Land, Labour and Rural Poverty: Essays in Development Economics* (New York: Columbia University Press, 1984); J. Kynch and A. K. Sen, "Indian women: Well-being and survival," *Cambridge Journal of Economics* 7 (1983); Devaki Jain and Nirmala Banerjee, *Tyranny of the Household: Investigative Essays on Women's Work* (New Delhi: Vikas, 1985).

25 For a particularly illuminating presentation and defense of this point of view see Veena Das and Ralph Nicholas, " 'Welfare' and 'well-being' in South Asian societies," American Council of Learned Societies/Social Science Research Council Joint Committee on South Asia (New York: SSRC, 1981).

26 One simple example of this relates to the issue of health perceptions of men and women. In a health survey taken in Singur, near Calcutta, in 1944—the year after the Bengal famine —the All-India Institute of Hygiene and Public Health reported certain re-

markable differences in the perceptions of ill health by men and women, e.g., widows having much lower perceptions of being ill than widowers. This can, of course, be a reporting hesitation, but that issue itself relates to a certain view of the legitimacy of particular perceptions. The situation, however, has changed over the years. The latest surveys in Singur do not repeat the same biases. Indeed, in neighboring Calcutta, women's self-perception of greater ill health is systematically reported. On these questions see my *Commodities and Capabilities,* cited in note 22, Appendix B.

27 Ester Boserup, *Women's Role in Economic Development* (London: Allen and Unwin, 1970).

28 On this question see Bardhan, *Land, Labour and Rural Poverty,* cited in note 24; and Pranab Bardhan, "On the economic geography of sex disparity in child survival in India: A note," mimeographed, University of California at Berkeley, 1987.

29 For example, Zarina Bhatty, "Economic role and status of women: A case study of women in beedi industry in Allahabad," ILO Working Paper, 1980; Maria Mies, *The Lace Makers of Narsapur* (London: Zed Press, 1982); M. F. Loutfi, *The Rural Women: Unequal Partners in Development* (Geneva: ILO, 1980). See also other empirical works referred to in my "Gender and cooperative conflicts," cited in note 23.

30 See, particularly, Mark R. Rosenzweig and T. Paul Schultz, "Market opportunities, genetic endowments, and intrafamily resource distribution: Child survival in rural India," *American Economic Review* 72 (1982); J. Berhman, "Intrahousehold allocation of nutrients in rural India: Are boys favoured? Do parents exhibit inequality aversion?," mimeographed, forthcoming in *Oxford Economic Papers.*

31 See Miller, *The Endangered Sex,* cited in note 24; Monica Das Gupta, "Selective discrimination against female children in rural Punjab, India," *Population and Development Review* 13 (1987). The discrimination against female children is particularly strong for *second* and *later* daughters, Das Gupta shows.

32 See N. Krishnaji, "Poverty and sex ratio: Some data and speculations," *Economic and Political Weekly* 22 (1987); Alaka Basu, *Culture, the Status of Women and Demographic Behaviour* (New Delhi: National Council of Applied Economic Research, 1988).

33 On this question see my "Gender and cooperative conflicts," cited in note 23. See also Bina Agarwal, "Social security and the family," in E. Ahmed et al. (eds.), *Social Security in Developing Countries* (Oxford: Clarendon Press, forthcoming).

34 Views differ as to whether or not this is the standard pattern. A contrary view has been presented by Samuel H. Preston: "It is clear that the frequency of systematically higher female mortality . . . declines monotonically as mortality levels improve" (*Mortality Patterns in National Populations,* New York: Academic Press, 1976, p. 121). On the opposite side, see, for example, S. Ryan Johansson, "Deferred infanticide: Excess female mortality during childhood," mimeographed, University of California, Graduate Group in Demography, Berkeley, California, 1983. Tim Dyson differentiates between a phase of mortality reduction and longevity expansion in which males benefit relatively more, and an equalizing process from which both sexes gain (see his "Excess female mortality in India: Uncertain evidence on a narrowing differential," mimeographed, London School of Economics, 1987). Certainly, the sex ratio in India has systematically fallen since the beginning of this century, reducing the number of females per 100 males from a reported 97 in 1901 to 93 in 1971. The sex ratio now seems to be slowly rising.

35 The motivation underlying the divisions is discussed in my paper "Africa and India: What do we have to learn from each other?," C. N. Vakil Memorial Lecture at the Eighth World Congress of the International Economic Association, December 1986; published in the proceedings, edited by Kenneth J. Arrow, *The Balance Between Industry and Agriculture in Economic Development,* vol. 1 (*Basic Issues*) (London: Macmillan, 1988).

36 China, however, is excluded from these categories. The Chinese experience has many special features, on which see Jean Drèze and Amartya Sen, *Hunger and Public Action* (Oxford: Clarendon Press, 1989), Chapter 11.

37 For a less hurried description of this alternative approach, see my "Gender and cooperative conflicts," cited in note 23, and Drèze and Sen, *Hunger and Public Action*, cited in note 36. The influences of variables other than outside activity (such as literacy and female education) on cooperative conflicts are also discussed there. One of the interesting cases is the state of Kerala in India, which has a female-male ratio higher than 1.03 and strong female advantage in life expectancy, similar to that in Europe and North America. Kerala is distinguished from the rest of India in having a long tradition of high literacy (including high female literacy), an active public health service, and a matrilineal inheritance system for some of the communities in the state. The big push toward high literacy and female education was initiated in parts of what is now Kerala by Rani Gouri Parvathi Bai, the ruling queen of Travancore (a "native state," formally outside British India), as early as 1817. These and related matters are discussed in *Hunger and Public Action*, Chapters 4, 11, and 13.

Land Reform and Family Entrepreneurship in East Asia

SUSAN GREENHALGH

AS A SOLUTION TO THIRD WORLD DEVELOPMENT PROBLEMS, land reform[1] is an idea whose time many observers believe has passed. In Latin America, Africa, the Middle East, and South Asia, land reforms implemented in the 1950s, 1960s, and 1970s have failed to produce the expected gains in equity and productivity.[2] While there is consensus that concentrated ownership promotes neither efficiency nor equity, and that in most settings small farms are better than large ones (Berry, 1984), due to insufficient "political will," most reforms have not been implementable. Even where land reforms are successfully carried out, experience has shown that, by themselves, they can accomplish very little. Since land reform programs are interdependent with other policy sectors, any given reform turns out to be no better than the policy environment in which it is carried out. Reflecting the current disillusionment with reform, John Montgomery has observed: "What happens in these policy areas [pricing, foreign trade, etc.] during and after land reform may be more important to underlying objectives of productivity and equity than the reform itself" (1984, p. 224).

What enthusiasm for land reform remains appears to be based largely on a handful of "success stories," all in the market economies of East Asia (and even these reforms have their detractors, e.g., Apthorpe, 1979). Irma Adelman, probably today's leading academic proponent of the view that land reform is a precondition for development benefiting the majority, bases her assessment of the positive economic and demographic consequences of land reform on the cases of Taiwan, Japan, and South Korea (Adelman, 1980, 1985; Adelman and Robinson, 1978).

The celebrated Taiwan reform of 1949–53 is credited with producing a major leveling of the distribution of income and wealth; large incentive-induced rises in agricultural productivity; measurable declines in unemployment due to intensification of labor inputs on the farm and nonmigrant transfer of surplus labor to off-farm jobs; and a stimulus to industrial development caused by rising demand for goods by the farm population and substantial net capital outflow from agriculture (Fei, Ranis, and Kuo, 1979; Ho, 1978; Koo, 1968; Lee, 1971; Thorbecke, 1979; Yang, 1970).

Scholars have been much more cautious in their assessment of the economic benefits of the 1945–52 mainland Chinese land reform. While agreeing on its critical political benefits for the regime (Perkins and Yusuf, 1984; Riskin, 1987; Shue, 1980; Wong, 1973), sinologists have downplayed its economic consequences. At most they have noted short-lived effects on distribution and a modest but important impact on development finance (Lippit, 1974; Moise, 1983; Perkins and Yusuf, 1984; Wong, 1973). The scholarly consensus is that only the political fruits of the reform were of lasting import; any economic or demographic outcomes were obscured or effaced by the process of collectivization that followed on its heels in 1955–56.

By contrast, the current rural reforms on the mainland, under way for a decade, have set off a developmental spurt not dissimilar to that witnessed on Taiwan in the 1950s and 1960s. The mainland reforms, which have returned the land to peasant family control (though not ownership), are hailed as a "second land reform" by the peasants. If, following the peasants, we consider the current reforms a sort of land reform, then the second mainland Chinese reform may sit alongside the Taiwanese, South Korean, and Japanese reforms, possibly as models, but certainly as cases worthy of re-examination. The first mainland reform also deserves another look, for it is possible that it produced significant changes in microlevel units that are not perceptible to the macroeconomic analyst.

Why have the East Asian land reforms succeeded in promoting sometimes rapid and equitable development, while reforms elsewhere too often have failed? To begin with, these reforms were successfully implemented, a notable feat in the post–World War II era. However, East Asia does not have a monopoly on "political will," as the implementation records of Kerala, Iran, and Egypt attest. A second hypothesis is that complementary agricultural policies, perhaps as important as the land reforms themselves, were different. While this argument is plausible for Taiwan and Japan, whose governments provided substantial assistance to farmers, it is less true of Korea, and even less true of the second mainland Chinese reform. Far from supporting the peasantry, the Chinese government now expects localities largely to finance their own social and economic development.

In this essay I advance a third hypothesis. I argue that the East Asian reforms generated relatively rapid economic growth and comparatively smooth demographic modernization in large part because of the micro-institutional context in which they were introduced. A similar argument could be developed for their equity effects; given space limitations, however, I do not treat distributional issues here.

Although the indigenous microsocial institutions of Chinese, Japanese, and Korean societies exhibit important similarities, the differences in such features as family structure are large enough to influence macrodemographic and economic outcomes. For this reason I confine attention to the two cases of Sinic culture, Taiwan and the Chinese mainland. Inclusion of Taiwan and the People's Republic within a single framework offers obvious analytic opportunities. A finding that indigenous microsocial institutions promote similar kinds of development in two such different policy environments would constitute a strong case for the resilience and contemporary importance of such institutions.

In capsule form the argument I will develop runs as follows: A major legacy of the Chinese imperial order was a society of profit-minded peasant entrepreneurs whose cultural toolkit contained a highly sophisticated set of socioeconomic mobility strategies. A second legacy was a highly (on the mainland) and relatively (on Taiwan) stratified class structure in which only a tiny minority could form demographically complete families capable of successfully pursuing mobility goals. The landless and land-poor majority faced constant threats to survival and shrinking capacity to maintain any family at all. By removing the class basis of peasant insecurity, the contemporary land reforms democratized family organization, making it widely available to the bulk of the population. While the land reforms redistributed the tangible means of upward mobility, at the same time they transformed the nature of the mobility ladder, removing status and profits from agriculture, and transferring them to commerce and industry. Where the post-reform environment provided certain conditions encouraging the expansion of family mobility strategies, the result was a virtual explosion of family entrepreneurship that, in turn, promoted a certain pattern of macroeconomic and demographic change. This pattern involved relatively rapid economic growth and diversification, rural–urban balance, and a comparatively rapid fertility decline in response to changes in the requirements for family socioeconomic success.

One problem in a study such as this is delineating the appropriate temporal frame for analysis. The difficulty lies in the fact, alluded to above, that land reforms are intertwined with other changes, so that the effects of the reform cannot be extricated from the effects of other policies and developments. The standard solution calls for restricting attention to short-term effects that occur, roughly, within the first decade. However, if we agree that social institutions might importantly mediate the effects of land reforms, then we must allow time, first, for the social-institutional changes to occur, and then for the institutionally mediated changes in economic and demographic behavior to work themselves out. Carving out a longer piece of historical time, of course, runs the risk of claiming too much for land reform. My solution is to consider the reform and its policy environment an indivisible package that produced a certain set of effects, some traceable to the reform itself, others attributable to the combination of reform and policy climate. Examining the impact of the reform alone would be informative but uninspiring: the most interesting and important consequences stem from the joint operation of the two.

A micro-institutional approach to the effects of the East Asian land reforms has few if any precedents in the literature. My ideas on the subject derive in part from four very disparate literatures that I have come to consider crucial to an appreciation of the long-term consequences of these reforms. These are the economic and political literature on the nature of traditional peasant society, feminist research on the effects of development on patriarchal family structures, economic and anthropological work on family entrepreneurship in developing countries, and, finally, sociological and anthropological studies of the role of the family in economic development. These literatures clearly take us some distance from land reform and agriculture. The shift is deliberate: in my view the developmental significance of the East Asian land reforms lies not in agriculture at all, but in commerce, industry, and the phoenix-like rise of the Pacific Rim economies.

This essay has six sections. The first elaborates briefly on these four bodies of literature in order to extract crucial pieces of the argument. The second then outlines that argument in as succinct a fashion as possible. The following three sections explore the economic and demographic consequences of the Taiwan land reform and the two mainland reforms. While the conventional effects—on distribution, productivity, and so on—are not unimportant, they are adequately treated elsewhere.[3] In this essay I concentrate exclusively on economic and demographic effects that are mediated through microlevel, largely familial, institutions. As family-level data from these time periods are scarce, the arguments developed here should be seen not as empirically substantiated conclusions, but as hypotheses for further testing. A concluding section spells out some implications for theory and policy.

Toward a micro-institutional view of the effects of the Chinese land reforms

Existing studies of the effects of the Chinese land reforms of the late 1940s/early 1950s are quadruply thin. (The following critique does not cover analyses of the current reforms on the mainland because they have not been treated as a land reform.) *Historically*, they tend to scrutinize only a narrow slice of post-reform time. Underlying this skimpy temporal coverage are the assumptions that preexisting social structures are irrelevant, and that post-reform changes occurring decades later are unrelated to the reform. *Sectorally*, existing studies concentrate primarily on the rural and agricultural, assuming that the urban, commercial, and industrial sectors remain relatively unaffected by the reforms. While some fine studies detail capital outflows from agriculture (e.g., Lee, 1971; Lippit, 1974), these works take a narrow (read: exclusively macro and formal) view of the channels through which land redistribution affects commercial and industrial growth. *Socially*, current research fails to notice the peasant nature of the host society, assuming that, as in modern industrial society, in traditional agrarian society economic behavior can be extricated from social structure. (This criticism does not apply to the work of political scientists, such as Shue, 1980.) *Causally*, in the existing literature the mechanisms through which land reform produces its effects remain relatively obscure. Much economic work consists of neat before-and-after or large-and-small (farm size) studies, which blithely ignore what happens to the messy institutional stuff of the society during the process of reform itself. In the implicit causal model generally adopted, economic effects flow directly from the reform and its constituent policies. Social institutions are conveniently black-boxed; the way they are stirred up by the reform has no effect on the nature of the economic outcomes.

A number of social science literatures challenge these assumptions. In the following sections I briefly review some of the arguments of four such literatures. These four were chosen because they contain kernels of a new, more institutionally sensitive approach to the problem.

The nature of peasant society

Research on traditional peasant societies highlights the intimate connection between economic and social life in such populations. The basic cell of both the economy and society is the peasant family; hence, this literature implies, the peasant family must form the starting point of any analysis of the effects of reforms introduced in such societies. Access to land, the economic foundation of peasant life, critically affects family social and economic organization. This research thus suggests that a land reform will profoundly affect microsocial structure and strategy, which in turn will affect later patterns of macroeconomic and, of course, demographic change.

While there is little disagreement among students of the peasantry on the close relation between peasant society and economy, authors are sharply divided on the question of peasant motivation. Some argue that peasant behavior is guided by a subsistence ethic operating in a moral economy whose cardinal rules are risk avoidance and safety first (Scott, 1976; also Chayanov, 1966). Others perceive a mobility ethic motivating economically rational peasant actors who are highly responsive to prices, supply and demand, and costs and returns (Popkin, 1979; Schultz, 1964).

Research on Qing imperial (1644–1911) and Republican period (1911-49) Chinese peasants suggests that this is a false dichotomy. Chinese peasants were motivated by both security and mobility goals; which goal dominated depended on the level of economic well-being, both that of the family and that of the time and place in which the family lived (Huang, 1985; Skinner, 1971).

Extended to the arena of land reform, these insights direct our attention to the distribution of security and mobility motives before the reforms—which boils down to the class structure—and the redistributive impact on those motives achieved by the reforms. More specifically, given the significance of land as both a security and mobility resource, we can hypothesize that the redistribution of land is in effect a redistribution of security and mobility goals. The extent of the goal transformation depends on the class structure before the reform and the amount of land distributed during the reform. The significance of whether most peasants are pursuing security or mobility will emerge below in the discussion of the literature on family entrepreneurship.

Effects of development on patriarchal structures

Feminist research on the effects of development on women highlights perhaps better than any other literature the critical importance of indigenous social and cultural institutions in shaping the consequences of such changes as commercialization, industrialization, and land reform programs.[4] The key finding is that, in many areas of the world, colonial and capitalist institutions did not simply destroy, but rather preserved, blended with, and built on precapitalist institutions in ways that were often inimical to women (Hartmann, 1976; Leacock and Safa, 1986; Nash and Fernandez-Kelly, 1983; Wellesley Editorial Committee, 1977).

The key study in the Chinese context is Judith Stacey's work (1983) on the consequences for mainland Chinese women of the 1945–52 land reform and succeeding policies adopted by the Chinese Communist Party (CCP). Stacey argues that by redistributing land to peasant patriarchs, whose social and economic power had been undermined by a century of agrarian crisis, the CCP land reform democratized family organization, making available to a large number of peasants the economic means with which to form the culturally ideal patriarchal families. By "neo-traditionalizing" social structure, the land reform reinforced traditional patriarchal structures and prerogatives, with deleterious consequences for women.

Stacey's study is particularly stimulating because it forces us to see that, from a social standpoint, what is redistributed during a land reform is not land so much as the rights land confers: patriarchal privilege and family formation. While Stacey is primarily concerned with the implications of a preserved patriarchal family for gender relations within the family, her argument can be extended to embrace effects of reform-induced changes on the socioeconomic structure and strategies of the family as a whole.[5] Given similarities between the late 1940s/early 1950s mainland and Taiwan reforms, this argument can also be extended to Taiwan. For both the PRC and Taiwan, we can hypothesize that the land reforms democratized family organization, vastly multiplying the number of units able to pursue whatever security and mobility strategies the post-reform environment permitted.

Family entrepreneurship

The literature on family entrepreneurship in developing countries casts doubt on the assumption that the effects of a land reform might be experienced only, or even primarily, by the rural and agricultural sectors. This literature underscores the *familial basis* of much Third World entrepreneurship, and the *rural bases* of many industrial and commercial entrepreneurs (Balmori, Voss, and Wortman, 1984; Strachan, 1976; Tripathi, 1981; White, 1974). For landed families with surplus to spare, the development of commercial and industrial establishments is often part of a deliberate diversification strategy aimed at spreading risks and expanding profits. Latin American (Balmori, Voss, and Wortman, 1984) and South Asian (Tripathi, 1981) case studies suggest that, in examining the economic and demographic effects of a land reform, we should follow members of landed families to the cities. There we are likely to find them running shops, factories, and even multi-company business groups.

Of particular utility in the Chinese context is Myron Cohen's (1976) work on petty tobacco farmer-entrepreneurs on Taiwan. Cohen's careful study of family dynamics shows how, when economic conditions permit, families with a secure land base develop explicit mobility strategies stressing demographic expansion, spatial dispersal, and sectoral diversification.

The entrepreneurship literature also indicates the importance of "getting the environment right." Families-qua-enterprises are vulnerable both to market fluctuations and to state manipulation and control. In mid-1960s Taiwan, studied by Cohen, the optimal environment seemed to be one in which a market was assured, and the state provided the necessary infrastructure, then went away.

The role of the family in economic development

While a previous generation of scholars viewed the family as an obstacle to economic development (e.g., Levy, 1949), in the past decade researchers have recognized the positive and causal role families may play in determining the nature and pace of economic development. This literature argues that families are most usefully viewed as intermediate "variables," which intervene between larger economic and political forces, on the one hand, and their demographic and economic outcomes, on the other. Among the best known in this context is Tamara Hareven's work (1982) on the role of "family time" in influencing the pace of industrial time in a late nineteenth/early twentieth century New England community.

The idea that the family might play a positive developmental role has only a handful of adherents among students of Chinese development. In the PRC context Stacey's book, introduced earlier, presents the first and most clearly articulated argument for the causal role of the family in constraining the policy choices of the Communist Party. In the literature on Taiwan my own research highlights most explicitly the contribution of strong corporate families to the island's record of rapid and equitable development (1984, 1988a; this theme is intimated, though not elaborated, in Myers, 1984). Those essays suggest that the family's net contribution was positive, especially during the initial stages of

Taiwan's industrialization in the 1960s and 1970s, but possibly also during the shift to more technology- and capital-intensive development in the 1980s. In the present essay I slightly reformulate this argument, paying more attention to the policy environment, and extend it to the PRC.

Together these bodies of literature suggest that the larger developmental significance of the Chinese land reforms is likely to emerge only if we broaden our temporal and sectoral frameworks and admit social, especially familial, factors into the domain of analysis. Incorporating insights from these bodies of research, in the next section I develop a micro-institutional, that is, family-mediated, view of the economic and demographic effects of these three reforms.

The argument

The argument I wish to advance is a complex one that, given space limitations, I can present only in skeletal form. Further work is needed to flesh out the details and to determine the extent of regional variation in applicability. The argument has three parts, dealing with the nature of the traditional socioeconomic order, the impact of the land reforms and post-reform environments on the peasant family, and the effects of family entrepreneurship on economic and demographic change.

The traditional socioeconomic order

A classic peasant society, late imperial China was a family economy, made up of countless small, independent production units, each with a family at its core.[6] "The family" was a highly complex organization:

> A Chinese family, especially in rural China, is far more than a group of related individuals. . . . It is a complex organization of family members, family property, domestic animals, family reputations, family traditions, and family gods. (Yang, 1945, p. 45)

Closer observation reveals that there was not one, but two families, one a trans-historical male fraternity concerned with inheritance and succession, the other a contemporaneous group of males and females concerned with daily subsistence and intergenerational reproduction. Although the Chinese considered both units their families, for analytic clarity we can call the first a descent line, and the second an economic family, or *jia*.

The descent line included male ancestors, living males, and male descendants, all united by the joint ownership of the ancestral estate, usually land and a house, but also other properties acquired over the generations. In exchange for "the gift of life," living members of the descent line had a sacred duty to preserve and, if possible, enlarge the patrimony for transmission to the next generation. While onward transmission of the estate ensured continuity of the family line into the future, and enlargement of the patrimony brought socially rewarded "glorification" of the family line, failure to transmit property to one's sons constituted a failure to live up to one's obligations that caused loss of face among kin and community alike (Yang, 1945). Thus, the dynamics of descent line perpetuation, coupled with the arithmetic of reproduction—the more sons, the more property one needed to accumulate in order to transmit the amount inherited from one's father to each heir—appear to have provided a built-in, morally sanctioned drive toward property accumulation.

Crosscutting the descent line was the *jia*, a corporate kin group of male descent line members and females related by blood, marriage, or adoption. A contemporaneous group, the *jia* was concerned with securing its subsistence and, whenever possible, improving its fortunes. Internally, the *jia* was stratified along lines of sex, generation, and age. Differentials of power and status notwithstanding, the *jia* was a relatively cohesive unit. Underlying this cohesiveness was a set of intergenerational "contracts," or mutual understandings, about the reciprocal rights and duties of each family member.[7] In this exchange parents were expected to provide their children with support before marriage; training; help in locating work and spouse; and, for sons, portions of the family property at the time of family division. In return, children were obliged to contribute to the family economy, and sons were also required to support their parents in old age. The strength of this package of rights and duties stemmed in part from the irrefutable rationality of its internal logic—namely, all investments deserve returns. Adherence to family expectations was also reinforced by the lack of "exit" available to unhappy individuals; as we shall see shortly, each family was a near-fortress unto itself. Strongly stressed by the culture, these rights and duties provided a glue binding family members together.

The family (*jia*), a relatively enduring kinship unit, was distinct from the household (*hu*), a less permanent residential unit made up of any family members (and occasionally also nonfamily members) who happened to live together at a given time. Because of the strength of familial obligations, members living elsewhere remained closely tied to their sending households through flows of material goods and shared expectations about fulfillment of future obligations.

If intrafamilial ties were strong, interfamilial connections were relatively weak. Although peasant families occasionally joined together to exchange labor, or share draught animals and farm tools, such mutual aid arrangements were generally seasonal, lasting only a matter of days or weeks (Shue, 1980, pp. 149–153). Given the presumed unreliability of those not bound by familial obligations, peasant families sought to keep as many activities, and especially the all-important economic activities, as close to home as possible. The result was a situation aptly described as "family atomism." Leon Stover explains it this way:

> If folk culture is family centered, it is because kinship is the only defense against mutual suspicion. Each family is necessarily thrown into competition with every other owing to the fixed availability of land within the Green Circle. . . . Villagers

> are acutely aware that any gain in property by one family is another's loss. (1974, p. 105)

Whatever the causes (Stover oversimplifies here), the effect was a thick social wall separating the untrustworthy world outside from the reliable circle of family and kin inside.

As it adapted to the vicissitudes of dynastic and market cycles, the *jia* developed a set of highly effective and remarkably resilient mobility strategies. Among these strategies four are particularly important for my argument. These are property accumulation (directed primarily at land), economic diversification (into nonagricultural undertakings), worker dispersal (ideally to higher central places), and family expansion (through the addition of children and formation of complex families). Each of these strategies defined a continuum, or ladder, from less to more ideal. Although they served poor families as well as rich (e.g., landless laborers were often forced to send members they could not support elsewhere to find work), in general there was a close and direct connection between class status (or access to land) and location on these mobility ladders.

Through frequent interaction with the market,[8] Chinese peasants developed an "economic rationality" of Schultzian proportions. Chinese peasants were not only attuned to supply and demand, costs and returns. They were also skilled managers accustomed to operating in a world where contracts, credit cooperatives, joint-share enterprises, and an array of other sophisticated economic institutions were commonplace (Cohen, 1970; Myers, 1972). One major legacy of the imperial period, then, was a society of economically relatively sophisticated, profit-oriented actors—in short, a society brimming with entrepreneurial potential.

A second, less fortunate legacy was a highly or, in Taiwan, and parts of the mainland (in particular the north), moderately stratified class structure, and thus unequal distribution of the ability to use these entrepreneurial strategies and skills. On the eve of the Chinese land reforms land in many areas was concentrated in the hands of a small landlord class, which alone among rural classes was able to pursue wealth and status through forming large, dispersed, diversified families. For the landless and land-poor, life was a constant struggle for security, at times even survival, in which problems of land shortage and lack of food were exacerbated by class exploitation. Economic failure was reflected in demographic failure and growing inability to maintain family life. The resulting "incomplete" families lacked the full complement of members and, consequently, were unable to fully deploy the strategies of accumulation, dispersal, and diversification.

Effects of the land reforms and post-reform environments on the peasant family

The land reforms of the 1940s/1950s and 1970s/1980s had three profound effects on this social order. One touched on security, while two involved peasant mobility.

First, through a variety of measures that undermined the power of landlords, and through the redistribution of land, itself an important safety net, the two early reforms *redistributed basic economic security* throughout the Taiwanese and mainland Chinese countryside. Previously extreme insecurity had forced large numbers of peasants to make do with incomplete or minimal family structures. By guaranteeing economic security to the peasant masses, these reforms democratized family organization, making the ability to form demographically complete families widely available to the bulk of the population.

Because land, a tangible asset, was a critical mobility resource, the land reforms also *redistributed the means for upward socioeconomic mobility*. At the same time, however, they *transformed the nature of the mobility ladder*. Through mechanisms both political and economic (detailed in the following sections), these reforms removed profit and status from agriculture, transferring them to commerce and industry.

The land reforms only distributed the *potential* for upward mobility. Whether newly endowed peasant families would be able to act on this potential and activate traditional mobility strategies depended crucially on the political and economic environment in which the reforms were carried out. In order for families to fully exercise their entrepreneurial skills, several conditions were necessary. These can be called the two guarantees and the four permissibles, or four encouragements. The two guarantees are respect for private ownership of property and the guarantee that families may control their economic affairs.

The four permissibles are lack of political restrictions on use of the four traditional mobility strategies outlined above. More specifically, these conditions exist when families may freely accumulate assets, freely diversify into any line of economic activity, freely migrate to wherever opportunities are bright, and freely determine their family size and structure. These activities may be politically *permissible*, but not easily *accomplishable* because of limits on available infrastructure, markets, or opportunities (for instance, inadequate transportation facilities may deter migration). Thus, we may say that these activities are to different degrees *encouraged* by economic conditions.

Applying these notions to the three Chinese reforms, I argue that the political-economic environment constructed after the first mainland reform stifled and eventually snuffed out peasant entrepreneurship. The context in which the second mainland reform has been carried out has been conditionally permissive, stimulating but also constraining the development of peasant entrepreneurship. Finally, the climate in which the Taiwan reform was introduced was both permissive and encouraging, and provided near-ideal conditions for the full flowering of Chinese family entrepreneurial activities.

Effects of family entrepreneurship on economic and demographic performance

Where the post-reform environment permitted or encouraged the development of family enterprises, these units fostered a distinctive kind of economic and demographic development, one marked by rapid economic growth and diversification, spatial balance, and relatively rapid fertility decline in response to changes in the costs and benefits of children.

The mainland Chinese view This argument takes on particular force because while it is relatively easy to support for "capitalist" Taiwan (Greenhalgh, 1984), it has recently been confirmed by events in "socialist" China. After suppressing sprouts of family entrepreneurship for 25 years, in the late 1970s China's leaders began to recognize the economic benefits of family-style development. In the past few years the view that family entrepreneurship can spur rural economic growth has been increasingly elaborated and now forms part of the official rationale for fostering the development of what is variously called the "individual economy," "household management," "household industry," the "peasant self-employed economy," "private enterprise," or the "family enterprise." Before setting out my own more formal version of the argument, it might be instructive to consider the nature of the arguments being advanced on the mainland. This discussion illuminates particularly well the strengths of the family economy—which probably stand out most sharply when contrasted to a collective economy—and it provides a rare glimpse into the implicit assumptions held by Chinese policymakers about the nature of the Chinese peasant economy.

According to Chinese writers[9] the family economy is a supplementary part of the larger socialist economy whose defining characteristics are as follows:

> The peasant self-employed economy is an economic sector in which laboring peasants use their private means of production to carry out production and management activities, operate as independent accounting units and shoulder sole responsibility for their profits and losses. It is of the nature of individual economy. (Wei and Zheng, 1985, p. 71)

Family enterprises are considered the main force behind the rapid development and diversification of the rural commodity economy. They are considered particularly effective in areas where large-scale, modern firms and supply and marketing cooperatives cannot effectively operate, because they "fill the blanks in the economic system." Their good economic results in turn elevate peasant incomes and raise tax revenues for the state. In addition to boosting revenues, family firms are said to facilitate development by absorbing and effectively utilizing surplus rural labor, pooling idle and scattered funds (accumulated since the introduction of the household responsibility system), accelerating the pace of rural town construction, and facilitating the popularization of science and technology.

The advantages of the family enterprise are traced to three features of its organization: incentive structure, entrepreneurial talent, and small size.[10] With respect to incentives, because family firms combine responsibility, authority, and reward, they are said to give play to peasant initiative. Heightened initiative, in turn, gives rise to enhanced efficiency and labor productivity. Since the self-employed economy directly advances peasants' material interests, peasants "do not grudge time and effort, make plans carefully, get up early and go to bed late, and do anything they can to reduce their expenditures and production costs" (Wei and Zheng, 1985, p. 69). Peasant economic diversification and townward migration are also explained as byproducts of newly released initiative:

> Their strong desire to continue prospering motivates them to divert capital and labor from traditional, unitary agricultural activities to other lines of production such as industry, commerce, construction, mining and transportation. (*Anhui Ribao*, 1985, p. 93)

The peasant economy also contains a great deal of entrepreneurial talent, the second factor cited by mainland writers to explain its vitality and contributions to rural development. As a *People's Daily* article put it, entrepreneurs are those who have a "strong sense of commodity." What this means is that they have a "concept of seeking material interest," which leads them to pay close attention to input, output, results, and accounting; and they have a "spirit of developing and blazing new trails," stemming from contact with market competition, which leads them to strengthen their information base, improve their technology, and so forth (Ma, 1987). Peasant entrepreneurs are particularly skilled at making much out of virtually nothing, a great advantage in China's resource-short countryside:

> They can build an enterprise by simply making use of a particular raw material, a special skill, a piece of information, or an able person. (*Anhui Ribao*, 1985, p. 93)

The third key to the success of the peasant family enterprise is its small size and scale of operation. Since small firms require little capital and simple technologies, they can establish themselves quickly and see quick results. Small firms have the advantage of organizational simplicity. With fewer administrative levels and unobstructed flows of authority, they are viewed as superior to large enterprises in both management and administration. These technical and organizational advantages of smallness lend family enterprises flexibility. As the saying goes, "small boats turn around quickly." When economic conditions change, they can adjust their products, labor force, capital flows, and even management methods.

These arguments about the benefits of family firms—in particular those highlighting incentive structures and accumulation motives, diversification and dispersal "strategies," and advantages of small size—are remarkably similar to those I developed elsewhere for Taiwan, suggesting that certain features of Chinese family organization have survived both capitalist- and socialist-style development. A major difference is that mainland Chinese writers believe that the family economy, while advantageous in the very early stages of development, will soon wither away as the economy develops. However, a close study of

Taiwan suggests that, given an encouraging environment, the family economy can continue to prosper long after land reform, promoting rapid and balanced growth well into the middle stages of industrialization. With respect to *why* families-qua-enterprises promote economic growth and diversification, the Chinese writers hint at the kinds of capabilities inherent in the family economy, but they contain no sustained analysis of these units that might explain their beneficial macroeconomic results. Furthermore, while exploring in depth family *economic* strategies, they largely ignore the rationale and effects of family *demographic* strategies. Finally, these writers are so busy lauding the virtues of *individual* peasant *entrepreneurs* that they overlook the critical role of peasant *families*, which provide the organizational springboards for entrepreneurial success. In this section I present a series of deductive hypotheses, based on wide reading of the sinological literature and studies of Third World family firms, that link microfamily organization to macroeconomic and demographic performance.[11]

An organizational view Given a relatively "free" environment in which families are allowed to pursue their interests and arrange their economic affairs, rural Chinese families equipped with some tangible assets tend to promote a pattern of dynamic economic growth and diversification, rural–urban balance, and a relatively rapid fertility decline in response to changes in child costs and benefits. As working hypotheses, I would maintain that these arguments are as true today as they were in late imperial China, and they are as true among "post-peasant" entrepreneurs in "post-rural" Taiwan as they are among peasant entrepreneurs in still very rural mainland China.

Several features of the family-qua-enterprise are critical to its *dynamism and potential for rapid growth.*[12] First, the Chinese family offers its core economic actors, traditionally adult males, a package of individual incentives and group insurance against failure that encourages the emergence of highly motivated, risk-taking entrepreneurs. Key individual incentives include promises of more assets after family division in return for more work before division and, at a later stage, salaries and rights to individual asset ownership. Second, the family is able to generate many strategic resources (labor, capital, information) from internal sources and family-based social networks. Reliance on family and networks to obtain inputs yields resource costs, increased reliability, and improved ability to respond rapidly when economic opportunities arise. Finally, through centuries of experience with markets, contemporary Chinese families come equipped with a toolkit that includes not only the motivation to increase profits and assets, but also a set of managerial skills and mobility strategies that helps them make maximal use of existing opportunities.

The *diversification of the rural economy* can be traced in part to the diversification strategies of peasant families as they seek to exploit new opportunities while insuring themselves against the risk entailed by relying on a single economic activity. Families often adopt an explicit strategy of intersectoral diversification—one son a farmer, another a merchant, a third a teacher—thus contributing to cross-industry diversification economy-wide. The peasant family promotes diversification not only because the strategy enhances family-wide security and mobility; it also allows efficient use of family labor, and it helps keep sons from breaking off to form new family units. Because the family labor force is quite diverse with respect to age, sex, skill, and experience, a strategy of diversification, with each "type" of labor allocated to the (culturally) appropriate task, enables the family to reduce waste and make maximal use of all the human capital resources available to it. Diversification also helps delay the breakup of the family unit by giving sons independent economic kingdoms where they can exercise initiative and develop skills relatively free of the control of the peasant patriarch.

Peasant family organization also fosters *urban–rural balance*, a synthetic term here implying a predominance of short-term over permanent migration, few symptoms of excessive urbanization (e.g., squatter communities, gross shortage of urban facilities), close links between city and country, and growing rural commercialization and industrialization. Family ties promote spatial balance in at least three ways. First, because dispersal is designed to serve *family* goals (as determined largely by the family head), it is at least theoretically a short-term affair, and it involves only part of the younger generation; the other part is needed at home to tend to agricultural work and familial obligations. Second, because dispersal operates in the context of continuing strong ties among related households, it promotes flows of labor, cash, commodities, technology, and information between city and countryside, thus helping to equalize the distribution of resources across the landscape. Because social and economic resources tend to concentrate in the cities, the flows from city to countryside are particularly crucial. Third, families play critical roles in rural commercialization and industrialization because, in part due to land reform, they own (or, in the PRC, control the use of) the bulk of the rural land and farm structures, in or on which shops and factories can be built.

Finally, in the post-reform environments family organization also promotes *fertility decline*. In the wake of a land reform one might expect two phases of fertility response. In the immediate post-reform phase fertility can be expected to rise as peasant couples respond to improved mobility opportunities in the traditional way—by having more children. As it gradually becomes clear that the locus of opportunities has shifted out of agriculture into industry and services, and that mobility in these sectors requires schooling, fertility can be expected to decline in response to the rising educational costs and falling labor benefits of children.

Although relatively large family memberships are advantageous to the use of accumulation, diversification, and dispersal strategies, the shift to lower fertility can occur because of the availability to Chinese families of alternative means of enhancing their demographic resources. In addition to having more children, they can also delay family division, thereby forming a complex (stem or joint) family. In contemporary environments the latter tactic is probably more appealing on grounds of direct cost and availability of prime-age family labor. As the educational, medical, and other costs of children rise (a crucial assumption is that they do), and education becomes a labor market necessity (a trend hastened by land reform) Chinese peasant families, ever attuned to cost–benefit ratios, respond relatively rapidly and reduce the number of their offspring.[13]

A third strategy for ensuring an adequate supply of labor and other resources is to cultivate kin and other social networks. Although the use of supra-*jia* networks greatly expands the number and geographical scope of available resources, these resources are less reliable than those obtained from within the family circle. Nevertheless, in an environment of rising child costs, the availability and ready exploitability of the network option is likely to make the decision to limit childbearing easier.

To lend substance to these hypotheses, I turn now to an empirical investigation of the changes wrought by the three Chinese land reforms. Since the changes I am interested in are not well captured in aggregate statistics, I rely primarily on microlevel sources, largely anthropological and other field studies and, where those do not exist, journalistic reports.

The Taiwan land reform

On the eve of the Nationalist land reform, the Taiwanese family was well off by Chinese standards. Fifty years of Japanese colonial rule had brought a green revolution to the island. Although most of the doors to mobility outside agriculture were closed to Taiwanese, and the fruits of the green revolution were reaped largely by Japan, enough of the agricultural surplus remained on the island to produce a gradual rise in living standards. The improvements in well-being were arrested only in the late 1930s/early 1940s when Taiwan was mobilized for war (Ho, 1978).

For the Taiwanese peasantry the colonial era was a period of family consolidation and expansion. In part because of the conservative social policies of the colonial rulers, and in part because of improvements in health conditions and living standards, mortality declined, fertility remained high or rose, household size increased, and marriage patterns shifted from less preferred to culturally ideal forms (Barclay, 1954; Pasternak, 1983; Wolf and Huang, 1980).

This overall picture of relative socioeconomic security concealed sharp differences in the circumstances of different classes. These differentials worsened in the hiatus between the departure of the Japanese in 1945 and the arrival of the Nationalist government in 1949. For the tenant and part-tenant farm majority, land rents in the 1940s took 50–70 percent of the harvest. Landlord demands for "iron-clad rents," security deposits, guarantee money, and advance payments further narrowed the margin of economic security. With lease contracts that were oral and of indefinite duration, landlords could evict tenants at any time, making tenant subsistence precarious at best (Chen, 1961; Yang, 1970).

Redistribution of security

The 1949–53 land reform was designed to guarantee political stability, and it achieved this goal largely by guaranteeing the social and economic security of tenant families, who formed 41 percent of the total in 1947 (Ho, 1978). The most important security provision was the 1949 regulation limiting rents to 37.5 percent of the standard yield of main crops. In addition to lowering and stabilizing rent payments, the rent regulations protected tenants against further landlord abuse by such provisions as universalizing written rental contracts, extending the period of tenant contracts, and giving tenants sole right to break contracts. Tenant and part-tenant security was further enhanced two years later when the government sold over half of the public farmland taken over from the Japanese to families who had tenanted it. Two years later in 1953 the Land-to-the-Tiller program destroyed the power of the landlords by compulsorily purchasing, at low cost, all land in excess of 3 *jia* (2.9 ha.) of medium quality, while further improving the security of tenants by allowing them to purchase the same land back from the government. These three measures brought immediate security benefits to the bulk of the Taiwanese peasantry. Seventy percent of tenant and part-tenant families enjoyed a reduction of rents to 37.5 percent. Seventy-five percent of tenant and part-tenant households, and 50 percent of all farm households, were able to purchase land during the reform. Since the number of landless laborers was extremely small to begin with, perhaps 4–5 percent of farm families (Chen, 1961), essentially only the landlords were left in insecure positions. And among the latter, only the small landlords were left in economically vulnerable positions; most large and middle-scale landlords had diversified out of agriculture long before the reform was implemented.

By redistributing economic security, the tripartite reform effectively democratized family organization by making the economic basis of family organization, a family estate, widely available to the tenant majority. Although there are few detailed studies of changes in family organization during this period, the spurt of housebuilding that occurred islandwide after the reforms suggests a process of family formation and renewal. Family histories I gathered in the late 1970s include cases of tenant families being reunited or delaying division, and tenant farmers marrying earlier around the time of the reform.

Redistribution and transformation of mobility

The Taiwan land reform did much more than provide security and the capacity for family formation to the tenant majority. More significant was its effect on family mobility. Because accumulation of land was the major route to prosperity, redistribution of the land in effect reapportioned the means of upward mobility.

What the reform gave, however, it also took away. For although land was the prime mobility resource *before* the reform, in the very process of reallocating it, the reform undermined its value as a mobility tool. Part of the explanation lies in the arithmetic: whereas before the reform a handful of wealthy families (2 percent in 1939) owned rather large estates of 10 or more *jia* (9.7 or more ha.), after the reform (in 1960) no farm family owned more than 10 *jia*, and only 3.2 percent of families owned 3–10 *jia* (Ho, 1978, pp. 351–352). Miniaturization of farm size was only one of the factors undermining the mobility value of land. Equally important was the well-enforced low ceiling on land ownership, which put it out of the running as a strategic resource for accumulation. Long before the Land-to-the-Tiller act was drawn up, landlords got wind of the new government's intentions and began to sell their land for any price it could command. As a result, between 1948 and 1952 the value of agricultural land dropped precipitously from an index figure of 100 to 38 (for high-grade land) (Chen, 1961, p. 310).

Probably because the land reform was followed by strong technical and organizational support for agriculture, the recognition that land could yield only limited mobility benefits came slowly to most Taiwanese cultivators. As population growth reduced farm sizes in the 1950s (from an average of 1.29 ha. in 1952 to 1.05 ha. in 1965; Thorbecke, 1979), and intensification of land use failed to stem the absolute decline in agricultural income in the 1960s, it became increasingly clear that the only way up was out of agriculture. When cities began to eat up their borders in the 1970s, peasants fortunate enough to own suburban land made fortunes by selling it. However, those were the lucky few. The government soon got wind of what was happening and developed stringent zoning regulations designed to preserve as much cropland as possible.

Although the reform changed the rules of the mobility game, it still gave peasants the means to get into the game. A farmer who planted vegetables or cash crops on his vestpocket farm could earn enough to invest in education for his children, a shop for his son, or even a small factory that could be housed in his living room and take advantage of surplus family labor. With so many families endowed with mini-estates, great potential existed for the widespread development of family mobility strategies. Whether family entrepreneurship would emerge, however, depended critically on the guarantees and incentives built into the post-reform environment.

The post-reform environment: Two guarantees and five encouragements

As any visitor to Taiwan today might suspect, the political and economic environment put in place at the time of the reform and further liberalized since has contained all the conditions identified earlier as essential to the full flowering of family entrepreneurship. Private ownership of property, of course, has been guaranteed, as has the right of families to manage their own economic affairs. Nor have there been political restraints on the use of the traditional mobility strategies. Rural families have been allowed to freely accumulate property (except land, which has remained subject to the ceiling set in 1953), diversify their economic undertakings, disperse their members to cities, and expand their demographic resources through having more children or delaying family partition and thereby forming more complex family structures. Especially since the late 1950s/early 1960s, when a series of policy measures transformed the basis of development from import substitution to export-oriented industrialization (Galenson, 1979), a seemingly unlimited global market for family labor and entrepreneurship has actively stimulated the expansion of family mobility strategies.

Family entrepreneurship and economic and demographic performance

Increasing use of family mobility strategies has contributed to the dynamism and rapid diversification of Taiwan's post-reform economy. Demographically, family strategizing has promoted urban–rural balance and briefly rising followed by falling fertility levels.

Economic dynamism Because the land purchase agreements signed in 1953 involved payments spread over ten years, in the first decade after the reform most Taiwanese peasants had few resources to invest in properties other than farm land, houses, and equipment. The landlords, who were compensated for their land in part with industrial stocks, were expected to shift easily into commercial and industrial activities. In fact, however, only the large landlords, who had already transferred many of their activities out of agriculture before the reform, were able to maintain their previous way of life. Small landlords, who comprised the vast majority, either turned to farming or started small businesses, many of which eventually failed (Yang, 1970).

Beginning in the late 1950/early 1960s, however, young people who had moved to the urban areas to find work had accumulated enough savings to invest in other kinds of assets. Because funds were limited they necessarily started small: a pedicab, a vegetable cart, a newspaper stall (Gallin, 1966). As business prospered, they moved up to taxicabs, trucks, and shops.

A decade later, as the costs of living and doing business in the cities climbed, these now-skilled, moneyed, and worldly young people began streaming back to their home villages to set up factories in their childhood homes. As a result, small-scale factories using primarily family land, labor, capital, and skills began to sprout all over the more densely inhabited western half of the island (Harrell, 1982; Hu, 1983).

Of all the literature on Taiwan, only ethnographic studies adequately convey the near-manic flavor of the acquisitional drive characterizing Taiwanese farmers-turned-petty-entrepreneurs. The desire to be the boss, to own a piece of property that can be built up and expanded, drives workers to labor 10 to 13 hours a day on their own volition simply to earn more money so they can establish their own businesses (Stites, 1982). Once established, the ambition is to expand as fast and in as many directions as possible. Although it is impossible to measure, there can be little doubt that this drive to accumulate, rooted in tradition but strengthened by the contemporary, opportunity-filled environment, has made a critical contribution to the dynamism of Taiwan's economy.

Economic diversification Often a deliberate strategy to spread risks and exploit new opportunities (Cohen, 1976; Wang and Apthorpe, 1974), diversification of family economies has increased rapidly since the early 1950s. The decline in agricultural profits that began on an islandwide scale in the late 1960s forced the vast majority of rural families to abandon agriculture as their main source of income. In one poor village in central Taiwan the proportion of income from farming plummeted from 95 percent in 1958–59 to 15 percent in 1978 (Gallin and Gallin, 1982). In two northern Taiwan farm communities I studied, between 1954 and 1978 the share of farm family income from nonagricultural sources grew from under 5 to 89 percent. Although agricultural planners have not, despite repeated experiments, found a happy solution to the problems of tiny farm size and low agricultural returns (described by Apthorpe, 1979; Huang, 1981; Sando, 1981), since the 1960s diversification has maintained farm family incomes at about 75 percent those of nonfarm families.

Family diversification strategies have also speeded the diversification of the economy as a whole. Partly in response to changes in family needs and assets (members, skills, tangible resources)—many of which were due to changes in the developmental cycles of the large baby-boom families of the 1950s—different sectors have developed in different decades. In the 1950s, when the baby boom generation was of school age, agricultural diversification strategies prevailed. As this generation reached adolescence in the 1960s, factory wage labor gained prominence. In the 1970s and 1980s, when the children began producing their own children, family-run factories became increasingly common.

Rural–urban balance A phenomenal increase in the short-term migration of workers and students has been a prominent feature of post-1949 Taiwan life. In my northern Taiwan sample (about two-thirds rural) the share of workers living outside the parental home rose from 16 percent in 1954–58 to 24 percent in 1964–68, 42 percent in 1974–78, and finally 45 percent in 1978. In the village on the northern tip of the island that I studied, fully one-half the members of interviewed families lived outside the village in 1978. In two central Taiwan villages one-third or more of registered inhabitants did not live in the village in the late 1970s and early 1980s (Gallin and Gallin, 1982; Thompson, 1984).

Family dispersal strategies have played an important role in the urban–

rural balance for which Taiwan's economy is justly famous. First, families have tended to disperse members and capital to areas where the economic opportunities are brightest, thus promoting a rational distribution of population relative to resources. In the 1960s, when opportunities were brighter in the cities, Taiwan's farm families sent large numbers of younger members into urban jobs. In many rural areas movement to the city after graduation became a routine phase in the life cycle of young men and women (Sando, 1981; Thompson, 1984). Then in the 1970s, as the costs of land, labor, and capital in the cities skyrocketed, many returned to the countryside to go into farming or start a business (Gallin and Gallin, 1982; Thompson, 1984). This willingness to go where—and only where—the work is has fostered a relatively successful urbanization process in which migrants have been rapidly absorbed into urban labor markets, and symptoms of overurbanization common in much of the Third World have been largely avoided (Chang, 1984).

Second, because dispersal operates in the context of continuing ties between migrant and sending households, it promotes flows of labor, cash, commodities, and information between city and village. The flows from city to countryside have been particularly noteworthy. In three villages studied in 1971–72, a minimum of 8 to 24 percent of total family income was earned by workers living elsewhere (Wang and Apthorpe, 1974). In my northern sample fully 43 percent of total family income was earned by dispersed workers in 1978. In some areas researchers contend that flows of remittances from family members in the cities have prevented the collapse of farming villages (Huang, 1981). Urban–rural family ties have also helped to ensure the continued operation of family farms even when many family workers are urban-based, and to infuse new technology and skills into the countryside, providing the means for the development of rural services and industries (Hu, 1983; Wang and Apthorpe, 1974).

Finally, family ownership of rural land and farm buildings, traceable in part to the land reform, has played a critical role in rural industrialization. While much cultivated land is zoned for agriculture and cannot be converted to industrial use, families have simply added factories onto or inside their farm houses, and used the courtyard and other open space for storage of raw materials and equipment.

Fertility decline To take advantage of the economic opportunities in their environment, Taiwanese families needed a critical mass of family members. Anthropological research suggests that they achieved their desired family sizes in different ways at different times. In the 1950s high fertility probably contributed most to large families. In the 1960s and 1970s, even while dispersal shrank the size of the co-residential household, delayed division appears to have increased the size of the family (*jia*) (Chuang, 1972; Gallin and Gallin, 1982). Here we are concerned primarily with the effects of family expansion strategies on fertility.

In the immediate post–land reform period, when peasants apparently were responding to improved mobility opportunities in agriculture, high fertility was probably the main contributor to large families. The total fertility rate peaked at 7.04 children per woman in 1951, and remained above 6.00 until 1959. With the shift of mobility opportunities away from agriculture toward occupations requiring education, however, the costs and benefits of children moved in a direction favoring lower fertility. Ever sensitive to economic stimuli, rural Taiwanese couples reduced their fertility levels with remarkable speed. For the island as a whole total fertility fell from 6.06 in 1958 to 5.35 in 1963, 4.33 in 1968, 3.21 in 1973, 2.71 in 1978, 2.16 in 1983, and 1.7 in 1988 (Population Reference Bureau, 1988). Fertility levels of rural couples remained about 10 percent higher than the islandwide average throughout this period (DGBAS, 1986; Ministry of Interior, 1975).

The Taiwan Value of Children study suggests that cost considerations, including the costs of education, were the major factor convincing parents to have fewer children (Wu, 1977). And surveys from the 1950s indicate that land reform played a critical role in raising educational aspirations. In 1952 families had to be reminded, cajoled, and threatened to obey compulsory school attendance laws. By 1959, however, every rural family "strongly desired that all children should have at least elementary education." Schooling was now considered not only economically feasible, but also necessary as preparation for a job "higher" than farming (Koo, 1968, pp. 105–106). Ethnographic research indicates that throughout the post-reform decades education has been widely held to be the surest route up and out of agriculture (Gallin, 1966, pp. 124–125; also Chen, 1977; Pasternak, 1972; Sando, 1981). Although the land reform can claim only some of the credit for the rural fertility decline, its indirect role that operated by transforming the mobility structure and promoting the universalization of education was clearly crucial.

One of the world's "model" land reforms, the Taiwan reform is famous for creating an agricultural success story of increasing agricultural productivity and improved rural distribution. While this assessment is not incorrect, a view incorporating the role of the family suggests that it is a partial one that places the emphasis in the wrong sector. The most important effect of the Taiwan land reform was to force resources out of the rural and agricultural sectors into the urban, commercial, and industrial sectors of the economy. The transition was so smooth because it was mediated by a family accustomed for centuries to operating in both worlds and skilled at exploiting market opportunities for its own benefit. Because the reform was carried out in a political and economic environment that permitted and stimulated the expansion of family mobility strategies, the reform-environment package led to an explosion of family entrepreneurship that has been a motor force behind the rapid development of Taiwan into one of the postwar world's industrial success stories.

The first mainland Chinese land reform

The century preceding the Communist land reform was one of growing political and economic decay. Against the political backdrop of a collapsing Confucian state, persistent population expansion, intensifying landlord exploitation, destruction of peasant handicrafts, and a series of natural disasters combined to produce an agrarian crisis marked by growing landlessness and inability of the population to secure a basic livelihood.

In a peasant economy such as this, the familial basis of economic life ensured that the agrarian crisis would also be a family crisis.[14] As early as the beginning of the nineteenth century, demographic pressures had forced peasants to lower their aspirations from achieving upward mobility to simply securing basic subsistence (Ho, 1962, p. 273). By the early twentieth century the margin of subsistence had become so thin that small fluctuations in the market could sink families into debt, while natural calamities could threaten their very survival (Fei and Chang, 1945; Fei, 1939; Yang, 1959). While a tiny minority of landlords and rich peasants were able to achieve the ideal of large, diversified families, for a growing majority of the landless and land-poor, persistent economic stress produced demographic distress in which peasants were unable either to sustain their current families or to create new families. Symptoms of family failure included high infant mortality, low fertility, late marriage, frequent divorce, early family partition, and shrinkage of size due to sale of secondary (i.e., female) members and loss of young men to the ranks of the floating, unemployed population (Lang, 1946).

Redistribution of security

Given this setting of disintegrating social and economic life, the most significant aspect of the 1945–52 land reform was not so much the redistribution of land as the redistribution of economic security, and with it the basis for family life. Several provisions of the reform, which was carried out in the north and east from 1945 to 1949, and elsewhere from 1950 to 1952, brought basic economic security to those who before had known none. The cancellation of debts and the refunds of rent deposits and excess rents wiped out years of unpaid obligations. The rent reductions, by 25 percent of current rent or to 37.5 percent of the main crop value, gave tenants a chance to rebuild their family economies on a new, more secure foundation. The redistribution of the land and assets of landlords gave tenants and poor peasants small parcels of land that could serve as safety nets when times got hard. Perhaps most important, the land reform and the concurrent "anti–local bully campaign" guaranteed the elimination of the landlord class and its armed protectors. In the late 1940s the Party allowed landlords and local thugs to be dealt with seriously and, often, violently. An estimated 800,000 landlords lost their lives. Many of the others fled to the cities, emptying the villages of a major source of peasant economic insecurity.

The elimination of landlordism, coupled with the distribution of landlords' assets, provided the basis for a democratization of family life. Families previously on the brink of demographic collapse now had small amounts of cash, grain, and land to support a nucleus of family life. In addition, the land reform included many provisions specifically designed to put families torn apart by the chaos of civil war back together again and ensure their integrity as units of rural production (Stacey, 1983). For example, class status was assigned on a family basis and land was distributed to the family head according to the number of family members. Families were also reunited by the reform: unemployed migrants were required to return to their home villages to claim a share of land, elderly vagrants were rejoined to their families, and wives who had deserted their husbands were encouraged to return and give marriage another try. The family-based economy was preserved by such means as distributing plots to those who had actually tilled them, and consolidating geographically dispersed plots into single family holdings.

Redistribution and transformation of mobility

Land was not only a source of security; in the mid-twentieth century Chinese countryside, its accumulation was also the primary means of upward mobility. Thus, the redistribution of land entailed also the reallocation of the *potential* for upward mobility to the majority of the peasants. *Fa jia zhi fu*—start a family and make a fortune—was the slogan of the day, and peasants responded enthusiastically to the promise that they would be able to secure and expand their livelihoods. Looking back over 30 years of socialism, Fujian peasants in the early 1980s remembered the immediate post–land reform years as a time of prosperity and hope (Nee, 1984). In fact, from the Party's perspective, newly endowed peasants responded a bit too well to the incentives created by the land reform: many peasant cadres and activists began to neglect their duties or even abandon their posts in order to devote their energies to rebuilding their family economies (Shue, 1980).

While the land reform gave peasants reason to think about upward mobility, it also destroyed the viability of land as a basis for upward mobility. For one thing, the amount of land redistributed was too small to provide the basis for much family economic advancement. Because the land reform was not intended to produce a major reshuffle of agricultural assets, the scope of redistribution was relatively modest. The land reform program of 1950–52 involved 43 percent of cultivated land distributed among 60 percent of the rural population, or an average of 1.7 *mou* (.11 ha.) per person (Wong, 1973). The redistribution of tools and draught animals was even more modest. On a per capita basis each peasant entitled to reform assets received .017 head of oxen, .24 tool, .16 house, and 53 *cattles* (23.5 kg.) of grain (Wong, 1973, p. 164)—hardly enough to guarantee security, let alone mobility. The miniaturization of the village land probably made it clearer than it had ever been when the land was concentrated in a few hands that the total amount of land available was so limited that no one could prosper through agriculture. Insufficient land, coupled with the absence of complementary measures supporting agriculture, left the majority of peasants hard pressed to make ends meet. Some were forced to sell their land. These parcels were bought by rich or well-to-do middle peasants who had recovered from the reform and begun expanding their economies. The result was a process of re-polarization, which threatened to undo the gains of the reform.

Perhaps more important than this *economic* message conveyed by the size of the parcels distributed was the *political* message, carried by sometimes violent means, that landlordism and the accumulation of land would be forbidden by the new regime. In the violent phase of the reform in the mid- to late 1940s, landlord property was forcefully confiscated and many landlords were killed. With land ownership a political liability, the price of land was sharply devalued. Beginning in the mid-1940s landlords tried to sell their property to friends or relatives, gave it away to those they trusted, or, failing those means, destroyed it before it could be expropriated. As a result, the value of the land plummeted between the late 1940s and early 1950s.

The obvious message was that families wanting to improve their socio-economic status would have to find an arena other than agriculture and landownership in which to do it. However, families had little chance to act on this insight, for this message was superseded by another one that took away their ability to own any kind of productive property.

The post-reform environment: No guarantees and one permissible

In the years following the land reform all but one of the conditions hypothesized as necessary for the full emergence of family entrepreneurship were either eliminated or drastically restricted.

The first guarantee to be removed was that of private ownership of the means of production. Even before the land reform had been completed, peasants were skeptical about the Party's continued support for the peasant family economy. Because the construction of a socialist society had been part of the Party's line when it came to power, even peasants who had received title deeds had good reason to fear that their newly acquired land would eventually be taken away. In some areas only usufruct certificates were issued, fanning fears about the future (Yang, 1959). Naturally, uneasiness about the Party's intentions acted as a strong disincentive to investment of either labor or capital on the land. Similar problems occurred in the area of credit, leading to severe shortages of credit and capital. Even those with money to spare were afraid to lend it for fear that they would never be repaid (in the campaign against usury that had just concluded, many debts had been canceled) or that they would be labeled "feudal exploiters" and targeted in future campaigns.

In the years following the reform, of course, peasant fears proved to be correct. Only two years after the completion of the reform the leadership became impatient with moderate policies, such as the promotion of mutual aid teams, which had failed to check the tendency toward polarization. By late 1953 top leaders had decided to collectivize agriculture. Collectivization proceeded through lower- and then higher-stage producers' cooperatives, in which land was collectivized, then accelerated with the formation of the super-collectivized communes in 1958.

With the formation of lower-stage producers' cooperatives the peasants lost their guarantee that the family would be the unit of economic decisionmaking. This unit was raised to a co-op management committee in the producers' cooperatives, further elevated to the commune during the Great Leap Forward of 1958–61, and then lowered to the production team in 1962, at which level it generally remained until the "second land reform" in the late 1970s.[15]

Opportunities to pursue traditional mobility strategies were also closed off or drastically restricted. In the early 1950s the industrial and commercial sectors of the economy were severely disrupted, eliminating markets for peasant handicrafts and generally discouraging diversification strategies (Yang, 1959, pp. 165–166). After collectivization, peasant energies were channeled primarily into agriculture and construction of basic infrastructure. Only in their spare time were members of the rural collectives allowed to work in sidelines and on private plots. Furthermore, since private marketing was severely restricted, sideline activities could enhance family consumption, but not family income.

Dispersal strategies were also discouraged and eventually forbidden. In the early 1950s opportunities for short-term migration to urban jobs were drastically restricted by severe urban unemployment (Yang, 1959). By the late 1950s the wave of peasants moving to the cities to escape poverty and turmoil in the countryside had become so vast that the regime introduced a set of measures designed to eliminate rural-to-urban migration altogether.

Accumulation strategies were, of course, also off limits because of the ban on private ownership of productive property. In the immediate post-reform environment the only mobility strategy that was not administratively curtailed was family expansion. Until the early 1970s families had substantial control over demographic decisions affecting family size and structure. In the absence of private sector economic opportunities, however, the benefits of demographic expansion to peasant families were much less obvious.

Recapitulating, the first land reform made basic economic security widely available and largely democratized family organization. Although it redistributed the potential for upward mobility and began to change ideas about the road to prosperity, the post-reform environment provided none of the guarantees and few of the conditions that traditionally stimulated the development of family entrepreneurship. Instead, the entrepreneurial potential created by the reform was snuffed out by the mid-1950s.

The second mainland Chinese "land reform"

The collectivization of Chinese agriculture did not work out as Mao had planned. Even after the unit of accounting was shifted back to the production team in 1962, the collective system stifled peasant initiative, constrained income-earning opportunities, and inhibited the efficient use of resources (Lardy, 1985). In 1979 the central leadership made the startling announcement that rural consumption levels in the late 1970s were no higher than they had been in 1957, the year before the Great Leap Forward. The revisionist view of Western scholars is that among the major legacies of collectivism were alienation and apathy, as the state removed the peasants' control over their lives and condemned expressions of self-interest as capitalist (Friedmann, 1987). Class struggle campaigns that descended with unsettling unpredictability not only failed to improve rural living standards; they also victimized growing numbers of peasants, bringing bitterness and cynicism in their wake (Chan, Madsen, and Unger, 1984).

Determined at last to remove the blight of rural poverty, in December 1978 the post-Mao leadership introduced the first of a series of agrarian reforms designed to invigorate the rural economy by arousing the initiative of the masses. Touching virtually every aspect of rural life, the reforms introduced since then have dismantled the collective system, replacing it with one more closely resembling the pre-Communist system in the scope given to family initiative. The most dramatic of these reforms was the household responsibility system, which shifted the unit of economic decisionmaking from the production team back to the peasant family. This reform was heralded by the peasants as a second land reform, and, from the viewpoint advanced in this essay, for very good reasons. The reform package also increased farm prices, encouraged diversification and specialization, reopened free markets, permitted the hiring of labor, and promoted the flow of peasant funds into private productive assets and enterprises (for reviews of these reforms see Griffin, 1984; Perry and Wong, 1985; Riskin, 1987).

Privatization of security

The first land reform had greatly improved the security of the peasantry by eliminating the exploitative class structure. Collectivization had further strengthened security guarantees by providing a collective safety net that members could fall back on in times of need. In fact, security was probably the main contribution of the collective to peasant welfare. Recent interviews by Western scholars reveal that peasants were willing to tolerate collective organization for so long because of the promise of economic security that it held out (Butler, 1985; Unger, 1985). (Even this modest promise, however, was sometimes broken, most disastrously during the Great Leap Forward.) After 25 years of collective security, however, what the peasants craved most was the opportunity for mobility, the chance to enjoy a few comforts in their own lifetimes and to create brighter prospects for their children's lives. Although the solution adopted—privatization of security as well as production functions—was experienced by demographically incomplete families as a loss of security, evidence available suggests that most Chinese were able to fill the gaps in their family structures, and worries about security were gradually dispelled (Unger, 1985–86). Indeed, once the private production system was introduced, it appears to have taken little time for the majority of formerly security-conscious peasants to get on with the business of improving their family fortunes.

Redistribution and transformation of mobility

The responsibility system redistributed the means of upward mobility by taking the land of the collective and parcelling it out to individual households to manage. Of the many forms of responsibility system introduced in the late 1970s and early 1980s, *da bao gan*—contracting everything to the family—became the most popular and was adopted by the vast majority of rural households by 1982. In early 1984 the leadership extended contracts for ordinary crops to more than 15 years, and contracts for other types of projects (e.g., development of orchards, forests, or wasteland) for even longer. While the state has made it clear that ownership of land will remain vested in the collective, it has guaranteed long-term usufruct rights to the land, in some cases issuing certificates to this effect, thus ensuring peasants that their investments will be returned.

While providing a tangible means for climbing the first rung of the mobility ladder, at the same time this second land reform reiterated the message conveyed by the first: that the road to mobility lay outside agriculture. Three decades of at times rapid population growth had reduced the amount of cultivable land available to only 2.70 *mou* (.18 ha.) per (rural) person in the early 1980s. While collective management of the land had probably helped to conceal the increasingly unfavorable population/land ratio, the parcellization of land out to individual families, combined with the loss of any economies of scale they may have enjoyed under the system of collective production, probably brought home to the peasants the sharp limits on income to be made from agriculture. Regulations forbidding private ownership of land further reduced incentives to stay down on the farm. Post–responsibility system measures further facilitated this shift of mobility strategies out of the agricultural sector. Among the most significant was a document issued in early 1984 that allowed peasants to transfer their contracted land to others. This document opened the way for a few efficient farmers to work all the land and for most to leave agriculture altogether.

The post-reform environment: Two (conditional) guarantees and three permissibles

The reforms introduced along with and after the responsibility system have gradually created an environment providing all but one of the conditions promoting peasant entrepreneurial activities. However, because of the leadership's insistence that the private economy remain a subordinate sector within the larger socialist economy, all of these guarantees and encouragements have been hedged about with many limitations and conditions.

Since the late 1970s guarantees of peasant control over property and economic decisions have been introduced and gradually strengthened. Although peasants are not allowed to own land, they have been guaranteed the right to own and accumulate other forms of assets, including farm equipment, motor vehicles, shares in enterprises, commercial and industrial establishments, and so forth. These rights are protected by law; if they are threatened peasants may lodge complaints with the government or people's courts.

Families have also gained substantial, though far from total, control over their economic affairs. They control most production and marketing decisions, and they determine the distribution of their labor to different productive and reproductive tasks. Team leaders and village and township authorities continue to play important roles, however, controlling the allocation of contracts, access to critical inputs and new opportunities, and levels of taxes and special fees (Oi, 1989; Unger, 1985–86).

In the reform environment of the 1980s three of the four traditional mobility strategies—diversification, dispersal, and accumulation—have again become politically permissible. Strategies of diversification and dispersal have been officially promoted or tolerated. Since the late 1970s rural families have been encouraged to "develop the local economy in an all-round way." While they are to ensure the production of grain, "specialized" and regular households are also to diversify into cash crops, agricultural sidelines, commerce, small-scale industry, construction, and transportation. Because such shifts absorb surplus rural labor, the migration of peasants into small towns and cities has been encouraged, and the short-term movement of the "floating population" into major metropolises has been tacitly accepted.

Through slogans such as "let some get rich to lead all to common wealth" and stories of successful 10,000-yuan households, the state has made it clear that peasant accumulation is in favor in the 1980s. Yet there are also limits, however ill-defined, to the level of wealth that is permitted. For example, a 1986 document on rural work states: "Although we must encourage . . . specialized households to become well-off. . . , we will not tolerate their 'becoming rich and influential families' " (Xinhua, 1986, p. K7).

In contrast to economic control, which has been partially ceded to families, demographic control remains in the hands of the state. The expansion of family demographic resources via increased fertility has been administratively restricted in the rural areas since the early 1970s. Under the one-child policy, introduced concurrently with the key economic reforms, rural couples are permitted to have only one or, in "difficult" circumstances (the most common being lack of a son), two children. Although the policy was slightly relaxed in 1984 and 1988, and regulations vary from place to place, it appears that in most parts of the country it remains difficult for peasant couples to have more than two children, whatever their sex.

Although the post-reform environment has severely restricted the use of one of the traditional mobility strategies and placed ambiguous but real limitations on the other conditions for entrepreneurship, given the pent-up demand for family autonomy and mobility created by decades of collective life, even these limited guarantees and encouragements can be expected to stimulate peasant entrepreneurship.

Family entrepreneurship and economic and demographic performance

This relatively limited stimulus produced a flood of entrepreneurial initiative and talent that, one surmises, surpassed the wildest expectations of the reforms' designers. Family entrepreneurial strategies in turn stimulated dynamic economic growth and rapid diversification. The demographic accompaniments were moderate migration and improved urban–rural balance, and fertility levels higher than the regime would like but low relative to rural Third World standards. Regional differences in policy environment doubtless produced regional differences in these patterns of economic and demographic performance. Limitations on space and data, however, preclude our exploring these regional variations here.

Economic dynamism At the heart of the rapid economic growth in the Chinese countryside lie millions of peasant families seeking to improve their incomes and accumulate private wealth. The tremendous dynamism of the family economy can be seen in the growth rates of the activities in which it is most active. Between 1978 and 1984 agriculture (crop cultivation) grew by 6.7 percent a year, but animal husbandry grew by 9.4 percent and sidelines grew by 18.6 percent (Riskin, 1987, p. 291). Of all undertakings, peasants see factories as the most direct route to prosperity and, where a market is available, invest substantial resources in developing them (*Far Eastern Economic Review*, 1985). By 1985 over three-quarters of the village and town enterprises were privately owned (Agricultural Publishing House, 1986, p. 147). The Chinese press reports that peasants who have no enthusiasm for township enterprises are very active in developing family factories (Zhao, 1985) and come "streaming in to register as shareholders" when joint-stock enterprises are established (*Shaanxi Ribao*, 1985).

Peasant families have been accumulating not only industrial assets, but also residential, commercial, agricultural, transport, and service properties. The most important asset of a peasant family is its house, and the reforms set off a surge of house building, often on illegally occupied land (Bernstein, 1984). Every year between 1980 and 1985 peasants built an average of .2 to .3 room per family (Agricultural Publishing House, 1986, p. 288). In agriculture, after an initial drop in mechanization due to the responsibility system–induced shrinkage of plot size, there has been a marked increase in mechanization, as cropland has been concentrated in the hands of large-scale, specialized farmers working the land with modern machinery (*China Daily*, 1985). At the end of 1985 individual peasants owned 89 percent of the small tractors, 62 percent of the medium- and large-size tractors, and 61 percent of the motor-driven threshers (Agricultural Publishing House, 1986, p. 258). Some farmers have even bought light aircraft, presumably for spraying their crops. Although land remains publicly owned, overeagerness to make a profit has led peasants in some areas to indiscriminately occupy it and to buy and sell land in violation of the law (*Fujian Ribao*, 1986). By far the most common nonresidential asset is transportation equipment. In the city of Wenzhou, a development model in Zhejiang famous for its household workshops, by mid-1986 individuals had acquired 790 trucks and 9,970 freight vessels (*China Daily*, 1986).

Economic diversification The reforms have also stimulated the revival of peasant diversification strategies. The rationale is captured in this saying: "There is no stability without agriculture, no vitality without commerce, and no way to become rich without industry." The optimal strategy is to have enough family members of different sexes and ages to be able to have fingers in every sectoral pot. A family in the Pearl River community of Wantong was able to achieve just this: The family head and his wife managed the family farm, the eldest son used his hand-tractor to transport goods, and the three daughters worked in a local fireworks factory. In their spare time family members engaged in putting-out work and sidelines, especially bee raising. In 1982 their net income from all these activities was 14,000 yuan. In Wantong in 1982, 10 percent of the households had achieved this multiple undertaking ideal (*Far Eastern Economic Review*, 1983).

Since the reforms released great pools of underutilized labor, formerly unemployed family members such as children, the elderly, sick, and disabled have found productive employment in the family economy. Between 1978 and 1985 the average number of able-bodied and semi-able-bodied workers per agricultural household increased from 2.27 to 2.95 while the number of dependents fell from 2.53 to 1.74 (Agricultural Publishing House, 1985, p. 235; 1986, p. 288). Many of these new workers are employed in the "courtyard economy" raising grapes and earthworms, weaving carpets and baskets, and so on (Guo, 1985). Others work in family-run workshops and factories.

Family diversification strategies have been an important force behind the rapid diversification of the rural economy. Whereas in the late 1970s the great bulk of the rural labor force was in agriculture, by the mid-1980s much of the work force had shifted to other sectors. In the highly developed areas of Guangdong and Jiangsu, by 1985 60–80 percent of the labor force had moved into sidelines, industry, or services (*China Daily*, 1985). Diversification has also occurred within agriculture. Between 1978 and 1984 the share of crop production in the value of gross agricultural output fell from 67.8 to 59.0 percent, while the share of sidelines (including factories) rose from 14.6 to 20.9, that of animal husbandry increased from 13.2 to 14.3, and that of forestry rose from 3.0 to 4.1 percent (Agricultural Publishing House, 1985, p. 285). Largely because of state measures ensuring the adequacy of grain production, the share of grain in total crops fell by only 2 percentage points, from 80.3 to 78.3, while the share of the more lucrative industrial crops rose from 9.6 to 13.4 percent (ibid.).

Rural–urban ties Although peasants have responded enthusiastically to the freeing of restrictions on migration, in part because of continuing constraints on permanent change of residence, migration has been generally short-term and circular in type (Goldstein and Goldstein, 1985). Spatial flows of people have been accompanied by spatial flows of economic resources such that both urban and rural areas have benefited. After two decades of policies that so exacerbated urban–rural differences as to create a dual economy (Perkins and Yusuf, 1984) or two social systems (Whyte and Parish, 1984), it will take a great deal of time to produce anything resembling rural–urban balance. In a few short years, however, the peasant family economy has substantially increased rural–urban ties and reduced the economic backwardness of the villages.

One reason China has avoided massive rural-to-urban flows lies in the creation of a great many rural jobs by family factories and workshops. This pattern of labor utilization is known as "leaving the land but not the countryside" or "entering factories but not cities." Equally important is the fact that migration is part of a family dispersal stratagem in which the "dispersed" members retain close ties and obligations to the members at home who determine overall family economic strategy. As a result, migration is generally short-term and, to the extent the migrants succeed, accompanied by flows of remittances, gifts, information, and even technology.

Available data suggest that after migration restrictions were lifted, peasants lost no time in activating dispersal-cum-remittance strategies. Although aggregate demographic data are lacking, household income data indicate that the proportion of income from remittances and two kinds of state subsidies grew from 6.9 percent in 1978 to 10.7 percent in 1980, then fell slightly to 9.6 percent in 1985 (unfortunately it is impossible to separate the remittances from the subsidies; Agricultural Publishing House, 1985, pp. 235–236; 1986, p. 289). Observers in southern Guangdong in the early 1980s reported a large exodus of the young to Hong Kong, followed soon after by large flows of remittances and gifts to the villages (Chan, Madsen, and Unger, 1984). A survey of 28 villages indicates that by 1982–83 a majority of men in 21 of them had found ways to earn money outside the community. While workers in richer villages generally found jobs in cities, those from poorer localities often performed farm work in rich villages (Unger, 1985–86).

Increasing movement of peasants to the cities has contributed to urban and rural growth alike. Despite serious strains on urban transport and housing facilities, the cities have benefited from the rural-born explosion of micro-scale service establishments, the availability of lower-cost construction labor, and even the influx of peasant capital (Zheng et al., 1985). By 1984, for example, peasant capital had built 289 restaurants, 98 department stores, 38 wharves, and more than 20 hotels, some over 20 stories high, in Guangzhou alone (*Guangzhou Ribao*, 1984). Peasant skills and capital have flowed as well into small cities and rural market towns, promoting their development as new foci of rural social and economic life. Skills and resources acquired in cities and towns have also flowed back to spur the development of the countryside. For example, the factories of Wenzhou, the development model in Zhejiang, virtually all employ returned urban migrants as technical specialists.

Fertility decline After decades of being forced to merge their family economies with their neighbors', Chinese peasants appear eager to cut off ties with other villagers and narrow the circle of economic decisionmaking to their immediate families. In most areas that have been studied by Westerners, peasants have not only been averse to farming collectively (Unger, 1985–86),[16] they have also taken a privatistic attitude toward welfare, "not noticing" when their neighbors' security was threatened in the first years of the reforms (Nee, 1985; Unger, 1985–86).

In order to take advantage of the new opportunities, peasant families have needed to round out their memberships, forming complete families in which the division of labor can be fully developed. Countless newspaper stories of families with large labor forces entering the ranks of 10,000-yuan households have daily publicized the links between large families and large incomes (recall the Wantong family described above). In some cases the task of demographic completion has entailed the recombination of whole family units that had formally divided in the lean years of collectivism. For most, however, the option of merging family units has probably been unavailable or unrealistic. For the majority, completion of the family has meant only addition of the proper number and types of children.

Inasmuch as fertility has been administratively restricted since the implementation of the reforms, the "natural" pattern of rising followed by falling fertility witnessed on Taiwan has not occurred on the mainland. Instead, after 1979, when the one-child policy was introduced, the total fertility rate (TFR) fell to about 2.0 in 1984–85 before rising to 2.8 in 1987.[17] These fluctuations were due to a combination of economic and population policy changes (Greenhalgh, 1989). Upward pressure on fertility can perhaps also be seen in the persistent preference for more children than the current policy allows. Stated fertility aspirations have been depressed by the massive birth control compaigns of the past 20 years, but in the 1980s the modal preference appeared to be two children, one son and one daughter (Whyte and Gu, 1987). It seems that no amount of pressure—educational, economic, or administrative—has been able to push these preferences lower, and peasants have gone to great lengths, concealing pregnancies, moving to cities to give birth, and so forth to have the minimum number and sex of children they consider essential.

Thus, while the regime has reaped tremendous economic benefits from allowing the family economy to develop, the demographic price-tag has been a fertility level higher than that implied by the official target of about 1.2 billion at the turn of the century. However, while the 1986 total fertility rate of about 2.4 is substantially higher than the official goal, by the standards of other Third World countries at similar levels of development it is remarkably low. Furthermore, from the family economy perspective developed here, it is difficult to concur with the official view that higher fertility (say, two or three children rather than one) translates directly into lower per capita income. The family strategy view also suggests that the current phase, marked by pressure from couples for more children than they are allowed, will in due time give way to another phase in which the great majority of couples realize that the road to mobility lies through lower fertility and higher education to jobs outside agriculture. Regardless of whether Chinese leaders accept the family economy arguments, the record to date suggests that a fertility level in the area of 2–3 may be a pill they will have to swallow until modernizing forces, in particular industry and education, more thoroughly penetrate the Chinese countryside, bringing lower fertility in their wake.

Like the Taiwan land reform 30 years earlier, the second mainland reform has been introduced in an environment giving families considerable elbow room to control their economic destinies and to reactivate traditional mobility strategies. Again like the Taiwan reform, probably the most significant effect of the mainland reform is the consequent shift of resources out of the rural and agricultural sectors into the urban, commercial, and industrial sectors of the economy. However, while the family economy has shown remarkable growth potential, it remains constrained by policies that send ambiguous messages about its future, and by cadres who help themselves to the new wealth by collecting exorbitant fees and levies. While peasant families appear to have overcome the limitation on fertility by simply defying the one-child policy, in periods such as 1986 and 1989 excessive taxation and retrenchment policies designed to slow the pace of reform dampened the growth of private enterprises. Whether the second land reform on the mainland will eventually produce the kinds of economic and demographic change seen on Taiwan depends to an important extent on the kind of post-reform environment China's leaders decide to create.[18] If they can reconcile the family economy with socialism and maintain a policy environment with certain basic guarantees and encouragements, China's leaders may reap the benefits of family entrepreneurship for decades to come.

Conclusions

This essay has raised far too many issues to resolve in a brief conclusion. Here I simply highlight a few theoretical issues, policy implications, and areas for future research.

If my central hypothesis is correct—that, by changing the security/mobility structure, the land reforms and accompanying policy environments removed the fetters on family entrepreneurship, and this in turn promoted rapid economic growth and demographic modernization—then this analysis supports the argument of Hareven (1982), Stacey (1983), and others that the family can play an active, even causal role in economic development. As for the nature of that role, I have stressed primarily the *benefits* of the family economy. However, significant *costs* also exist, and need to be built into the account. With respect to timing and phasing, Taiwan's experience suggests that families-qua-enterprises can promote development at least through the advanced labor-intensive stage of industrialization and possibly also into the capital- and technology-intensive stages. Thus, analysts in the PRC may want to rethink their conclusion that the family economy will soon wither away, giving rise naturally to a cooperative economy in which production units are owned by several families, neighborhoods, or even whole villages. That is likely to happen only if the role of the private economy is politically circumscribed.

The Taiwan–mainland China comparison invites one to posit many similarities, contrasts, and tentative conclusions, which, unfortunately, I cannot indulge in here. Most important, this comparison has shown how a single set of indigenous social institutions can survive, even be preserved by, two entirely different political-economic policy environments, and then emerge to play a somewhat similar role in later development. From this vantage point the two societies appear far more Chinese than "socialist" or "capitalist."

The answer to the question posed at the outset—why have the East Asian land reforms been uniquely successful?—contains fewer policy lessons for other countries than one would hope. The answer suggests not that the reforms or their policy environments were unique, but that the micro-institutional climate in which they were introduced was unique, and that this institutional context played a determining role in the economic and demographic outcomes. Thus, transferring Taiwan-like reforms to other sociocultural settings is not likely to produce Taiwan-like economic and demographic outcomes.

The lesson here is not that other countries should try to create institutions like those of the Chinese. The message, rather, is that countries should build on what they have. For one thing, it may be impossible to destroy some indigenous institutions—at least this is what a quarter century of mainland Chinese efforts to "cut off the capitalist tails" suggests. But there is more to it than that. While "Confucian" culture may have few parallels as a promoter of capitalist development, it is likely that all institutional complexes contain some features conducive to modern economic and demographic change. An institutionally sensitive policy process is one that first figures out what those strong points are, then selectively reinforces them, harnessing them to the achievement of larger economic and demographic goals.

This essay has left many areas for future work. For example, my treatment of the policy environment has been very crude. Future work should specify much more precisely the effects of different political and economic conditions on familistic entrepreneurship. Because of limits on space and available data, I have largely ignored regional differences within China. Future research should explore those differences, examining the extent to which gradations of "Confucian" culture and variations in contemporary policy may combine to produce regional differences in family entrepreneurship and economic and demographic change.

With respect to the development of institutional approaches generally, this essay represents a modest contribution. Family organization shaped the effects of the postwar land reforms in many more ways than I have been able to describe. Even more important, the family is but one of many institutions in the Chinese cultural complex that colored the way the reforms were designed and implemented, and the nature of their results. The real challenge for institutional analysis is to demonstrate the integrity and developmental role of a whole institutional complex, a task that requires an explicitly comparative, or cross-cultural, approach. Especially in the Chinese case, the exceptional continuity and complexity of the traditional system, coupled with its resilience to date, suggest the possibility that East Asian development patterns may remain forever distinctive. However, this hypothesis, while intriguing, will require a great deal of comparative and historical work to support it.

Notes

I am grateful to Martin K. Whyte for his comments on an earlier draft of this essay.

1 In this essay land reform is defined as the wide redistribution of ownership or usufruct rights to land from landlords or collective production units to individual peasant families. This definition embraces the East Asian reforms of the 1940s and 1950s as well as the

current set of rural reforms on the mainland, in particular the production responsibility system. Land reform is intertwined with, but analytically distinct from, the policy environment in which it is introduced. This essay employs these two concepts—land reform and policy environment—but does not address the broader issue of agrarian reform.

2 The successes and failures of Third World land reforms are carefully assessed in FAO, 1979 and FAO and ILO, 1976.

3 The effects of the Taiwan reform are analyzed in Fei, Ranis, and Kuo, 1979; Ho, 1978; Koo, 1968; Lee, 1971; Thorbecke, 1979; and Yang, 1970. Economic consequences of the late 1940s/early 1950s mainland Chinese reform are explored in Lippit, 1974; Perkins and Yusuf, 1984; Wong, 1973, among others.

4 The advantages of this body of literature stem from the abundance of empirical studies of women and development from many areas of the Third World.

5 Stacey also relies on a "moral economy" or "reaching back toward tradition" argument. My own work emphasizes instead the economic rationality of the preservation of certain features of traditional family organization.

6 This generalized picture of the traditional socioeconomic order is based largely on the work of anthropologists and historians. Among the key sources are Cohen, 1970; Fei, 1939; Fei and Chang, 1945; Rozman, 1981; and Yang, 1945.

7 For elaboration see Greenhalgh, 1985.

8 On the mainland about 30 percent of farm produce was marketed in the nineteenth century (Perkins, 1969). This proportion was undoubtedly higher on Taiwan (Myers, 1972). The degree of peasant "economic rationality" probably varied from region to region depending on the development of markets.

9 This discussion is based on a limited review of articles published in Chinese newspapers and economics journals between 1984 and 1987. A more comprehensive review would probably produce an argument that is more complex but whose main lines are similar to the one presented here.

10 This organizational analysis is my addition. The Chinese sources simply give laundry lists of various characteristics of family enterprises.

11 Key sources on Chinese family firms include DeGlopper, 1979 and Mark, 1972. Among works on Third World family enterprises the most useful include Benedict, 1968; Strachan, 1976; and Tripathi, 1981.

12 The following hypotheses are developed and empirically examined in Greenhalgh, 1988a.

13 This argument is elaborated in Greenhalgh, 1988.

14 This idea owes much to Stacey's (1983) stimulating discussion.

15 In the 1970s some localities attempted to raise the unit of accounting to the brigade level.

16 Available evidence suggests that peasants in wealthier units where collective organization of labor had been beneficial to individual households were less enthusiastic about the reforms of the late 1970s/early 1980s (Zweig, 1985).

17 These figures, based on official surveys, probably underestimate the level of fertility.

18 It is unlikely that the pace of economic and demographic change experienced by Taiwan since the early 1950s could be matched by the mainland. Taiwan began the postwar period with a much more modernized economy and society; it was extensively influenced by foreign, especially American, development models; and it enjoyed the benefits of a global trade boom in the 1960s. These and other differences ensure that the two parts of China will continue to develop in divergent ways.

References

Adelman, Irma. 1980. "Economic development and land reform," *American Behavioral Scientist*: 437–456.

———. 1985. "A poverty-focused approach to development policy," in *Development Strategies Reconsidered*, ed. John P. Lewis and Valeriana Kallab. New Brunswick, N.J.: Transaction, pp. 49–65.

———, and Sherman Robinson. 1978. *Income Distribution Policy in Developing Countries: A Case Study of Korea*. Oxford: Oxford University Press.

Agricultural Publishing House. 1985. *China Agriculture Yearbook 1985*. Beijing: Agricultural Publishing House.

———. 1986. *China Agriculture Yearbook 1986*. Beijing: Agricultural Publishing House.

Anhui Ribao. 1985. "Anhui governor speaks on rural enterprises," in *China—Political, Sociological and Military Affairs*, 5 July, pp. 87–106. Washington, D.C.: Joint Publications Research Service.

Apthorpe, Raymond. 1979. "The burden of land reform in Taiwan: An Asian model land reform re-analyzed," *World Development* 7, no. 4/5: 519–530.

Balmori, Diana, Stuart F. Voss, and Miles Wortman. 1984. *Notable Family Networks in Latin America*. Chicago: University of Chicago Press.

Barclay, George W. 1954. *Colonial Development and Population in Taiwan*. Princeton, N.J.: Princeton University Press.

Benedict, Burton. 1968. "Family firms and economic development," *Southwestern Journal of Anthropology* 24, no. 1: 1–19.

Bernstein, Thomas P. 1984. "Reforming China's agriculture," paper presented at the American Council of Learned Societies/Social Science Research Council Conference "To Reform the Chinese Political Order," Harwichport, Mass.

Berry, R. Albert. 1984. "Land reform and the adequacy of world food production," in *International Dimensions of Land Reform*, ed. John D. Montgomery. Boulder, Colo.: Westview, pp. 63–87.

Butler, Steven B. 1985. "Price scissors and commune administration in post-Mao China," in Parish, 1985, pp. 95–114.

Chan, Anita, Richard Madsen, and Jonathan Unger. 1984. *Chen Village: The Recent History of a Peasant Community in Mao's China*. Berkeley: University of California Press.

Chang Ming-cheng. 1984. "Economic adjustment of migrants in Taiwan: Recent findings and policy implications," paper presented at a conference on Urban Growth and Economic Development in the Pacific Region, Institute of Economics, Academia Sinica, Taipei, 9–11 January.

Chayanov, A. V. 1966. "Peasant farm organization," in *The Theory of Peasant Economy*, ed. Daniel Thorner, Basile Kerblay, and R. E. F. Smith. Homewood, Ill.: Richard D. Irwin, pp. 29–269 (orig. publ. 1925. Moscow: Cooperative).

Chen Cheng. 1961. *Land Reform in Taiwan*. Taipei: China Publishing Company.

Chen Chung-min. 1977. *Upper Camp: A Study of a Chinese Mixed-Cropping Village in Taiwan*. Taipei: Institute of Ethnology, Academia Sinica.

China Daily. 1985. "Trend toward large, specialized farms develops," 6 February, p. 1.

———. 1986. "Zhejiang city model of household industry growth," 10 July, p. 1.

Chuang Ying-chang. 1972. "T'ai-wan nung-ts'un chia-tsu tui hsien-tai-hua te shih-ying" (Adaptation of rural families in Taiwan to modernization), *Bulletin of the Institute of Ethnology* 34: 85–98.

Cohen, Myron L. 1970. "Introduction," in *Village Life in China*, ed. Arthur H. Smith. Boston: Little, Brown, pp. ix–xxvi.

———. 1976. *House United, House Divided: The Chinese Family in Taiwan*. New York: Columbia University Press.

DeGlopper, Donald R. 1979. "Artisan work and life in Taiwan," *Modern China* 5, no. 3: 283–316.

DGBAS (Directorate-General of Budget, Accounting and Statistics), Executive Yuan, Republic of China. 1986. *Statistical Yearbook of the Republic of China, 1986*. Taipei: DGBAS.

Far Eastern Economic Review. 1983. "Letter from Wantong," 3 November.

________. 1985. "China's fear of flying," 5 December, pp. 64–65.

Fei, John C. H., Gustav Ranis, and Shirley W. Y. Kuo. 1979. *Growth with Equity: The Taiwan Case.* New York: Oxford University Press.

Fei, Xiaotong (Hsiao-t'ung). 1939. *Peasant Life in China.* London: George Routledge and Sons.

________. 1983. *Chinese Village Close-up.* Beijing: New World Press.

________, and Chih-i Chang. 1945. *Earthbound China: A Study of Rural Economy in Yunnan.* Chicago: University of Chicago Press.

Food and Agriculture Organization (FAO). 1979. *Review and Analysis of Agrarian Reform and Rural Development in the Developing Countries Since the Mid-1960s.* Rome: FAO.

________, and International Labour Organization (ILO). 1976. *Progress in Land Reform: Sixth Report.* New York: United Nations.

Freedman, Ronald. 1987. Personal communication, 30 November.

Friedmann, Edward. 1987. "Maoism and the liberation of the poor," *World Politics* 39, no. 3: 408–428.

Fujian Ribao (*Fujian Daily*). 1986. "Fujian congress session stresses land control," in *China—Daily Report,* 14 February, pp. 01–02. Washington, D.C.: Foreign Broadcast Information Service.

Galenson, Walter (ed.). 1979. *Economic Growth and Structural Change in Taiwan: The Postwar Experience of the Republic of China.* Ithaca, N.Y.: Cornell University Press.

Gallin, Bernard. 1963. "Land reform in Taiwan: Its effects on rural social organization and leadership," *Human Organization* 22, no. 2: 109–112.

________. 1966. *Hsin Hsing, Taiwan: A Chinese Village in Change.* Berkeley: University of California Press.

________, and Rita S. Gallin. 1982. "Socioeconomic life in rural Taiwan: Twenty years of development and change," *Modern China* 8, no. 2: 205–246.

Goldstein, Sidney, and Alice Goldstein. 1985. "Population mobility in the People's Republic of China," Papers of the East-West Population Institute, no. 95. Honolulu: East-West Population Institute.

Greenhalgh, Susan. 1984. "Networks and their nodes: Urban society on Taiwan," *The China Quarterly* 99: 529–552.

________. 1985. "Sexual stratification: The other side of 'growth with equity' in East Asia," *Population and Development Review* 11, no. 2: 265–314.

________. 1988. "Fertility as mobility: Sinic transitions," *Population and Development Review* 14, no. 4: 629–674.

________. 1988. "Families and networks in Taiwan's economic development," in *Contending Approaches to the Political Economy of Taiwan,* ed. Edwin A. Winckler and Susan Greenhalgh. Armonk, N.Y.: M.E. Sharpe, pp. 224–245.

________. 1989. "Fertility trends in China: Approaching the 1990s," Working paper no. 8, Research Division. New York: The Population Council.

Griffin, Keith (ed.). 1984. *Institutional Reform and Economic Development in the Chinese Countryside.* Armonk, New York: M.E. Sharpe.

Guangzhou Ribao (*Guangzhou Daily*). 1984. "Guangzhou peasants pooling over 200 million yuan to run tertiary industry," in *China—Political, Sociological and Military,* 4 April, pp. 136–137. Washington, D.C.: Joint Publications Research Service.

Guo Shougui. 1985. "Tentative exploration of present rural surplus labor outlets," *Shanxi caijing xueyuan xuebao* (*Journal of Shanxi Finance and Economics College*) 6: 10–11.

Hareven, Tamara K. 1982. *Family Time and Industrial Time: The Relationship Between the Family and Work in a New England Industrial Community.* Cambridge: Cambridge University Press.

Harrell, Stevan. 1982. *Ploughshare Village: Culture and Context in Taiwan.* Seattle: University of Washington Press.

Hartmann, Heidi. 1976. "Capitalism, patriarchy, and job segregation by sex," in *Women and the Workplace: The Implications of Occupational Segregation,* ed. Martha Blaxall and Barbara Reagan. Chicago: University of Chicago Press, pp. 137–169.

Hinton, William. 1983. *Shenfan: The Continuing Revolution in a Chinese Village.* New York: Random House.

Ho, Samuel P. S. 1978. *Economic Development of Taiwan, 1860–1970.* New Haven, Conn.: Yale University Press.

Ho Ping-ti. 1962. *The Ladder of Success in Imperial China: Aspects of Social Mobility, 1368–1911.* New York: Columbia University Press.

Hu Tai-li. 1983. "My mother-in-law's village: Rural industrialization and change in Taiwan," Ph.D. dissertation, Department of Anthropology, City University of New York.

Huang Shu-min. 1981. *Agricultural Degradation: Changing Community Systems in Rural Taiwan.* Washington, D.C.: University Press of America.

Huang, Philip C. C. 1985. *The Peasant Economy and Social Change in North China.* Stanford, Calif.: Stanford University Press.

Jacobs, J. Bruce. 1985. "Political and economic organizational changes and continuities in six rural Chinese localities," *The Australian Journal of Chinese Affairs* 14: 105–130.

Koo, Anthony K. C. 1968. *The Role of Land Reform in Economic Development: A Case Study of Taiwan.* New York: Praeger.

Lang, Olga. 1946. *Chinese Family and Society.* New Haven, Conn.: Yale University Press (repr. 1968. New York: Archon).

Lardy, Nicholas R. 1985. "State intervention and peasant opportunities," in Parish, 1985, pp. 33–56.

Leacock, Eleanor, and Helen I. Safa (eds.). 1986. *Women's Work: Development and the Division of Labor by Gender.* South Hadley, Mass.: Bergin and Garvey.

Lee Teng-hui. 1971. *Intersectoral Capital Flows in the Economic Development of Taiwan, 1895–1960.* Ithaca, N.Y.: Cornell University Press.

Levy, Marion J., Jr. 1949. *The Family Revolution in Modern China.* New York: Octagon.

Lim, Edwin, et al. 1985. *China: Long-Term Development Issues and Options: A World Bank Country Economic Report.* Baltimore: Johns Hopkins University Press.

Lippit, Victor D. 1974. *Land Reform and Economic Development in China: A Study of Institutional Change and Development Finance.* White Plains, N.Y.: International Arts and Sciences Press.

Lipton, Michael. 1985. "Land assets and rural poverty." World Bank Staff Working Paper no. 744. Washington, D.C.: World Bank.

———. 1984. "Family, fungibility and formality: Rural advantages of informal non-farm enterprise versus the urban-formal state," in *Human Resources, Employment and Development,* Vol. 5: *Developing Countries,* ed. Samir Amin. New York: St. Martin's Press, pp. 189–242.

Ma Encheng. 1987. "Let a large number of peasant entrepreneurs spring up," in *China—Daily Report,* 13 February, pp. K26–K27. Washington, D.C.: Foreign Broadcast Information Service.

Mark, Lindy Li. 1972. "Taiwanese lineage enterprises: A study of familial entrepreneurship," Ph.D. dissertation, Department of Anthropology, University of California, Berkeley.

Ministry of Interior, Republic of China. 1975. *1974 Taiwan–Fukien Demographic Fact Book, Republic of China.* Taipei: Ministry of Interior.

Moise, Edwin E. 1983. *Land Reform in China and North Vietnam: Consolidating the Revolution at the Village Level.* Chapel Hill: University of North Carolina Press.

Montgomery, John D. 1984. "Prospects for international action," in *International Dimensions of Land Reform,* ed. John D. Montgomery. Boulder, Colo.: Westview, pp. 221–232.

Myers, Ramon H. 1972. "Taiwan under Ch'ing imperial rule, 1684–1895: The traditional economy," *The Journal of the Institute of Chinese Studies of the Chinese University of Hong Kong* 5, no. 2: 373–409.

———. 1984. "The economic transformation of the Republic of China on Taiwan," *The China Quarterly* 99: 500–528.

Nash, June, and Maria Patricia Fernandez-Kelly (eds.). 1983. *Women, Men and the International Division of Labor.* Albany: State University of New York Press.

Nee, Victor. 1985. "Peasant household individualism," in Parish, 1985, pp. 164–190.

Oi, Jean C. 1989. *State and Peasant in Contemporary China: The Political Economy of Village Government.* Berkeley: University of California Press.

Parish, William L. (ed.). 1985. *Chinese Rural Development: The Great Transformation.* Armonk, N.Y.: M.E. Sharpe.

________, and Martin King Whyte. 1978. *Village and Family in Contemporary China.* Chicago: University of Chicago Press.

Pasternak, Burton. 1968. "Some social consequences of land reform in a Taiwanese village," *Eastern Anthropologist* 21, no. 2: 135–154.

________. 1972. *Kinship and Community in Two Taiwanese Villages.* Stanford, Calif.: Stanford University Press.

________. 1983. *Guests in the Dragon: Social Demography of a Chinese District, 1895–1946.* New York: Columbia University Press.

Perkins, Dwight. 1969. *Agricultural Development in China, 1368–1968.* Chicago: Aldine.

________, and Shahid Yusuf. 1984. *Rural Development in China.* Baltimore: Johns Hopkins University Press.

Perry, Elizabeth J., and Christine Wong (eds.). 1985. *The Political Economy of Reform in Post-Mao China.* Cambridge, Mass.: Harvard University Press.

Popkin, Samuel. 1979. *The Rational Peasant: The Political Economy of Rural Society in Vietnam.* Berkeley: University of California Press.

Population Reference Bureau. 1988. *World Population Data Sheet.* Washington, D.C.

Riskin, Carl. 1987. *China's Political Economy: The Quest for Development Since 1949.* Oxford: Oxford University Press.

Rozman, Gilbert (ed.). 1981. *The Modernization of China.* New York: Free Press.

Sando, Ruth Ann E. 1981. "The meaning of development for rural areas: Depopulation in a Taiwanese farming community," Ph.D. dissertation, Department of Anthropology, University of Hawaii.

Schultz, Theodore W. 1964. *Transforming Traditional Agriculture.* New Haven, Conn.: Yale University Press.

Scott, James C. 1976. *The Moral Economy of the Peasant: Rebellion and Subsistence in Southeast Asia.* New Haven, Conn.: Yale University Press.

Shaanxi Ribao (*Shaanxi Daily*). 1985. "Resolving issue of rural, small town enterprise funds," in *China—Political, Sociological and Military,* 24 October, pp. 20–21. Washington, D.C.: Joint Publications Research Service.

Shue, Vivienne. 1980. *Peasant China in Transition: The Dynamics of Development Toward Socialism, 1949–1956.* Berkeley: University of California Press.

Skinner, G. William. 1971. "Chinese peasants and the closed community: An open and shut case," *Comparative Studies in Society and History* 13, no. 3: 270–281.

Stacey, Judith. 1983. *Patriarchy and Socialist Revolution in China.* Berkeley: University of California Press.

Stites, Richard. 1982. "Small-scale industry in Yingge, Taiwan," *Modern China* 8, no. 2: 247–279.

Stover, Leon E. 1974. *The Cultural Ecology of Chinese Civilization: Peasants and Elites in the Last of the Agrarian States.* New York: Times Mirror.

Strachan, Harry W. 1976. *Family and Other Business Groups in Economic Development: The Case of Nicaragua.* New York: Praeger.

Thompson, Stuart E. 1984. "Taiwan: Rural society," *The China Quarterly* 99: 553–568.

Thorbecke, Erik. 1979. "Agricultural development," in Galenson, 1979, pp. 132–205.

Tripathi, Dwijendra. 1981. *The Dynamics of a Tradition: Kasturbhai Lalbhar and His Entrepreneurship.* New Delhi: Manohar.

Unger, Jonathan. 1985. "Remuneration, ideology, and personal interests in a Chinese village," in Parish, 1985, pp. 117–140.

________. 1985–86. "The decollectivization of the Chinese countryside: A survey of twenty-eight villages," *Pacific Affairs* 58, no. 4: 585–606.

Wang Sung-hsing. 1976. "Family structure and economic development in Taiwan," paper presented at a Social Science Research Council Conference on Anthropology in Taiwan, Portsmouth, New Hampshire.

———, and Raymond Apthorpe. 1974. *Rice Farming in Taiwan: Three Village Studies*. Taipei: Institute of Ethnology, Academia Sinica.

Watson, Andrew. 1983. "Agriculture looks for 'shoes that fit': The production responsibility system and its implications," *World Development* 11, no. 8: 705–730.

Wei Daonan, and Zhixiao Zheng. 1985. "On the peasants' self-employed economy," *Jingji Yanjiu* (*Economic Research*) 3: 66–71.

Wellesley Editorial Committee (ed.). 1977. *Women and National Development: The Complexities of Change*. Chicago: University of Chicago Press.

White, Lawrence J. 1974. *Industrial Concentration and Economic Power in Pakistan*. Princeton, N.J.: Princeton University Press.

Whyte, Martin King, and William L. Parish. 1984. *Urban Life in Contemporary China*. Chicago: University of Chicago Press.

———, and S. Z. Gu. 1987. "Popular response to China's fertility transition," *Population and Development Review* 13, no. 3: 471–493.

Wolf, Arthur P., and Chieh-shan Huang. 1980. *Marriage and Adoption in China, 1845–1945*. Stanford, Calif.: Stanford University Press.

Wong, John. 1973. *Land Reform in the People's Republic of China: Institutional Transformation in Agriculture*. New York: Praeger.

Wu, T. S. 1977. *The Value of Children: Taiwan*. Honolulu: East-West Population Institute.

Xinhua. 1986. "Plan of the CPC Central Committee and the State Council for rural work in 1986," in *China—Daily Report*, 24 February, pp. K1–K9. Washington, D.C.: Foreign Broadcast Information Service.

Yang, C. K. 1959. *A Chinese Village in Early Communist Transition*. Cambridge: MIT Press.

Yang, Martin M. C. 1945. *A Chinese Village: Taitou, Shantung Province*. New York: Columbia University Press.

———. 1970. *Socio-Economic Results of Land Reform in Taiwan*. Honolulu: East-West Center Press.

Zhao Tianzhen. 1985. "An investigation on transforming town and township enterprises into peasants' joint-stock enterprises," *Jingji Guanli* (*Economic Management*) 5: 22–23, 12.

Zheng Guizhen, Xian Liu, Yunshu Zhang, and Jufen Wang. 1985. "Analysis of the mobile population in Shanghai," unpublished manuscript, Fudan University, Population Research Institute.

Zweig, David. 1985. "Peasants, ideology, and new incentive systems: Jiangsu Province, 1978–1981," in Parish, 1985, pp. 141–163.

Futures of the Family in Rural Africa

JACK GOODY

I NEED TO BEGIN AN ESSAY ON FUTURES with a warning about predictions. Even those social scientists who work with harder data and tighter theories, such as economists and possibly demographers, find the immediate future difficult to foretell, let alone the distant prospects of two decades. In the study of the family, with the one exception of census material, the national data for Africa are of outstandingly poor quality because of the general lack of registration, of welfare services, and of efficacious bureaucracies. The aggregate figures for age at marriage, rates of divorce, and so on, are difficult to use for most serious purposes; even the contemporary picture is harder to sketch than in other parts of the developing world. For many aspects of the family one is largely dependent upon case studies that have often only a limited application to wider data because of the range of variation we are dealing with, the result of differences in past structures and in the impact of present influences. I have no commanding knowledge of the range of current studies, which are carried out more and more at the local level and in countries where the circulation and even the printing of such information has often become increasingly difficult. My remarks are based largely on West Africa, in particular on Ghana, to which Esther Goody and I have returned many times over the last four decades, including extended visits to some rural areas during the last two years. My comments on future trends have been much influenced by the new forms of "salaried farming" we found there, side by side with the neo-traditional modes of production and reproduction.

Parameters

In looking at what family forms might look like several decades hence, we have to take into account some basic parameters for the study of social interaction in Africa. While we can discern certain general trends, especially at the level of population growth (that is, in factors affecting mortality and to a more limited extent fertility), a great deal of differentiation is often concealed by the use of aggregate statistics. This differentiation derives from two sources: first there are the earlier differences, for example, those between centralized and noncentralized, patrilineal and matrilineal, societies; second there are the graded effects of (largely) externally stimulated changes, in education, employment, the distribution of wealth, and rural or urban residence. Each of these affects (and in turn is affected by) family life, but even radical economic changes do not necessarily influence every aspect of the situation in any direct and immediate way.

Recall the case of England and the simple variable of household size. Family life certainly underwent significant changes during the agricultural and industrial revolutions. But the mean household size continued at a figure of 4.75 from the late sixteenth century until the first decade of the twentieth century, and its decline was even preceded by declines in France (below 4.0 in 1880) and Sweden, England being "somewhat more resistant to the economic, social structural and demographic changes" (Laslett, 1972: 139). If such a relatively straightforward variable as mean size of household did not drop until 100 years after the beginning of the industrial revolution, we should be wary of anticipating a rapid change in those countries in which little industrial activity has yet emerged.

But economic activity in this sense is only one of the factors affecting the structure of family life in Africa today. Indeed in many rural areas, the neo-traditional ones, relatively little has happened to the system of agriculture that might affect the domestic domain, even though improved crop varieties, water supplies, transport, and health services have made life easier, as has access to imported iron and cloth paid for through the cultivation of some cash crops. There is still an important neo-traditional sector where many aspects of family life remain identifiable with earlier practice. What has dramatically affected the situation is the increased migration, between rural areas but especially between town and countryside, or between one country and another, a function of the greater mobility associated with improved transport and with more open political systems. A second factor is the spread of education and the new employment situations created by an expanding government bureaucracy and service sector. Third, national politics has offered opportunities for widening and maintaining breaches in family relations that might otherwise have been papered over; the state itself has sometimes put forward its own definitions, for example, of marriage (for Uganda see J. Goody, 1986: 167–168); but, most importantly, in many areas it has failed to provide the reasonably stable conditions that might have allowed continuing and self-sustaining rural development to take place. Fourth, religious influences can have a very strong and immediate effect on family life, whether in the form of loosening existing norms (which are often supported by supernatural sanctions) or of spreading new ones, for example, close marriage in the case of Islam or distant marriage in some forms of Christianity. In Christianity, we also find an emphasis on monogamy, direct inheritance, and transfer between spouses—practices that exist only as tendencies in Islam, though the latter has a more open attitude to positive checks on fertility. In Africa, as distinct from Eurasia earlier, religious changes are now rarely effected on a national, or even regional, basis, but in a more "voluntaristic" way, creating internal differences of a sectarian kind (including new sects) in which the search for cures or salvation tends again to divide as much as unite families. In such world religions and their offshoots, as in political parties and trade unions, brotherhood acquires a different dimension, cross-cutting kin groups, although such networks already existed in the case of age-grades, "secret" societies, and similar associations.

African background: General factors

In any analysis of Africa's future we have to bear its past constantly in mind. A consideration of change at any level, familial or agricultural, must start by setting aside simplistic schemes that divide the world, like Gaul, into three parts (or worlds), or worse still into two: opposing modern to traditional, capitalist to noncapitalist, industrial to preindustrial. While each of these dichotomies may be of some use in limited contexts, one is not going to get very far by way of analysis, application, let alone prediction, without recognizing certain basic features of the socioeconomic system of the African continent.

One critical feature is environmental: whether we are dealing with the forest, the savannah, or the Sahel. Although historically farming everywhere consisted largely of shifting cultivation by means of the hoe, forest agriculture was certainly more productive except for livestock (and the Sahel less so with the same exception), whereas under cereal production in the savannahs only a limited surplus was normally possible and the use of land was "extensive." Compared with Eurasia, population was relatively sparse, kin ties were widely dispersed (giving individuals some protection against localized shortages), kinship rights and duties were complex (e.g., giving individuals a general claim on areas of jointly held land), and the opportunities for full-time specialization were restricted by technology and by demand. Most domestic groups, even of craftsmen, engaged in agriculture, usually constituting farming groups of limited size. In the 1950s the average among the LoWiili in northern Ghana was one of the highest reported, namely 11.1 (J. Goody, 1972), but the average size of household (or perhaps houseful) for African countries in the aggregate United Nations census material varied between 3.5 and 5.0, rural Tanzania averaging 3.9 (1958) and urban Zaire 4.0 (1958).

The larger households usually consisted of male relatives and their wives, often polygynous, who comprised the work group; over large areas of the continent women carried out many of the basic agricultural as well as the household activities. It is important to remember the range of these tasks both in the past and the present, bearing in mind the restricted use of nonhuman energy in the absence of livestock trained for ploughing or for traction, and the difficulties in raising water and in transporting goods in the absence of the wheel. For the near future such technological restrictions remain important factors in the organization of rural labor, despite the fluctuating importation of some motor vehicles and cycles, at a heavy cost in foreign exchange. It is not of course the absence of the material products alone that matters but of the whole social organization that produces and uses these techniques. Employing the technology presents some problems; but the organization of its production and the maintenance of the plant involve the greatest difficulties, even in the case of relatively simple products such as the ox-drawn plough; and such difficulties in turn lead to a heavy and fragile dependency upon outside supplies and upon outside funding, sources that are subject to constant interruption, both for external and for internal reasons. The fragility of this situation means that many families constantly have to adapt to conditions that change first in one direction, then in another. This high level of uncertainty, both political and economic, forms the background against which family strategies and individual decisions have to be taken. At the same time it militates against the success of any elaborate development projects that aim to be self-sustaining.

Modern technology and the differentiation of labor tend to decrease the need for cooperation between neighbors, while increasing the cooperation (or dependency) required at more inclusive levels, especially at those of the firm and the state. The traditional social system in Africa required a high degree of cooperation between members of the same household ("family") and lineage (or other wider grouping of kin or community). That is to say, the absence of specialized services and of centralized welfare institutions meant considerable dependence upon kith and kin. Orphanhood, widowhood, and divorce, while they were always personal tragedies (despite the assumptions of some European family historians and demographers), were not faced alone. Nor were there, in general, alternative sources of welfare. While in certain regions chiefs had an obligation to supervise those suspected of witchcraft, government traditionally played a radically different role than it began to do, at least in small measure, with the division of Africa among colonial powers some 100 years ago, and especially with postindependence moves toward social development. Nor did ecclesiastical or secular charity have any great part until the advent of Christian missions. While Muslims made some provision for the deprived in the form of the Friday *saddaqa* and other institutionalized gift-giving, there was not the sustained surplus to set up trusts or foundations of the kind widely distributed throughout Europe and Asia. The difference is profound and longstanding. Already in the Sumerian code of Ur-Nammu (end third millennium BC), provision was made for protecting the orphan, the widow, and the poor from ill treatment and abuse (Kramer, 1954: 41). Over time such provisions expanded. In the virtual absence of such arrangements in Africa, the kin group was of correspondingly greater importance, as it continues to be today (Sabelli, 1986), although often on a reduced scale, especially since it is clear that the current economy has difficulty in supporting the strong, let alone the weak.

It is this continuing interdependence, although less strong than before, that supports women and children while their husbands are away working; that contributes to the return fares of migrants in distress; that helps toward the expenses of schooling—acts that in the short term at least reinforce the interaction and the mutual dependencies. In this sense extended ties of kinship (what some mean by the vague phrase "extended families") are critical aspects of contemporary Africa. And while in Tunisia today, as in China and India earlier in this century, such dependencies may be rejected by the very "modernizing" young who have so often benefited from them, they are essential to situations in which the community can neither supply nor afford other forms of welfare. In other words such ties need to be preserved (or better, left alone), even though agricultural and other experts may see themselves as frustrated in their efforts at development by the absence of the same kind of individualized tenurial systems that exist in the West. One cannot easily have the one with the other, and a too heavy dose of "individualizing" could reduce a community to a moral desert populated by people in distress. No growth will compensate for such an outcome; morality and welfare are intrinsic components of all development, of all economic activity.

Differentiating factors: Migration

Migration and mobility

One of the most notable changes that have taken place in Africa, in the colonial period and after, has been the widespread growth of towns. Family life has been affected in several ways. First, the average household size of town dwellers has tended to be less than in the country, though the urban figures are often difficult to assess because of the more complex patterns of residential, production, consumption, and reproductive groups. Second, the growth of towns has been largely the result of migration, hence physically separating the members of kin groups and often of elementary families, and throwing an increased burden on the rural populations who raise children but often lose their help as adults, a fact not necessarily conducive to family limitation. From this particular standpoint the rural economy bears part of the cost of urbanization, but looked at from the standpoint of a particular family, it is their own members (mainly the departing ones) who "benefit" from the movement to the towns.

In precolonial times mobility generally took the form of:

(1) the migration of cultivators to new areas because of soil depletion, drought, or war;

(2) the movement of long-distance traders, some of whom established a residence along the way;

(3) the regular and intermittent movements of nomadic herders (Stenning, 1959);

(4) the enforced movement of captives, sometimes very considerable in extent (Meillassoux, 1986);

(5) the movement of conquering groups.

While movements of the last two kinds have diminished, the first two have greatly increased. In addition the division of labor has meant increased movement of various kinds. In the preindependence phase migrants were largely unskilled laborers in search of work in mines, on roads and railways, and in cash-crop farming. Some migrated permanently; most went and returned, often seasonally.

Such forms of labor migration, based on the movement of individuals rather than of families, still continue. Indeed migration has spread to partially schooled women, the young often going for employment and prostitution, the older ones for market trading. But education has also brought with it the mobility, largely out of the countryside, that bureaucratic and business jobs require. This migration has taken on an international aspect (Weil, 1986), which has been very marked between the more prosperous countries and their poorer neighbors. However, relative position may change over a short period; the earlier pull of the cocoa economy of Ghana in the 1950s has been replaced by the subsequent mass migration of Ghanaians and many others to Nigeria as a result of the oil boom.

Economics and migration

African states differ considerably in the buoyancy of their economies. At present, Côte d'Ivoire, Kenya, Malawi, and Nigeria are relatively prosperous, while Ghana, Guinea, Mozambique, Tanzania, Uganda, and Zambia are relatively depressed. Shifts in prosperity have taken place over the short term. Twenty years ago, at the time of Independence (which everywhere raised "great expectations," some of a "cargo cult" variety), Ghana and Uganda were among the better-off nations. Economic policy and political instability have been mainly responsible for the change in their fortunes. In other cases, there have been resource and infrastructural reasons, such as the discovery of oil in Nigeria. From a socioeconomic point of view, these longer term trends have been largely unpredictable.

The results, however, have been important at the level of the family, since the poorer nations have been the providers of migratory labor for the better-off, although such migration takes place largely within the groupings of anglophone and francophone nations, with these metropolitan languages providing the basic *linguae francae* for the market place as well as for social contacts.

This migration has involved both skilled and unskilled labor. Nigeria attracted large numbers of Ghanaian workers of both kinds, just as South Africa has for its surrounding territories, at least for unskilled labor. Any migration of highly trained workers involves a substantial loss to the exporting country that has educated them at great expense. This loss may be partly made up by remittances from abroad, which, while less significant than in the case of India, are of considerable importance to the local economy. On the other hand, it does mean that the institutions of the poorer nations tend to be proportionately less well staffed than in the richer states, except in those sectors where they are able to attract foreign workers through outside aid.

Given the different marriage ages and conjugal responsibilities of men and women, migration, whether to the town or to another rural area, is largely male. When such migration is long term, as in Southern Africa, it leads to the peripheralization of men in the family. In Lesotho women conduct much of the farming, even the ploughing (Murray, 1981). Socially and psychologically, the result is a kind of matrifocal family of the West Indian type, also found in certain lower class groups in England, the United States, and elsewhere (Smith, 1956; Alexander, 1977). Long-term migration of this kind fills a particular niche in the development of large-scale mining in Africa, and, in the more successful enterprises, one would anticipate its replacement over time by resident labor. In any case only part of the rural population is involved, and given the importance of the male role in many rural activities (not necessarily agriculture) and the increasing propensity of women to migrate, it seems unlikely that there would be any general move to a West Indian type of family structure.

The predominance of males in African towns contrasts starkly with the predominance of females in early modern Europe, where women would go to work, often in domestic service, to accumulate a dowry before making a late marriage. In Africa, the dowry is virtually nonexistent, except in a weak form in some Islamic areas; it is rather men who initially used this work opportunity to accumulate the bridewealth, which was sent to the in-laws—and which their fathers or maternal uncles would have provided, had the young men been living and working at home. Apart from those with secondary education, however, there seems to be a significant increase in the number of girls going to the town, temporarily in the first place, to trade sexual services, further complicating the relations between town and country.

Bureaucratic "employment"

The movement to towns has clearly not been a function of industrialization, since so little has taken place, but rather of increased activity in trade, commerce, the service sector, and above all of employment by the state.

One of the most radical changes over the last 30 years has been this growth in "employment," mainly by the state or under state auspices. Initially, when wages and salaries were high, such activity led to a profound split between the farming and nonfarming worlds, the latter being composed largely of that part of the rural population that had attended school or gone in search of unskilled work. The schooled tended to move out of the countryside permanently, the unschooled temporarily.

With the change that has followed the oil crisis, the increasing political instability, and the diminishing amounts of "aid" in absolute terms, all wages have taken a downward turn and a considerable proportion of the urban population and of the employed more generally have reactivated their rural or agricultural ties and begun to farm, sometimes with the help of hired labor. While some produce from this kind of farming finds its way to the market, much of what does is used to pay for the inputs necessary for subsistence.

One of the consequences is that, unlike the situation in urban areas, every person (including many young children) is not only employed, but is likely to be a producer as well as a consumer. This applies even to schoolteachers and administrators. As food prices have taken an increased part of the salaries of wage-earners, the incentive to grow for oneself is strong, and there is an equally strong incentive for full-time farmers to produce more, where the market is accessible. But if employed personnel in the rural areas find it possible to produce some of their own food, this is not always the case in the towns, so that the increasing food demand of the burgeoning urban population leads to the raising of prices of any surplus the rural areas may produce (for Southwest Nigeria, see Adepoju, 1975).

The result is a form of agriculture very different from neo-traditional practice on the one hand and from that of "big farmers" on the other. While this is not the place to expand on that topic, the implications for family structure are important. Such a return to the land has not led to the reconstitution of the larger production units of "traditional" farming among this segment of the population. Employment, even when it is a matter of poorly paid, socialized, or moonlighting jobs, individualizes the budget, and people are reluctant to return to a system in which the young were supposed to pool their cash incomes with their elders—and to draw upon a common pool for marriage payments and the like.

While kinsfolk do help one another, especially children and their parents, households and budgets of the employed are largely distinct and people "beg" and hire the labor of outsiders rather than depend on that of other kin, from whom they are often living separately. One reason for this individualization is that farmwork has to be fitted in with the salaried job, and the job imposes different adjustments and different timetables on each person. Schoolteachers are free during the holidays, which are increasingly adapted to agricultural schedules; many office workers interpret each day in a flexible fashion. The bureaucracy runs at half speed so that it can feed itself.

Such a situation is increasingly recognized by governments. In Ghana in 1985 a farm institute, formerly devoted to training in mechanized techniques, was now taking contingents from the army and police force to train them in forms of nonmechanized agriculture that did not require heavy investment. In other words it was assumed these employees would farm as well as carry out their duties in the police and army. "Salaried farming" had moved out of the moonlight into the sunlight.

Relations between town and country

While the focus here is on the rural family, I have discussed migration and employment because it is becoming increasingly difficult to disentangle the rural from its urban counterparts for the following reasons.

(1) Since most urban migrants have arrived relatively recently, they still have strong links with the countryside. Where land constitutes a fairly open resource and where kin groups and localities continue to exercise joint claims, those leaving for the town are not excluded from returning to claim farmland unless they literally "sell-up"; sale of agricultural land is still very rare either because it was traditionally prohibited or because it is a common resource under government control (whether socialist or not).

(2) Since 1975 there has been an even greater incentive to hang onto or acquire joint claims to land because the increasing shortage of food, locally produced or imported, has led town dwellers to turn to farming themselves, as an alternative to purchasing food. This has meant some blurring of the boundaries between rural and urban as well as between wage employee and subsistence worker.

Family and marriage

Let us first consider the African family in terms of a series of specific variables, namely household size and structure, age at marriage, the nature of marriage (plural or monogamous), divorce, the treatment of the old, parent–child relationships, widowhood, wider kinship ties (such as intrafamilial exchange), houseforms, and sexuality. I will then refer briefly to some demographic variables that are hardly separable from the above and that are subject to some better measurements.

Marriage

Let me begin by summarizing the numerical data on marriage in Africa. The average age at which women marry varies regionally from below 17 to about 22. The lowest figures occur in West Africa, the highest in East, and the figures seem to have remained stable over time, with the possible exception of a later age developing in Kenya. The male age at marriage varies but is considerably later than the female.

The differential ages for males and females are linked to the rates of polygynous marriage and of widow remarriage. In Dorjahn's survey of polygyny in the continent (1959; see discussion in J. Goody, 1976), the mean percentage of married men who were polygynists was 24.7 percent in East Africa, 33.8 percent in the Western Sudan, and 46 percent on the Guinea Coast, with a mean for sub-Saharan Africa of 35.0 percent.

The rate of polygyny is in turn related to household size, which tends to be larger in the West than in the East. Size is affected by polygyny not only because of the larger number of wives and children per married man, but also because the greater extent of plural marriage means that marriage for men is delayed so that young men continue to farm with their fathers for a longer period, being unable to set up on their own without a wife. The East–West African contrasts are summarized in the following table.

	West	**East**
Marriage of women	earlier	later
Polygyny	more	less
Household size	larger	smaller

By and large, later marriage for both sexes is connected with simpler households, while early marriage is associated with more complex ones. The mechanical reasons are fairly obvious. Given the age of marriage of most women in contemporary Africa (as in northern India), they are hardly in a position to set up separate households, even if their husbands are considerably older. Or rather they may set up separate cooking units (hearths) but within a more complex dwelling group in which many other tasks are carried out cooperatively. Such is the situation, for example, among the LoDagaa of northern Ghana.

Two conditions under which a separate residence becomes possible for the young wife (an only wife) are when the husband's mother is a widow and when domestic servants are available. In rural Africa there was no tradition of domestic service, apart from slaves, who, if female, tended to become co-wives. While one aspect of the widespread practice of noncrisis child fosterage lay in the provision of domestic services, this was rarely the major feature of a custom that in any case took the form of the transfer of kin rather than the employment of unrelated servants (E. Goody, 1973, 1981; Isiugo-Abanihe, 1985).

Polygyny is widely practiced in many regions of the world, but in Africa it displayed a very different pattern from that in Asia (at least in the major societies of that continent) as far as rates are concerned. In Asia fewer than 5 percent of marriages are polygynous, even where it is allowed. In Africa, in the recent past, approximately one-third of all unions were plural.

The results for household structure are significant. Widespread polygyny is possible because of a high differential marriage age, with women marrying early and men late (although it is also the case that primary sex ratios are more evenly balanced than in Eurasian populations); in extreme cases (e.g., the Konkomba of northern Ghana) men married between 35 and 40, preempting mates in the cradle. Indeed the competition is such that there has recently been a tendency among this group to preempt the child of a pregnant woman, increasing the differential betrothal age if not the differential marriage age. At the same time the opportunities to escape from such arrangements are greater (albeit fewer for women than for men), mainly by leaving the neo-traditional rural sector.

While migration to towns means that initially a man is often separated from his wife, this has not substantially lowered polygyny rates, although the marriages may take a different form, even in some rural centers. A survey of Accra carried out in 1971–72 showed that of 100 married males, 39 were married polygynously; this figure is higher than for some other African cities (e.g., Abidjan, 15) but others (e.g., Odienne, 38) come close. At the same time it is higher than the overall rate for Ghana itself, which according to the 1960 Census postenumeration was 26, and low in comparison with Guinea (38) and Dahomey (31) (Brass et al., 1968). Polygyny in the urban setting seems to have increased rather than declined, but it takes a different form, as René Clignet (1970) showed for the Ivory Coast. Each of the several wives of one man may have a different home, rather than live in different parts of a single compound, as is the case among rural populations. Only in crowded slums and in planned resettlement villages have I seen two wives living in the same room, squalor indeed from a rural standpoint where, as elsewhere, people can live more spaciously if with fewer "services."

This fact reminds us that to take the norms of the most highly educated group and project these on to the future is a failing of many planners; it is a practice that would have led to very mistaken predictions in the case of European marriage and the family where, in certain significant ways, it was the practices of lower groups that tended to become the later norm (for example, regarding the disappearance of dowry, smaller households, fewer servants, etc.).

Note what is happening to polygyny in Africa. Here the highly educated (university) group displays the lowest rates with 19 percent of marriages being polygynous, according to a survey published in 1978; those with no education had 28 percent, while those with secondary and primary education had 48 percent and 65 percent respectively. The same was true when marriage patterns were measured in terms of wealth ("areas of residence" being the index). "In the high status area, 26 percent of male marriages were polygynous, in the poorest area about the same, but in the middle area the figure rises to 49 percent" (Aryee, 1978: 371). The nature of this distribution makes trends difficult to predict, especially because, for obvious reasons, norms of monogamous marriage, incorporated in the teachings of most Christian churches and the expectations of most modernizers, are internalized by women more readily than by men. But while this fact has an effect on household structure (as I have suggested), it has less effect on the nature of conjugal groups since it does not affect the question of a plurality of women attached to a single male.

I have touched upon the size of coresidential farming groups. Let me turn to consider a number of factors affecting their structure.

Postmarital residence patterns

In earlier Africa, the location of the marital home depended on the system of descent groups. In patrilineal societies, the marital home was the husband's home. In matrilineal societies the location varied, although virilocal residence was still often the case, but usually for a shorter period of a woman's life. Urbanization appears to increase the tendency for the marital home to be defined by the husband and his work, as was also the case in Europe; on the other hand, there urbanization might increase the number of those rural situations (perhaps 10–20 percent or more of the cases) where a son-in-law was called to live with his wife in a filiacentric union, more especially where she was the heir to the farm but also because of the need for his labor (the later the age of marriage, the more pressing the need to find someone to take over the farm, or at least the heavy labor). In parts of Mediterranean Europe (Greece and Cyprus), however, a woman's dowry might include a house or apartment, but its location, under urban conditions, would nevertheless tend to be defined by the man's job. Neither of these situations is to be found in Africa, for reasons connected with land holding and social differentiation.

Household size

One of the simplest variables from which to obtain information on African families is mean household size, which forms a constant of most census records. But the results are frequently difficult to use, since they are usually based on the concept of the hearth (cooking unit), which may be embedded in wider groupings, namely, the farming group, and on the concept of the houseful or dwelling group. Indeed the activities of consumption and production may well extend beyond any boundary-maintaining units through the exchange of food and of labor. Wider kinship relations continue to be important in many contexts, including, in many areas, the "exchange" of children (E. Goody, 1981). The existence of these relations makes calculations based on the value of children difficult to assess, since one's own children are not the only relevant consideration. Particularly in matrilineal societies, maternal uncles assist in putting their sisters' children through school. While in many contexts there is an increasing concentration upon the conjugal family, this is not the case with the inheritance of property in matrilineal societies (e.g., Asante cocoa-farms), where the reduction of oral custom to written law has had the effect of conserving the model, and to a large extent the practice, of uterine transmission—that is, to a full brother and then to a sister's son.

Let us take the average size of the "domestic group" in developed countries to be 3, and averages in developing societies to be around 5 (Mogey, 1978: 3). Households are generally smaller in towns than in the country. Nelson Addo's survey of Southeastern Ghana showed an average of 3.5 against 4.8. In the towns 27 percent of households were female-headed, and these were larger than the male-headed households (4.1 and 3.3 respectively), a fact related to the matrilineal system (1974: 351). Many migrants are single, and even the married, especially the illiterates, come without wives and children; indeed, among the employed, marriage formalities and responsibilities are increasingly avoided by women as well as men. Average households in towns grew in size between 1960 and 1966, whereas those in the villages dropped. This may be due to migration but it is also the case that the richer and better educated have larger households, not only because of servants but also because of kin and the responsibilities for their education. Social mobility has a halo effect. Such extended households, even the more inclusive set of relatives living close together, provide the support needed for having more children because of the childcare functions offered by non-childbearing women, either grandmothers or daughters.

The problems with regard to the number of children are several; under present conditions (which seem likely to change only slowly) one does not know how many will survive and one does not know what sex they will be. Both questions are now responsive to medical science, but a resolution of the second is held back by ethical considerations. The matter is greatly complicated, however, by the likelihood that children who have attended school will leave the village and hence deplete the working group, the structure of which depends not only on numbers and sexes but also on age.

Retirement

Except in the employed sector there has been little of the "early retirement" characteristic of some European regions. That is to say, most elderly people continue to live in rural households as "heads," not as dependents. This was so with 87 percent of men aged 65 and over in the Ghanaian census of 1970. Even 46 percent of elderly women were heads of their own households (Ware, 1983: 28). This is not to say that control is exclusively vested in the senior generation; even in rural areas of a neo-traditional kind, the role of those who have been to school and who participate in or are able to manipulate the "modern" sector is fully recognized. Nevertheless the idea of handing over control finds little favor, except in some pastoral communities where cattle have long been passed on *inter vivos*.

Kin terminology

There is one aspect of the influence of education on family behavior that is of some relevance to the changing family in Africa. Since the languages of instruction and of political action are basically either English or French, individuals constantly have to translate or employ European terms to explain words in their maternal tongues. In this way "brother" gets differentiated into "brother" and "cousin" (or sometimes "cousin-brother") and the terms for father and mother in similar ways. In other words, every school child learns an "individualizing" rather than a "classificatory" terminology in a high-status context. Such usages emphasize the nuclear family (or rather the direct line of filiation) as against the wider ties of lineage or kindred, and therefore tend to set limits to the obligations of relatives to one another.

Community size

The size of local communities, although often difficult to measure since many consist of dispersed settlements, has an effect on family life. Clearly size in turn is affected by political, ecological, and population factors. In some cases, usually among noncentralized peoples, compounds are widely dispersed over the countryside, rendering difficult the provision of centralized services. Among politically centralized peoples and in forest or desert environments, communities tend to be nucleated. Some of these communities constituted "towns" of limited size, but towns and villages are very much smaller than in India or elsewhere in Asia. The consequences for "development" are not always recognized. But it does mean that even in traditional groups, ties of kinship tend to be quite widely scattered, not simply at the level of the clan but also among close relatives.

Demographic factors affecting the family

Population growth

The rate of growth of the population in Africa is now the highest in the world, even though it is generally recognized that, unlike the case in other major areas of the globe, this has not yet reached its peak. In demographic terms this rate of growth has been largely the result of decreased mortality, but the increase in fertility is also of considerable significance.

I have elsewhere (1971, 1975) tried to relate this rate of growth to the socioeconomic background of the continent, with its relatively low levels of technology and production and its relatively open set of agricultural resources. I have also suggested that the demographic consequences of this system of production manifest little of the discrimination against female children that we find in Europe and especially Asia; "stopping rules" indicate no sex preference (J. Goody et al., 1981a and 1981b) nor do hospital attendances. This is what we would expect from a system of marriage transactions involving high bridewealth, where what is sent out for a son's wife has to be brought in by a daughter. The very high child dependency ratios that rapid population growth entails do not appear to have substantially altered this situation, even though marriage payments are becoming of less importance in the employed and urban sectors. In any case the labor of girls is still essential not only in the house but on the farm and in the market. They do not require an endowment, and if they are given education, the main costs are usually borne by the state.

Fertility

There seems little likelihood of any decline in fertility in the near future. Indeed quite the contrary. In an examination of the proximate determinants of fertility in Africa, Bongaarts et al. (1984) distinguish between those that effectively exert an upward pressure and those that tend to reduce fertility:

- Fertility-enhancing trends: shortening of breastfeeding and postpartum abstinence; decline in pathological sterility.
- Fertility-reducing trends: rise in age at first union; higher prevalence and effectiveness of contraception. (p. 533)

The authors conclude that there is no evidence for any reduction in fertility in Africa; if anything there is an upward trend due to the slackening of social taboos on postpartum abstinence and breastfeeding, while pathological sterility is declining; at the same time the adoption of contraceptive practices is very limited and the age of first union is not substantially increasing. Indeed fertility and growth are highest in Kenya, one of the most prosperous of African countries. That country's total fertility rate rose from 6.6 births per woman in 1950–55 to 8.1 in 1975–80, and this situation appears to presage a rise for the rest of sub-Saharan Africa, given adequate subsistence from farming or international aid.

Bongaarts et al. remark that "Large declines in fertility will not occur until these traditional patterns of reproductive behavior are modified" (p. 535). The main obstacles to change are high levels of illiteracy and of infant and child mortality, the large numbers of children desired, the high prevalence of sterility, and the lack of access to health and family planning services. Several of these factors, sterility, mortality, lack of health care, do of course reduce fertility, the implicit argument being that if the risks associated with childbearing are reduced, the need for so many births will decline as well. But in itself that could simply mean adjusting the size of the family at any one moment to what it was before. Given that education (literacy) in Africa has a limited effect on the reduction of fertility, we are left with "the desire for large numbers of children," which was the problem from the beginning.

Undoubtedly the high levels of fertility are related to high levels of mortality. The problem has been that in Africa, as in early nineteenth century England, there has been a decrease in the mortality of children without any immediate adjustment in terms of fertility. Indeed fertility is increasing; the number of children desired is high.

Discussions about the reversal of wealth flows suggest that when children cease to be regarded as an asset, then fertility will drop. Since asset has to be defined in general terms, the hypothesis is difficult to test, especially in polygynous families where it is the number of a woman's children, not a man's, that is significant, and especially where costs and benefits may be shared between reproductive units; indeed the state itself takes on the bulk of expenditure on education, a fact that family strategies have taken advantage of. The decreasing availability of land, and economic downturn more generally, may reverse this situation in the longer term, since population is increasing so rapidly and since the state has found itself shouldering burdens it cannot possibly sustain. But what is critical is the reaction of families to the changing situation (since states would find it difficult to intervene on the Chinese family planning model) and here neither recent European nor Asian experience is of much help.

Checks on fertility

The positive checks to population growth are war, famine, and other conditions that limit the span of natural life. The preventive checks are the postponement of marriage, postpartum prohibitions on intercourse, continence, contraception, and other means that, intentionally or not, prevent the rise in population.

In Africa, the positive checks were undoubtedly of greater importance in the precolonial period. These adverse factors were relieved by the political control exercised by colonial powers, by the advent of transport systems that could alleviate local famines, by the improvement of health, and by developments in agriculture, especially the availability of higher yielding varieties of rice, maize, and sorghum. Looking back to the nineteenth century, another factor was the greater availability of cheap iron, principally for hoes and machetes, but also in the form of guns, and hence for the capture of animal protein, although the wild has become a rapidly dwindling resource.

As for preventive checks, while marriage was often postponed for men, women were married early; indeed this imbalance was intrinsic to polygyny. Polygyny necessarily means not only delay for men but some "continence" for individual wives, unless one supposes that polygynously married men are sexually more active than monogamous husbands. So it is not surprising to find that plural marriages are less fertile than others. While we may wish to count this form of marriage as a check on growth, it is not intentionally so. From the standpoint of the actor, the aim is quite the opposite. The same may be true of the lengthy postpartum taboo. It has been argued that in tropical forest areas a protein-deficient diet encourages a long postpartum break from intercourse; the postponement of childbearing, in turn, encourages men to take another wife polygynously. A long postpartum taboo is also found in societies where monogamy prevails but there is clearly a logical link with polygyny. It seems likely, however, that, if intention can be ascribed, the postponement of childbearing was aimed at maximizing the holding of children at any one time; hence, like polygyny, it may have limited childbirth but not family size.

Sterility

As is well known, across Central Africa from Kenya to the Cameroons runs a belt of very high sterility that has substantially reduced the capacity for childbearing. To the extent that this condition is subject to medical control, there is obviously potential for further increases in fertility. As we have seen the same is true for the decline of the postpartum sex taboo with the weakening of public sanctions against shorter birth intervals, and for polygyny, the future lessening of which could have the same result.

Mortality: Health and disease

Enormous advances have been made in tropical medicine over the last four decades, including the provision of hospitals, doctors, and clinics throughout the African continent. Such services suffer from the same economic and political considerations as the rest of the social system but are characterized by the great expenditures that individuals are prepared to make to maintain the health of their offspring and other kin. While state provisions have often diminished in recent years, they have been supplemented by private payments (often under the counter). At the same time health, like famine, is an area in which foreign agencies—governmental, religious, and humanitarian—are most prepared to help, and this aid is a continuing factor in the support of such facilities.

Given the dramatic improvement in curing widespread diseases, such as yaws and leprosy, it would be wrong to stress only the failures. Nevertheless the worldwide growth of chloroquine-resistant strains of malaria, the disease that has been the prime cause of infant mortality after diarrhea, is an obvious cause for concern, and a factor likely to militate in favor of continuing high rates of fertility. On the other hand the work of European historians that suggests a connection between high rates of infant mortality and the lack of child-orientation or of a sentimental attachment to children in the premodern period (Ariès, 1965; Stone, 1977; Shorter, 1975) receives no support from the African experience and hence in itself provides no basis for prediction about the course of family life.

The attention of the outside world has been yet more dramatically caught by the advent of AIDS, generally supposed to have reached endemic proportions in parts of Africa. In terms of family structure, its impact may well be greater than that of malaria since it touches the adult rather than the infant population; current projections for Africa indicate that it may radically affect both agricultural and urban productivity.

Models of family structures

For the colonial period it was possible to look at the African family in a given ethnic group as relatively undifferentiated in its structure. This is no longer true: many ethnic differences still exist, but they are complicated by class and national distinctions. While the use of countrywide aggregate statistics tends to lead to a new "homogeneity" in respect of developments, the actual situation requires us to elaborate a number of models in order to offer even an approximate framework for analysis. Such models have to take account of three levels:

(i) "class" (or hierarchical differentiation),

(ii) "ethnicity" (local cultural differentiation),

(iii) "national" differences, relating to political policies, economic pressures, and religious and educational factors.

"Class"

Considerable scrutiny has been given to elite families in Africa, especially in terms of the strength or jointness of conjugal bonds as compared with kin ties and the attention devoted to children as compared with parents. It is undoubtedly true that those who first attained positions in the new system of occupational roles, eligibility to which was largely through education, continue to try to reproduce their status in their offspring. The strategies involved include emphasizing the use of the official language (French or English) in the home and the selection of appropriate primary schools, often private. But at the secondary and university levels most investment comes from the state; in this group the children's labor that has been forgone is not a material consideration but in any case can be made up with more children, possibly borrowed from kin or, less usually, with hired labor.

In contrast to the earlier post-independence period, the position and composition of this elite group has changed. Economic factors have tended to bring senior government employees closer to what one might call the salaried class, which has altered their strategies of gaining a livelihood as well as of social reproduction. Political changes have ensured that the military, professional politicians, and economic intermediaries constitute a much greater element of the "upper" echelon, better at effecting decisions and at reaping benefits than the university-trained professionals who served as earlier role models. Economic and political instability has radically affected the lifestyles of all these elements, whose fortunes ebb and flow with considerable rapidity, leading in many cases to partially enforced emigration or exile. In the early 1980s, the two countries providing the most political refugees to Britain were, first, Iran, followed closely by Ghana. Other African countries have acted both as receivers and providers of refugees. As a result, many families are physically split (as with labor migration), often as a deliberate strategy; it is important to maintain one foot abroad, even in the case of families in the military.

Two other "class" categories have already been discussed in the context of family strategies. The clerical or salaried class largely consists of government employees, who in many cases now constitute the salaried farmers of the rural areas. The neo-traditional farmers and herders adapt or modify strategies to deal with the increase in market activity, the "loss" of personnel through education, and the advent of new crop varieties and techniques, but they continue to operate in the "traditional" sphere and are less affected than others by the surrounding uncertainties. A fourth consists of the urban working class, an amorphous group with many continuing connections with the rural areas from which they came in the course of the rapid build-up of the town populations. Their demographic behavior is often influenced by their continuing access to the resources of the countryside, but in general the size of households is lower.

"Ethnicity"

So far I have concentrated upon general factors affecting the segment of the rural African population that has received some education and that has some employment, usually the inhabitants of small townships rather than of the villages themselves. The neo-traditional village families are affected by the migration of men and women to the towns, by the imposition of taxes, and by the fluctuating prices of foodstuffs as the result of the changing demands of towns, of some nonfarming rural workers, and of boarding schools. There are two sources of variation that we need to take into account, now and in the future: the differences in the traditional structures; and the differential impact of vectors of change.

Among the traditional factors are differences in family structure. Some African groups practiced matrilineal inheritance (of property), succession (to office), and descent (to social groups); others patrilineal; some both concurrently; and yet others displayed an absence of descent groups (i.e., were bilateral) but transmitted property and office from parents to children. Each of these systems has changed but they retain certain earlier characteristics, the strength of which depends upon position in the new hierarchy and the impact of external factors. Take the matrilineal case.

Matrilineal inheritance and succession persist under a great variety of changing circumstances. While it may be that such systems initially developed under conditions of early hoe-farming, where the woman's harvesting role in hunting and gathering societies may have led to her assumption of farming tasks, they are found today under a variety of socioeconomic conditions, by no means the least complex, as the examples of the Nayar of South India and the Asante of West Africa show. The Asante, who formed an important military state with rich forest and mineral resources, became early cash-crop farmers when they adopted cocoa at the end of the last century. While Christian missionaries and legal politicians have made efforts to instigate a change to a European mode of reckoning, with transmission from husband to wife to children, the Asante persist in passing down much landed property through the uterine line. Matrilineal "family farms" are the typical units of much cocoa production (Hill, 1963). Some observers report an increasing tendency to transmit property to the wife and son (e.g., Sarfoh, 1974: 137), that is, within the conjugal family. But early records and standardized myths suggest that some transmission to sons, at times in secret, was always a feature of such systems. Typically men inherited from men, women from women, but there now appears to be an increasing emphasis on the passage of "gifts" from husband to wife, reflecting a widening of conjugal ties and a corresponding narrowing of lineage ones (Oppong, 1974). The transmission of property to wife and sons, which was insisted upon by the early Basel Missions, for example, as a way of getting rid of matrilyny and instilling a Christian marriage, meant that even brothers would be excluded; hence the sibling group was likely to split economically before this event occurred (Poh, 1975: 64).

It was not only in matrilineal systems that there was traditionally no transmission of property between husband and wife, but it was there that the continued attachment of the wife to her own kin was most noticeable. It was members of the "lineage," whether matrilineal or patrilineal, rather than husband and wife (members of different lineages by definition), who inherited a man's property, brothers first (uterine or agnatic), then either sister's sons or sons, depending upon the dominant mode of reckoning. In the case of women, daughters (and sometimes sons) inherited from their mothers. In order to maintain the system men inherited from men, women from women.

The greatest strain appears in matrilineal societies, not so much because of the fate of the widow, but because of the children. The system came under early pressure from European legal ordinances as well as from representatives of the Christian churches (de Graft-Johnson, 1974; Woodman, 1974), who regarded it as their duty to press for provision for widows and orphans. While they had some limited success, the matrilineal family has displayed great powers of persistence even under the new circumstances of cash-crop production, and this mode of social organization has even been incorporated into the written laws of the new Ghanaian nation, thus giving it renewed life. The shift from oral custom to written law has had some conservative influences on family life.

"National" differences

The political force most likely to influence family structure is the general instability of national regimes. The frequent overthrow of civilian governments by military personnel, the slow return to some form of civilian rule, then the repetition of the cycle—it is difficult to see this pattern changing in the near future, although it has been much less marked in some territories than in others. Given this situation, we are likely to find little continuous government interference with family life, even at the level of promoting a population policy. Meanwhile family strategies will tend to be defensive in kind, placing little reliance on the prospect of state help as far as pensions and other benefits are concerned. Even in those countries where centralized civilian control has managed to fend off military intervention, the sums needed for effective intervention in welfare are either not there, or else have been whittled away by inflation. Consequently the family has an important part to play in protective strategies, even in those cases where it has ceased to be a productive or residential unit.

Education has some relationship with fertility in Africa, but no overall pattern emerges. As I have suggested, "education given" (putting children into school) may have a positive effect, looked at from the standpoint of the working group. But this may be the case even with "education received" (the children of the schooled). In Ghana, for example, an inverse relation appears to exist between education and fertility (Page, 1975: 55). Female literacy rates are higher there than in Africa as a whole, 34 percent in 1970 for females aged 6 and over as compared with 16.3 percent for the whole continent (UNESCO Statistical Yearbook, 1972: 47).

Rural schooling currently suffers from a lack of intensity. As a result of the changes in food prices in relation to teachers' incomes, the amount of time teachers spend in school is often much less than official timetables would suggest. At times boarding schools have been unable to provide sufficient food for the pupils, who therefore have to be sent home. Even if attendance figures appear to be increasing, the quality (and even quantity) of education is certainly suffering. Indeed in the Gambia, as elsewhere, there appears to have been some falling off in schooling over the past decade (Weil, 1986: 296). In any case, the less children participate in school, the more they are available to contribute to the family.

Educated women generally postpone marriage and then establish separate households when they do so. This is partly because they tend to work in towns, where residence patterns are necessarily different if only because of housing. Partly too, participation in full-time education is seen by most women as incompatible with marriage, given the nature of the commitment required from students and of the accommodation available to them. Nevertheless schooling brings boys and girls together outside the house so that the family has less control over the age at sexual intercourse (as distinct from marriage) and the choice of partners. Schoolgirl pregnancies are not uncommon, even if marriage is sometimes postponed. In fact education in itself puts off marriage only for those who attend the senior classes of secondary schools with a view to proceeding to tertiary education. But whether or not schooling influences the age of marriage, there is no clear association between fertility and education in Africa, either given or received (Timur, 1977).

Discussion: Decisionmaking under uncertainty

Fluctuating economic and political conditions mean that many processes of decisionmaking have to take account of uncertainty on a large scale. While a process of unilineal development did seem possible to actors and observers at the time of independence, such predictability was short-lived. Even before that period, many parents had placed some children in school, keeping others on the farm—a matter of insurance to them and their families, rather similar to the technique of families elsewhere of putting one member into the church and another into the army (or in some places the ruling party).

What makes this process more relevant at the present day (and in the "foreseeable future") is that there is little stability even in the career one chooses, or has been directed to, for the following reasons:

(a) the prices of crops, machinery, imports, and labor fluctuate enormously, leading to the necessity for shifting between one crop and another;

(b) salary levels are profoundly affected by inflation and by different political regimes;

(c) goods come and go depending on loans from the World Bank and other international agencies, meaning that at one moment sudden imports of maize may sell more cheaply than the local product, then stop altogether. It is the same with the ingredients for soap, or with iron. When these disappear one has to resort to an earlier technology that depends less upon outside help. Roads of an almost freeway standard revert to worse than dirt tracks, since the tarmac gets holed so unevenly. Then a subvention from outside may return the situation to what it was. Moreover such fluctuations are endemic, even without the additional complication of civil war.

There seems an important reason why the kinds of conditions that are developing in Africa should lead to smaller decisionmaking groups (from the economic standpoint) than are found in many Asian societies, at least in the middle and upper strata. In Chinese law, at least in contemporary Taiwan, and in Hindu law, familial property may continue to act as a focus for joint concern long after the death of the head; looked at from a particular standpoint, partition, at least of the basic productive resources, is delayed and it has to be undertaken formally.

In Africa, land did not fall under the control of domestic groups in the same way. An individual had a right to farm lineage land not being used, so that only rights of ownership were limited, especially in regard to powers of alienation. Division was related to use.

When wage employment or even the cultivation of cash crops expands, the impact is different in these two systems. In Asia, the "joint undivided family" often retains a unity with regard to property and possessions; for example, in the Gujarati village of Nandol, the chairman of the council, a farmer, and his brother, a university professor, pooled their incomes in one bank account, for they had not "divided." This would, I think, be unheard of in Africa. That is not to say kin do not help one another. It has often been remarked that rich men have to support many relatives. But they do so, generally, in terms of help and handouts. Affairs are rarely managed jointly.

Domestic groups in Africa do hold other forms of property and do cooperate. But land rights like land usage are "extensive" rather than "intensive" and there are fewer inhibitions on fission. Of course we find large farm groups that cooperate closely. But when the situation changes and the reason for such cooperation disappears or diminishes, then it is easier for fission to occur. Again, such fission does not mean that cooperation of a more diffuse kind does not take place. But individualistic economic practices seem to emerge more readily, as in the case of the "salaried farmers" to which I have referred.

It should not be thought that these "individualistic" attitudes or means of organization are necessarily more conducive to development; that is a Western European myth. They may facilitate worker participation in industrial activity just as they certainly facilitate and are facilitated by migration, especially rural migration. But it is the "undivided" families of Taiwan that are better able to raise capital and provide support for entrepreneurial investment and management.

A second point has to do with the variety of social forms found in Africa, in contrast to the relative standardization of societies influenced by the Great Traditions associated with world religions or with imperial rule. As a consequence there exist a whole range of adaptations to change, which are often concealed in aggregate figures. For example, if averaged out, differences in the distribution of the institution of fostering would be reduced to a mean rate for the country or region as a whole. Differences in the indigenous practices of "ethnic" groups do matter in a variety of ways, not only to the actors but also in their implications for the family's adaptation to development. For example, the existence of fostering anticipates and encourages the more recent practice of sending children to stay with relatives to learn new trades or to enter new jobs (E. Goody, 1981).

Conclusions

I have tried to survey a wide range of factors affecting the future of the family in Africa. In conclusion, I would draw attention to the following points.

(1) The diversity of preexisting social forms and the differential impact of external pressures.

(2) The common features related to the nature of an agricultural system with relatively open access to resources and its relationship to migration and the growth of population, now more rapid than elsewhere in the world.

(3) The effects of increasing population on land and upon limited material resources.

(4) The lack of a "joint family organization" in the Asian sense, which renders fission at the domestic level an increasing possibility.

(5) The corresponding strength of the wider kinship ties of lineage, weakening under changing modes of livelihood but still important for welfare purposes.

(6) The effects of mass schooling and mass bureaucratic employment on the economy, with the development of "salaried farming" as a new mode of adjustment, based on a reduced household.

(7) The concomitant existence of neo-traditional farming.

(8) The reaction of all types of "farm families" to the high degree of uncertainty in all spheres of social life and the effect of such uncertainty on decisionmaking and activity within the household.

Note

The author is much indebted to Esther Goody for her contribution to this essay, and to Dafna Capobianco for help with the bibliography and word-processing.

References

Adepoju, A. 1975. "Some aspects of migration and family relationships in South West Nigeria," in C. Oppong (ed.), *Changing Family Studies,* Legon Family Research Papers, no. 3, University of Legon, Accra.

Addo, N. 1974. "Household patterns among urban and rural communities in south-eastern Ghana," in C. Oppong (ed.), *Domestic Rights and Duties in Southern Ghana,* Legon Family Research Papers, no. 1, University of Legon, Accra.

Alexander, J. 1977. "The role of the male in the middle-class Jamaican family: A comparative perspective," *Journal of Comparative Family Studies* 8: 369–389.

Ariès, P. 1965. *Centuries of Childhood.* New York: McGraw-Hill.

Aryee, A. F. 1978. "Urbanization and the incidence of plural marriage: Some theoretical perspectives," in C. Oppong et al. (eds.), *Marriage, Fertility and Parenthood in West Africa.* Canberra: Australian National University.

Bongaarts, J., et al. 1984. "The proximate determinants of fertility in sub-Saharan Africa," *Population and Development Review* 10: 511–537.

Brass, W., et al. 1968. *The Demography of Tropical Africa.* Princeton: Princeton University Press.

Clignet, R. 1970. *Many Wives, Many Powers.* Evanston, Ill.: Northwestern University Press.

de Graft-Johnson, K. E. 1974. "Succession and inheritance among the Fante and Ewe," in C. Oppong (ed.), *Domestic Rights and Duties in Southern Ghana,* Legon Family Research Papers, no. 1, University of Legon, Accra.

Dorjahn, V. 1959. "The factor of polygamy in African demography," in M. Herskowitz and W. Bascom (eds.), *Continuity and Change in African Cultures.* Chicago: University of Chicago Press.

Goody, E. 1973. *Contexts of Kinship: An Essay in the Family Sociology of the Gonja of Northern Ghana.* Cambridge: Cambridge University Press.

———. 1981. *Parenthood and Social Reproduction.* Cambridge: Cambridge University Press.

Goody, J. 1971. *Technology, Tradition and the State of Africa.* Cambridge: Cambridge University Press.

———. 1972. "The evolution of the family," in P. Laslett (ed.), *Household and Family in Past Time.* Cambridge: Cambridge University Press.

________. 1975. "Population, economy and inheritance in Africa," in R. P. Moss and R. J. A. R. Rathbone (eds.), *The Population Factor in African Studies.* London: University of London Press.
________. 1976. *Production and Reproduction: A Comparative Study of the Domestic Domain.* Cambridge: Cambridge University Press.
________. 1983. *The Development of the Family and Marriage in Europe.* Cambridge: Cambridge University Press.
________. 1986. *The Logic of Writing and the Organisation of Society.* Cambridge: Cambridge University Press.
________, et al. 1981a. "On the absence of implicit sex preference in Ghana," *Journal of Biosocial Science* 13: 87–96.
________, et al. 1981b. "Implicit sex preferences: A comparative study," *Journal of Biosocial Science* 13: 455–466.
Hill, P. 1963. *The Migrant Cocoa-Farmers of Southern Ghana: A Study in Rural Capitalism.* Cambridge: Cambridge University Press.
Isiugo-Abanihe, U. C. 1985. "Child fosterage in West Africa," *Population and Development Review* 11: 53–73.
Kramer, S. N. 1954. "Ur-Nammu Law Code," *Orientalia,* series 2, 23: 40–51.
Laslett, P. 1972. "Mean household size in England since the sixteenth century," in P. Laslett (ed.), *Household and Family in Past Time.* Cambridge: Cambridge University Press.
Locoh, T. 1982. "Demographic aspects of the family life cycle in sub-Saharan Africa," in *Health and the Family Life Cycle.* Geneva: World Health Organization.
Meillassoux, C. 1986. *Anthropologie de l'Esclavage.* Paris: Presses Universitaires de France.
Mogey, J. 1978. "An essay on irregularities in family research," in C. Oppong et al. (eds.), *Marriage, Fertility and Parenthood in West Africa.* Canberra: Australian National University.
Murray, C. 1981. *Families Divided: The Impact of Migrant Labour in Lesotho.* Cambridge: Cambridge University Press.
Oppong, C. 1974. *Marriage Among a Matrilineal Elite.* Cambridge: Cambridge University Press.
________. 1983. *Paternal Costs, Role Strain and Fertility Regulation: Some Ghanaian Evidence.* Geneva: ILO.
________. 1985. "Reproduction and resources: Some anthropological evidence from Ghana," in G. M. Farooq and G. Simmons (eds.), *Economic Analysis of Fertility Behavior in Developing Countries.* New York: Macmillan.
Page, H. 1975. "Fertility levels: Patterns and trends," in J. C. Caldwell (ed.), *Population Growth and Socioeconomic Change in West Africa.* New York: Columbia University Press.
Poh, K. 1975. "Church and change in Akwapem," in C. Oppong (ed.), *Domestic Rights and Duties in Southern Ghana,* Legon Family Research Papers, no. 1, University of Legon, Accra.
Sabelli, F. 1986. *Le Pouvoir des linéages en Afrique: la réproduction sociale des communautés du Nord-Ghana.* Paris: L'Harmattan.
Sarfoh, J. A. 1974. "Migrant Asante cocoa-farmers and their families," in C. Oppong (ed.), *Domestic Rights and Duties in Southern Ghana,* Legon Family Research Papers, no. 1, University of Legon, Accra.
Shorter, E. 1975. *The Making of the Modern Family.* New York: Basic Books.
Smith, R. T. 1956. *The Negro Family in British Guiana: Family Structure and Social Status of the Villages.* London: Routledge and Kegan Paul.
Stenning, D. 1959. *Savannah Nomads: A Study of the Wodaabe Pastoral Fulani of Western Bornu Province, Northern Region, Nigeria.* London: Oxford University Press.
Stone, L. 1977. *The Family, Sex and Marriage in England, 1500–1800.* London: Weidenfeld and Nicolson.
Timur, S. 1977. "Demographic correlates of women's education: Fertility, age at marriage and the family," unpublished paper.
Ware, H. 1983. "Female and male life-cycles," in C. Oppong (ed.), *Female and Male in West Africa.* London: Allen and Unwin.
Weil, P. 1986. "Agricultural intensification and fertility in The Gambia (West Africa)," in W. P.

Handwerker (ed.), *Culture and Reproduction: An Anthropological Critique of Demographic Transition Theory.*'' Boulder, Colo.: Westview.

Woodman, G. 1974. ''The rights of wives, sons and daughters in the estates of their deceased husbands and fathers,'' in C. Oppong (ed.), *Domestic Rights and Duties in Southern Ghana,* Legon Family Research Papers, no. 1, University of Legon, Accra.

COMMUNITY AND GOVERNMENT

Social Organization and Ecological Stability under Demographic Stress

GEOFFREY MCNICOLL

IN SOME LATE REFLECTIONS ON AGRARIAN CHANGE in Southeast Asia, the geographer Karl Pelzer (1968: 272–273) described the encroachment of wet-rice farming by Javanese migrants on formerly swidden areas of Sumatra, based on his observations over several decades:

> [P]arts of South Sumatra that for several thousand years had carried the imprint of shifting cultivation had now undergone a drastic change involving complete removal of the second-growth forests; the leveling, bunding, and, where necessary, terracing of the land; and, finally, construction of irrigation works which included dams, water distribution headworks, and a system of canals. . . . A new feature of the landscape was the presence of large rice mills to handle the rich harvest of an area that had become in a relatively short time span a new rice granary capable of supporting some four hundred to five hundred persons per square kilometer. . . .

Forest clearing does not necessarily lead to these large and continuing economic gains. There are new potential sources of instability: terraces require substantial labor allocated to maintenance—the equivalent of investment to cover capital depreciation—and assurance that upstream water run-off remains under control. The loss of species diversity makes for greater plant disease and pest problems. Thus vulnerability may be increased along with productivity, and an indefinite maintenance commitment taken on. In the same article, Pelzer (ibid., p. 276) recalls a contrasting Philippine instance, involving migrants from the severely eroded—in large parts virtually treeless—island of Cebu to the Philippines' land frontier in Mindanao:

> In 1940 I visited pioneering Cebuano maize cultivators on the hills facing the north coast of Mindanao in the provinces of Misamis Oriental and Occidental. Returning in 1951 I found that they had already ruined their first homesteads and moved into

> the interior where they were "eating the forest and the soil." It will only be a matter of three to four generations before large parts of Mindanao will be a replica of Cebu.

More recent evidence of the pace of denudation in Mindanao suggests that two generations would have been a more accurate forecast (see Ooi, 1988).

Rural population growth and its resulting intensification of settlement and agricultural activity can thus transform a landscape in ways that lead to permanent gains in economic productivity, subject to upkeep being met; or they can set up processes that lead to a fall-off in productivity, perhaps ending in abandonment of the land. The question posed in this essay is: what kinds of factors make the difference in the economic-ecological outcomes?

In very broad terms the answer is clear enough. One set of factors influencing the outcome describes the resilience of the ecosystem under human impact—how difficult is it to raise its productivity without endangering ecological stability? (I am using "stability" in a limited, nonpurist sense: the amenity-value of the environment and any wilderness characteristics it had retained may be drastically altered.) Ecosystems, of course, are also in flux for reasons that have nothing to do with the local human impact. Such underlying processes are usually too gradual to be relevant in considering that impact, but not always so.

A second group of factors influencing economic-ecological outcomes describes the nature and intensity of human activities impinging on the ecosystem. In a stylized picture, individuals or households, in pursuit of their economic welfare, seek to slew off the negative effects (including environmental effects) of their economic activity onto others—just as they seek to arrogate positive effects of others' activities to themselves. They may do so individually or collectively, directly or strategically (through a political system). They are, however, variously constrained from acting in their rank economic self-interest by common or codified law, by intervention of government authority, by social pressures (backed by sanctions or threat), or by their acculturation to conservative customary practice.

Ultimately, economic-ecological outcomes under conditions of population growth are determined by the interactions of these two groups of factors. For exposition, however, it is convenient to start by discussing the factors separately. I take most of my illustrations from alpine agrarian settings, particularly in South Asia, where the population-environment nexus is becoming fairly well documented and the time-scale of many relevant processes is foreshortened.

Ecological resilience

Agriculturalists, once taken by scholars as well as officials to be ignorant if not actually foolish, are now routinely assumed to be wise and rational; rarely are they seen as presumably they are: people like others, error prone, partially informed, culturally blinkered to varying degrees, and somewhat arbitrarily mixing short- and long-run considerations in their decisions. For ordinary people like these, it matters quite a lot how forgiving their environment is.

The case of alpine environments represents an extreme of unforgivingness, yet ecological outcomes there have ranged from prototypical disaster—some of the most striking contemporary instances of ecological degradation—to a fair degree of stability. These situations thus provide a rich store of empirical materials for a discussion of resilience.

The Himalayas, by most accounts, lie in the disaster domain. The case of highland Nepal is frequently described, but similar trends are reported in the Himalayan regions of India and China—notably, northern Uttar Pradesh, Himachal Pradesh, northwestern Yunnan, and western Sichuan. Erik Eckholm (1975: 769) concludes his bleak summary of Third World alpine situations, referring mainly to the Himalayas, with the prospect that "many mountain regions could pass a point of no return within the next two or three decades. They would become locked in a downward spiral from which there is no escape, a chain of ecological reactions that will permanently reduce their capacity to support human life." He cites Nepal's Planning Commission as foreseeing "the development of a semi-desert type of ecology in the hilly regions." Downstream, the lowlands will experience "devastating torrents and suffocating loads of silt" (ibid., pp. 764, 765). Eckholm's language and the urgency it conveys are echoed in numerous other reports, both scholarly and popular.

The immediate causes of degradation are generally held to be routine agricultural activities pushed to unsustainable intensities: grazing and fodder collection, tree-felling for timber and fuel, and expansion of cropland on potentially unstable slopes. The real culprits, in most of these accounts, are, in varying proportions, population growth and poverty; market penetration, "commoditization," and corporate self-interest; and ill-considered government land and forest policies. Among these, population growth is often the principal villain, albeit sometimes behind the scenes, operating either through a "tragedy of the commons" situation, in which each family's individual childbearing interests produce in aggregate an overexploitation of the environment, or, where those interests *are* consistent with ecological stability, through the inability of families to limit children to that desired number—the implicit and perhaps dubious premise of many family planning programs.

In the Himalayan case, one strand of recent research has called into question this set of explanatory relationships—what Jack Ives (1987) has called "the theory of Himalayan environmental degradation." The acceleration of population growth after 1950 did not apparently result in a major intensification of deforestation. Indeed, based on the fragmentary data that can be assembled, serious deforestation and erosion in the region were already evident two centuries ago. Virtually every linkage that casual observation has seemed to indicate—forest destruction by collection of fuelwood, for example—can be shown to lack secure empirical support. In the hill districts studied by D. A. Gilmour (1988), for instance, there appeared to have been little change in forest boundaries over the past century, although there was some decline in the density of tree cover. "The big linkage, population growth and deforestation in the mountains leading to massive damage on the plains, is not accepted" (Ives et al., 1987: 332).

Himalayan erosion, in this revisionist view, is primarily a long-term geophysical process, one that long preceded human settlement. Human impact can be additionally damaging—for example, by poorly engineered road construction and clear-cutting of forests—but can also be ameliorative. Terracing, in particular, is usually in the latter category, tending to slow rather than to exacerbate erosion.

It seems that we must recognize two superimposed processes of ecological change. One is the consequence of geophysical forces, elevating mountain ranges and eroding them. The other is the consequence of human activity in seeking to raise land productivity to accommodate steadily rising population numbers. For the plainspeople, the former should be the most worrying: the Ganges plain, it has been pointed out, already has a base of Himalayan silt to a depth of 100 meters or so; some collapsed terraces or road construction landslides in the mountains are scarcely the crux of the flooding problem. For the villagers in the hill country, the geophysical processes create the fragile background against which their agrarian economies must operate. But are they then, as many would have it, routinely devastating the landscape, "eating the forest and the soil" and moving on? Or are they more often retarding for a time the inexorable large-scale trends, their bunds and forest commons creating islands of at least temporary stability? And if (most likely) there are cases enough of both kinds, what makes the difference and does rapid population growth tip the balance toward the first? These questions will be taken up below.

Some further light on ecological resilience can come from the other extreme of alpine ecosystems, where stability at first glance is seemingly assured: the European alps. One difference here is of course the absence of a monsoonal climate, creating a somewhat less fragile landscape (although in that respect the Nepal Himalayas differ also from the western Himalayas). But agrarian pressures were not always as contained as they appear now under low fertility and net population exodus. Christian Pfister (1983: 297) remarks that from the sixteenth to the late nineteenth centuries, "the Hirtenland [Swiss highlands] resembled very much the overcrowded mountain ecosystems of the developing countries today." His description has a familiar ring to it: new crops and cultivation techniques permitting denser settlement, and both humans and livestock overstressing the natural environment. "The increase in the number of goats, and the greater fuel demands of the larger population, caused serious degradation of the mountain forests" with consequent floods, avalanches, and topsoil loss (ibid., pp. 293–294).

Alpine environments are not unique in the characteristics of interest for the present subject. An analogous situation in which "naturally" changing environmental conditions combine in uncertain proportions with effects of human activity to yield ecological degradation is that of desertification in the Sahelian region of Africa. A comparable array of ecological processes, involving climatic change rather than mountain formation and weathering, creates fragile environments in the semi-arid fringes that are easily overstressed by efforts to raise productivity. Patterns of transhumance, like hill-terracing, are variously seen as conserving or threatening ecological stability. Increasing human and animal densities are believed by many to promote desertification, by others merely to raise the numbers potentially at risk from it. (See Mortimore, 1989, and Hill, this volume.)

The pace of productivity improvements demanded by modern rates of population increase and heightened economic aspirations, and permitted by the transfer of agricultural technologies developed elsewhere, would strain the most resilient of ecosystems, let alone the fragile ones considered here. (Even in historical times, moreover, the introduction of new technologies could be severely disruptive. The introduction of potato cultivation into the hill agricultural systems of Europe in the eighteenth and early nineteenth centuries, as later in Nepal, represented a shock of that kind, allowing population densities to rise and often setting the stage for subsequent production crises.) Yet the outcomes are not uniformly disastrous, nor explicable in terms only of ecosystem parameters. Critical dimensions of the response—largely determining success or failure in environmental management—are found in social organization.

Social organization of environmental management

One dimension of the diversity of human societies is the demand placed by their individual members on the natural environment. Unlike the case with nonhuman species, little can usefully be said about human population–environment interrelations until details of technology, social organization, and culture are brought into the discussion. The concept of carrying capacity, for example, familiar in the ecological literature with reference to nonhuman species, can be meaningfully applied to human populations only when those details are specified. When they *are* specified, and notionally held constant, well-defined economic consequences of demographic change can be discovered and their implications for ecological stability gauged.

But the convenient device of holding other things constant yields only a partial derivative. Demographic change potentially has effects *on* technology, social organization, and culture, in turn giving rise to a new material, institutional, and cognitive context for economic-demographic processes and a possibly very different ecological outcome. It is the total derivative we are seeking.

Moreover, random events, small and large, undermine any strong determinism in the system's development. Hence by necessity we must be interested in the complexities of the social system on which demographic change is operating—and from which it is being itself modified through feedback effects of various kinds.

A persisting myth that lies behind many popular discussions of environmental degradation, usually implicit, is that there was some "traditional" social order under which stability was assured, which was then disrupted by external forces—such as the educational or public health advances that can rapidly improve rates of child survival, or the attractions of central planning. The old system of governance breaks down, and restraints on individual, corporate, and bureaucratic rapacity are loosened. The picture often comes replete with political message: arcadia was likely to have been communitarian or anarchosyndicalist as well as green, at least at the local level; the disruptions were capitalist or (recently conceded) Leninist.

That cases can be found which more or less follow such a pattern does not of course establish the pattern as typical, though it does suggest that we are not dealing with a mere straw man. Each element in the pattern, however, is dubious as a general proposition. Allowing for the truncation of evidence that results from the fact that utter ecological failure leaves little social residue, premodern societies probably differed and differ as much as modern societies in their treatment of landscape. There are some in which extreme ecological conservatism is virtually an aspect of religious belief—in which, for example, an extended reality peopled by spirits makes the status of the living and corporeal population no more than curator. There are others (recall Pelzer's Cebuanos in Mindanao) that would have nothing to learn from the worst contemporary cut-and-run despoilers of the environment aside from the technological prowess to permit a truly modern pace of exploitation. The majority of "traditional" societies lie somewhere in between, having mixed success in preserving a rough measure of ecological stability in the face of the evident interests of some members individually (and the evident capacities of the more powerful and enterprising of them) not to be bound by a social compact in this area or any other.

Constructing a plausible story out of these elements calls for filling out a description of local social realities: demography and economy, social organization and governance, and the cultural system. Doing so immediately shifts the locus of attention to precisely those realities. It is not that population growth is therefore ecologically unimportant; rather, its degree of importance is contingent on how the social system, broadly defined, responds to it. The persistence of population growth in the face of serious ecological degradation indicates an inadequate response.

There are many potentially pertinent elements of rural social organization. For a discussion of local environmental issues, however, the critical ones are those that define the territorial divisions of society and supply them to a greater or lesser extent with an instrumental capacity. The principal divisions of interest here are villages and local government areas. It is true that both of these tend to lose significance over the course of economic development, being supplanted by nonspatially based institutional forms. Moreover, even in premodern agrarian societies social relations cannot be exhaustively described in terms of those categories: class and kin are merely the most obvious of other principles of affiliation that call for attention. Nevertheless, the simple picture of rural society as a hierarchy of household/village/local government district/national government can serve adequately as a frame for this brief analytical foray.

Family-level spillovers

Consider a notional "society" composed of autonomous economic agents (households) grouped into territorial "communities" small enough to allow face-to-face contact among members—that is, small enough for "noticeability" in the sense of Mancur Olson's (1971: 45) treatment of the grounds for collective action. A simple analysis of environmental degradation in such a setting might distinguish four categories of inter-household transfers of costs, each providing a structural basis or explanation for the process of degradation.

(1) *Hardin-type transfers* Within a community, to the extent that one person's behavior generates negative external effects on others, there will often be some compensating transfer, offered or exacted, that on balance yields a rough measure of reciprocity. For population-induced degradation, reciprocal cost transfers are the prototypical case, made familiar in Garrett Hardin's (1968) classic article as "the tragedy of the commons." Each person added to the population of a community imposes demands on the environment, and beyond some limit, ceteris paribus, the productivity of that environment falls, perhaps drastically. Hardin's analogy is with overgrazing a pasture held in common, the original reference being to pre-Enclosure rural Britain. Implicitly a society, like the owners of the livestock, is seen as composed of members with more or less uniform power and capacity. The stylized "game" that models the situation is the Prisoner's Dilemma, in which each player has an immediate incentive to pursue autonomously self-interested behavior unless a binding agreement to act in the common interest can be reached—but if it can be, the outcome is superior for all.

Hardin's intent and that of numerous subsequent writers drawing on him was to illuminate contemporary ecological degradation in the rural Third World. Villages and villagers, in these accounts, breed themselves into poverty by competitive overexploitation of their common resources—typically, forests for fuelwood and fodder and local aquifers for well water. The Himalayan case again exhibits numerous studies offering Hardin-type diagnoses. (See, for example, Messerschmidt, 1987; Poffenberger, 1980.) Piers Blaikie et al. (1980: 214) describe the degradation cycle: "forest clearance for terraces, use of forest for compost and fodder to maintain production causing greater erosion, which in turn destroys terraces, necessitating forest clearances for new terraces. . . ." Each new household acquires draught and milch animals needing pasture and leaf fodder, and requires timber for construction and agricultural implements. The remaining tree cover and soil stability are threatened.

I earlier cited the criticisms of this scenario as a characterization of the region as a whole. A balanced account would allow also for positive responses to ecological threats, whether or not there was local culpability in their origin—or, indeed, any human agency at all. Human foresight and appropriate institutional design may ensure that the worst outcomes are avoided. Resource constraints can be eased, either by outmigration or by raising the productivity of the local economy; and negative feedbacks that cut down on rates of entry into the community—controls on births or on establishment of new households—can be devised.

Although it is usual to discuss this type of situation as one of reciprocal lateral transfers, there is evidently an important time dimension to it: a major part of the explanation for what might be seen as supine accession to increasing poverty, where that is the outcome, is in the divergence between the parental interests in family-building and the children's interests in inheriting less-diluted assets. Hence the category is not fully distinct from (4) below.

(2) *Interclass transfers* Few real societies are well described by a model that assumes the constituent families or households within a community to be relatively equal participants—similarly placed players in the Prisoner's Dilemma. The more common case is one in which many of the relevant cost transfers are effectively unilateral rather than reciprocal: the divisions by power or status (caste, for example) within the community are sufficiently strong for uncompensated transfers to become possible. (Identification of costs imposed by better-off sections of the society on the poor does not in itself entail a moral judgment. The lack of reciprocity may simply follow from the unequal shares in the economy, ethically blameworthy only under a radical egalitarianism. Moreover, there may well be other situations in which significant external costs are generated by the poor precisely because of the slightness of their stake in the economy at large.) The existence of face-to-face interaction, however, will often put limits on the extent of imbalance—for example, by imposing social obligations on the rich or, in a more hard-nosed interpretation, making it evident to them that some resources need to be allocated to buying off potential threats to a comfortable social order.

A feature of interclass transfers is that they are potentially reciprocal. Where those bearing the costs are able to organize themselves and acquire political voice, reciprocity may be established—either through the formal political process or through what Kenneth Boulding (1968) has called the "threat system." An alternative route that yields partial reciprocity is through clientage, to the extent that some of the poor find patrons among the rich, providing political or other services in return for crisis insurance. Some proportion of interclass transfers is internalized within such quasi-families, and, at least in the subjective view of the client, is balanced out in the relationship.

(3) *Intervillage transfers* Some spillovers extend across community boundaries, the spatial separation making it more likely that the transfers will be unilateral rather than reciprocal. Compensation here would depend on the power of the injured party or the existence of an overarching authority structure within which redress could be obtained.

Consider again the Himalayan case. The Kumaon Hills, the western Himalayan foothills in Uttar Pradesh, in the description of Sri Madhava Ashish (1979: 1058–1059), show a classic Ricardian frontier with the added feature of scope for (literal) spillover effects back to the core:

> The older and larger villages lay in the valleys where the soil was rich, the terraces wide, and water channels irrigated three full crops in the year. . . . A thousand to 1,500 feet above lay another line of villages, smaller, poorer, with narrower terraces, and with only mountain springs for water. Above them again came another line of dwellings, poorer still, having perhaps three crops in two years, many of them established as villages within living memory. This is the last line of villages at 6,500 to 7,000 feet altitude. When the population lower down increased to the point where the harvest yields no longer supported it, families migrated to start cultivation higher up, clearing more forests in the process.

The potential transfers here are between communities, specifically costs transferred downward on the alpine gradient. Streams are left depleted and polluted for downstream users; catchment areas are damaged, giving rise to more serious and more frequent flooding. Analogous spillovers can be identified in non-alpine environments. The case is Hardin-like, especially in its origins: the settlers in Ashish's top line of villages may well have come from those directly below. But once established in a more or less autonomous community, that origin and line of responsibility are immaterial. What now distinguishes the situation is the likely fact of nonreciprocity of the transfers.

(4) *Intertemporal transfers* Some unilateral transfers are characterized by the absence of even potential recourse by those on whom the costs are imposed. This is the case with intertemporal transfers. The intricacies of defining transfers over time—beyond readily measured assets and liabilities—usually argue for reducing them to net present-values, so that they either vanish (if they lie within a single family unit) or are treatable as quasi-lateral transfers. In this process, however, a great deal of the substance of the transfer problem is lost, and arbitrary and unexamined assumptions about the society, particularly to do with social discount rates and intergenerational relationships, are introduced.

Within a family, accidents aside, some degree of intergenerational balance of payments (broadly construed, including, for example, John Caldwell's "obligational flows") may be struck, though the point of balance is highly dependent on culturally prescribed standards, which are not immutable. But the relevant bargaining is conducted by one party only, setting short-run costs—of having children and of safeguarding assets—against imagined long-run recompense. Even Gary Becker (1988) acknowledges an intrinsic market imperfection here: that unborn children are in no position to contract with parents. Moreover, accidents—demographic and economic fortuity—are inevitable and are prominent determinants of family outcomes, doing most to undermine the value of a family-level decision calculus in just the kind of fragile environments that are of chief interest here.

In the society at large, analogous bargaining difficulties are more familiar—encapsulated in Joseph Addison's rhetorical question: what has posterity done for us? Cultural and institutional continuities, under stable circumstances, permit the bypassing of this question by "constitutional" means: the society *is* a continuing social form, replete with implicit intertemporal (intergenerational) transfers. In unstable conditions—say, with rapid economic change, population growth, or environmental degradation—an explicit awareness of intertemporal transfers is needed. The succeeding generation potentially benefits from its inheritance of physical and human capital from the parental generation, but it may be disadvantaged in per capita terms by being much more numerous—commonly, double the size—and it may similarly be disadvantaged by inheriting a degraded environment.

There are many conceptual problems here—in commensurability over time and between economic and environmental assets, in disentangling economics and ethics, and in taking account of the adaptation of preferences to realities—that to deal with would divert the course of this discussion. For the most part, especially if we consider a single community, these problems can be ignored. At the scale of community, the dimensions of the intergenerational transfers should be starkly apparent. For example, Norman Ryder (1975) has shown how the acceleration of population growth that occurs during the early phase of demographic transition is felt not as an increasing dependency burden to individual families but as a more rapid multiplying of the *number* of families: there can be few phenomena in a village more evident to all its members than establishment of new households.

Community containment

In discussing various categories of environmental transfer effects above, I have touched on some of the "remedies" that are routinely proposed to deal with them. I treat these here in some greater detail, first considering means that lie within the province of the local community—whether acting qua community or not—and then (in the section following) measures that call for local or higher level administrative action.

Village republics

Villages almost by definition have elements of corporateness. This feature is easy to exaggerate in scope and significance or to imbue with imagined idyllic qualities—witness the "village republic" (*gram raj*) of much early writing on Indian rural society—but in essence is a simple expression of the necessity or value of collective action in confronting certain contingencies. The contingency of environmental degradation is evidently one such, and could often be expected to elicit a village-level response working through existing or incipient corporate institutions.

Hardin's proposed solution to the tragedy of the commons was of this kind, "mutual coercion mutually agreed upon." (Hardin deals with the problem at the national level; from the standpoint of the locality the mutuality of the coercion may be hidden. The application here is to a single community treated as an isolate. Cf. McNicoll, 1975.) A rough participatory process is implied by "mutually agreed upon," justifying concurrence in submitting to regulation. The costs of the regulatory burden are presumably to be widely spread.

Ecological arguments for the prevalence of corporateness in fragile settings follow this line. Robert Rhoades and Stephen Thompson (1975), for example, trace the institutions of communal landownership and rotating leadership in Swiss, Nepalese, and Andean communities to the recurrent imperatives of dealing with negative spillovers. But are those cases of selective institutional survival, or is there a process that generates them? More broadly, is there any tendency for a social compact to emerge naturally as ecological crisis looms?

Some evidence that there is such a tendency comes from an important comparative village study in Andhra Pradesh by Robert Wade (1988). Wade explores the conditions under which local organizational initiatives are taken to cope with the collective threat of crop loss (through theft or livestock damage) and to control the allocation of canal water for irrigation. Both are public goods; more specifically, in Wade's usage, they are "organizational goods"—goods for which "people must organize to get the collective benefit" (p. 108). Within this single region, the villages that evinced corporate responses to such problems were those for which the potential benefit was greatest. The result is gratifying, if unsurprising, to economists, the more so since the internal social makeup of the village—its stratification of wealth and power and its intensity of factionalism—seemed to matter rather little in distinguishing among village responses.

These findings are far from conclusive for the problem at hand, and would be misleading if taken to suggest an automatic corrective response when ecological problems become serious enough. The chain of causation from individual behavior to environmental threat is often less obvious than in the instances dealt with. The actual damage may mount up more quickly and be hard to reverse. And the dismissal of social-structural factors occurs, as Wade acknowledges, within a series of culturally and historically similar populations: in a more diverse sample, the same arguments on benefits that supported collective action in some of these villages could well prevent or destroy such action. The findings do, however, point to significant possibilities for promotion of corporate institutional responses in confronting ecological problems.

A caveat to the quasi-determinism of tying the emergence of village corporatism either to specific details of agrarian ecology or to local institutional patterns is the observation derived from a slew of community development and similar extension programs that "success" is often associated with effective, charismatic local leadership—whether in the community or in the program. There is no reason to doubt that the same holds for the environmental domain. Indeed, T. B. S. Mahat et al. (1987), assessing reforestation potentials in Nepal, report that differences among panchayats in terms of forest trends, where success called for effectiveness in organizing protection and negotiating agreements with government agencies, were quite largely attributable to the qualities of particular leaders.

I should note here that the corporate village is not necessarily a good corporate citizen among others in the matter of intervillage spillovers, although corporate institutions may make pressures by or on behalf of outside threatened or injured parties more effective.

Notionally, villages are exhaustive of the rural population. Local corporate institutions do not of course have to be inclusive in that sense. In particular, nongovernment membership organizations are usually not so. Such organizations, even if we leave aside those concerned primarily with political mobilization, come in a profusion of forms in most agrarian societies—producer and marketing cooperatives, irrigation societies, rotating credit associations, and so on—sometimes contained within a village, often extending over several. These have been the subject of extensive research by Milton Esman and Norman Uphoff (1984), who point to the importance of their current and potential role as intermediaries in rural development. Those organizations that are tied to landholding would seem to resemble corporate villages in terms of environmental behavior. The readier option they have of extrusion of recalcitrant or nonconforming members may make them more cohesive; on the other hand, the readier option their members have of exit, if private pursuit of opportunity offered better returns, might arguably make them more fragile.

The irrigation societies of Bali are an extreme case of cohesion and efficacy: the locus of topographic control, working "like a cooperatively owned public utility with no necessary contact by state or local political bodies" (Netting, 1974: 33). The stability of Bali's rice terraces in the face of accelerating economic decline and rising political strife in the 1950s and 1960s, prior to the violent suppression of agrarian radicalism in Indonesia in 1966–67, speaks to the autonomous strength of those societies—a strength that would indeed be predicted by Wade's (1988: 111) criteria of the high net benefits to members of abiding by the rules and "the high degree of publicness of the goods produced."

The environmental effects of less tightly organized local organizations, lacking as firm a rationale as that imposed by a common watershed, are likely to range widely depending on the group's purposes and the ecological circumstances. However, the potential for the establishment or promotion of membership organizations specifically concerned with environmental issues is considerable, and the possibility of linking together locally based groups within a larger movement may offer a means to resolve the intergroup spillover problem alluded to earlier with reference to villages.

Erosion of territorial corporateness

It is a commonplace to observe that the vestiges of corporateness in village organization are eroding under the effects of modern communication and transportation, the enlarged scope for private enterprise, and the bureaucratic imperatives of the state. Labor and capital markets open up, patterns of commuting to nearby and even far-away towns develop, radio and television bring families directly into touch with a national culture, and the functions of village leadership are progressively usurped by government administration.

In writing on development, such changes are usually seen as contributors to as well as indicators of economic growth, bringing technology and opportunity to hidebound rural communities. Nigel Allan (1987) contrasts Nepal to the western Himalayas in that vein: "Nepal, almost bereft of roads, appears to be suffering from agricultural involution, the Geertz population-pressure model type. Any applicability of this model to Kalam [in Swat] is negated by modernization: the construction of roads, the integration into the larger polity and economy, and migration of any surplus population to the plains."

Whether such development incidentally serves ecological stability, however, seems very doubtful. While Hardin-type pressures are eased, so are the incentives for conservation and remedial action. Ashish (1979), in the study of the Kumaon Hills (Uttar Pradesh) cited above, argues that environmental degradation there could be attributed in part to that very "integration into the larger polity and economy." As the Kumaon agrarian economy fell behind, with tree-cover thinning and erosion becoming serious, attention did not shift to the Ricardian intensive margin (with investment in soil stabilization), or to efforts to halt demographic increase, but to employment opportunities in the plains. Human and draught animal populations could both continue to increase, supported by remittances from outmigrant family members and the grain supplies those remittances could purchase. Had this outlet not existed, Ashish speculates, "the natural consequences of over-exploitation of the environment would have, in the long run, pinned down population density at a level where the recuperative capacity of the environment balanced the amount of exploitation. But because of the possibility of migration and the resulting inflow of cash there are no forces inherent in the situation today to jolt local people into an awareness of what is happening" (ibid., p. 1060). A somewhat analogous picture is drawn by Michael Mortimore (1989) in discussing the effects of the 1970s Sahelian drought on villages in northern Nigeria: the major village response to ecological degradation was greatly increased short-term migration, permitting even some improvement in living standards as "farmers, denied their subsistence, ransack the ecosystem for alternative income" (p. 186). But in turn, this wider opportunity set works against collective remedial action on the local environment—to the extent such action might have been feasible. Conservation is labor intensive, a fact which, confined to the family level, justifies Mortimore's remark that "there is . . . an inherent contradiction in calling simultaneously for intensified, conservationary land use and for family planning" (p. 209).

Another component of this set of changes is the weakening of the village as holder of a local culture, maintaining customary rules of behavior and imposing sanctions for violations. Social pressures may have been exaggerated in depiction along with the conformity they supposedly engendered, but they would not have been inconsequential. Thus their erosion, just as it leaves a starker economic struggle, also removes a pattern of sanctions and taboos that may have been particularly important for ecological stability. In the longer term a new ideological support for conservation may emerge—the contemporary "green" movement, for example—but by that time much more damage may have been done.

Village realpolitik

I noted the tendency to romanticize the corporatist, collectivist qualities of "the remembered village" prior to its incorporation into the wider economy. A closer look, identifying the forces making for that supposed nonmarket comity, rapidly dissolves the vision. The micropolitics of collective action occupy the same space that the market later does, without the elements of dispassion and anonymity afforded by a price system. Conformism, mutual aid, patronage, and other such seemingly benign features of traditional village life can cover intense social pressures and outright coercion—at worst, in a mafia-style pattern of hierarchy and control by violence. Incorporation into the wider economy is typically found to have begun decades or generations ago; and in recurrent, face-to-face transactions the monetization of exchange brings less dispassion and anonymity than was supposed.

The main alternative to a corporatist social compact in Prisoner's Dilemma situations, one that figures prominently in much recent policy analysis and recommendation, is private ownership. Drawing on Ester Boserup (1965) and theorists of property rights, it is argued that the essential problem of the commons is the inappropriability of gains. As pressures on the commons rise, notably by population growth, so do the benefits of appropriation. Historically, as Boserup and many others have detailed, the tenuous land rights of the shifting cultivator are transformed into the usufruct and later freehold of the agriculturalist. Rights are claimed by planting perennial crops or by marking boundaries, or they are awarded by a customary or formal process of legal entitlement. The mechanisms vary, as does the nature of the rights, but the proposition is now well supported that institutional change works in the direction of a strengthening of property rights and their lodging in individuals rather than communities. The commons are enclosed. Secure property rights in turn provide the incentive for investment in productivity improvements by the holders. The entire process—population growth inducing technological advances and maintenance or even improvement in per capita production—is seen by many as a reason for some degree of complacency about rapid population growth in rural settings.

While privatization creates an incentive structure favoring improvement in economic productivity, it gives no similar assurance of ecological stability. Families in a fragile environment cannot be assumed to be conservationist: as economic optimizers, indeed, they may be relatively unconcerned with negative spillovers from their activities. Madhav Gadgil (1985: 1909) sees "a gradient of cultural practices of ecological prudence, from being strong in productive and stable environments to being much weaker in harsher and/or less stable habitats." Private property owners may have as much incentive to transfer their costs to others as they do to raise their own economic productivity. Privatization merely lowers the level at which benefits are internalized and costs (if possible) externalized, leaving unspecified the presence and efficacy of a supervening social order. Thus something more is required to resolve the problem.

At the national level, the obvious requirement, highly imperfect though the reality may be, is a system of law that provides a stable and predictable framework for economic behavior, enforces limits on the cost-externalization option, and has procedures for dealing with torts. At the level of the village, an analogous structure is needed. Formally, the national system would provide it, extending down to the village. Under highly authoritarian regimes this may in fact be the dominant element of village governance. In the more common actual situation, village governance reflects local politics and the local distribution of economic power, with the national system wielded, and perhaps manipulated, by local interests. Those interests need have no long-term commitment to ecological stability: to the contrary, economic growth for them may be maximized by environmental asset-stripping. Thus the privatization route, while it offers the most rapid potential aggregate economic gains, will only in special cases also safeguard environmental conditions. Economic growth and ecological stability come into conflict and acquire ideological trappings.

Administrative containment

As the discussion thus far indicates, neither corporateness nor individual private ownership precludes a structure of incentives under which ecological degradation is tolerated or even rewarded. Corporateness in an agrarian system, where it is effective, pushes negative spillovers to the corporate border. Corporateness, however, seems an insecure and impermanent feature of a territorially defined community. It takes little to erode substantive cultural traditions to such an extent that they can no longer sustain the village as a distinctive social and economic entity beyond an assemblage of households and household economies—albeit a degree of symbolic continuity may be preserved. And ad hoc collective action to capture the economic gains available from cooperative response is likely to be at best weakly institutionalized and easily subverted—whether by hold-out free-riders, by smaller and more sharply defined alliances able to arrogate a disproportionate share of the gains, or by simple loss of interest and energy on the part of villagers and village leaders. Privatization, solving the appropriability problem, leaves unresolved the containment of negative spill-overs that now offer private benefits to those creating them even within the village.

In seeking to identify sufficient social-organizational conditions for ecological stability, we therefore need to examine the larger administrative system within which local communities (and other local organizations) are set. This system, comprising a political and legal culture and associated institutions and a formal government apparatus, is in a broad sense part of the "environment" of the community—together with the physical environment and the regional economy. It is subject to change as a facet of a country's political development or as a byproduct of national-level political conflict, most likely with scant regard for side effects on the matters that are here of concern. It is also, as I noted above, subject to being subverted or captured at its interface with the community and bent to the interests of local power holders.

We can broadly distinguish three types of administrative framework within which containment of local-level environmental spillovers is feasible, recognizing that most actual situations of effective containment would have elements of each. The first is a regime of authoritarian control, where ecological stability is accepted as a desired, albeit probably subsidiary or incidental, objective. The second, under a more minimalist state, is an effective legal system able to enforce a perceived national interest in long-run sustainability of ecosystem productivity and to offer redress of individual damage claims resulting from spillover effects through common-law processes or their equivalent. And the third is a system in which administrative authority is devolved to a level of local government where responsiveness and accountability coincide.

These three types do not necessarily translate into overall characterizations of rural polities, since administrative styles can vary somewhat across the spectrum of government interests. It is theoretically possible for there to be strict, centralized environmental regulation (or, at the other extreme, full devolution of responsibility for environmental management to local units) in an otherwise free-wheeling market economy. Divisions of this sort have existed between the administrative styles brought to bear on internal security and economic activity in a number of countries. Nevertheless, the likelihood is that a single style is dominant throughout the government, reflecting the country's administrative history and the regime's ideology and degree of responsiveness to popular will.

The realities of administrative performance often differ greatly from the formal design, so there is little point in devising recipes for administrative containment of ecological spillovers that are cast in terms of abstract models. Thus, in practice each of the three kinds of administrative framework mentioned above could fall far short of its intended performance. Authoritarian control is subject to top-level corruption, to subordination of environmental to shorter run economic or political interests, to policy and implementation errors introduced by the distance of decisionmaking from the situation on the ground, and to the magnified effects of sheer incompetence shielded by absence of public debate. Reliance on an impartial rule of law may be effective under certain regimes of customary law in periods of relative social stability, and also under stable constitutional government in a liberal polity; but hardly in the turbulence of political and economic change that has characterized most Third World countries over recent decades. In these latter situations the legal system becomes more an instrument or weapon in the political process than an independent arbiter. (See McEvoy, 1988: 226, who criticizes Hardin's spillover remedy for its implication of "an independent observer of nature and a neutral rulemaker somehow external to and immune from the interests being disciplined.") And devolution too, notwithstanding its current fashionability, may be tantamount to surrender of control of environmental outcomes to local-level special interests in a position to exercise influence over government program activity. In each of these cases, however, the effort to overcome such problems is inextricable from the development effort at large.

Administrative histories, from this perspective, show the stumbling from one structure to another, sometimes with ecological problems in mind, more often not. In a pattern of administrative reform that has been repeated in similar guises in many countries, Nepal in 1957 nationalized its private and village-controlled forests. Traditional rules of forest management were ignored and the incentives for local conservation removed, without the substitution of any comparably effective management regime. The outcome, now widely recognized, was a severe overexploitation of forest resources by local communities and corporate interests, for both of whom short-run gains rationally dominated. The perverse incentives were eventually removed in 1976 with Nepal's National Forestry Plan, entailing devolution of responsibility for forest protection and exploitation to the panchayat level (population around 5,000) and provision for long leases of forest area. (See Bajracharya, 1983.) The results of this policy are not fully in, though the evidence collected by Mahat et al., cited above, is promising.

Administrative structure and change can have substantial influence over ecological trends without being directly related to natural resources or the environment. The scope for freedom of economic maneuver clearly is set by government—given that we include the case in which that scope is enlarged by influence of economic agents *over* government. The processes of political development are promoted or impeded by administrative action. The same is true for the legal order, and for management of the difficult transition that many states are engaged in from customary and religious law to codified civil law. The space for local organizational initiatives on environmental (or, for that matter, all other) issues is defined by the administrative system. Yet rarely are the environmental implications of these structures or the side effects of changes in them given explicit consideration. Indeed, one often has the impression that substantial reforms of administrative systems are entered into with rather slight exploration of consequences, for quite narrow or partisan reasons—for example, to help mobilize political support for the government.

The arbitrariness of such reforms is lessened in practice by the continuity of administrative institutions: the tendency for new administrative regimes to revert de facto to long-established patterns of behavior (for instance, in the style of relations between local officials and community leaders) because that is the least-effort means of conducting necessary routine business. Those old patterns, however, while they may have served well the incidental purpose of environmental preservation under low-pressure demands, are not necessarily well suited for the high-pressure demands that are generated by contemporary technological and demographic change. Indeed, just as patterns of social organization may fortuitously be well or poorly suited to economic management under rapid technological change or to acceptance of fertility regulation under fast-declining mortality, so may the pattern of local administration, for equally fortuitous reasons, help or hinder the cause of environmental stability. Constrained though the feasible range of reform may be by this weight of institutional inheritance, the systematic search for improvements in how the administrative regime deals with environmental considerations warrants far greater attention than it gets.

Ecological futures under nonterritorial governance

A theme of this essay is the changing scope for local solutions to ecological problems. Encroachment of the urban world and the consumer economy, the widening of the labor market, increased temporary and permanent outmigration, and the weakening of formal and informal local authority structures are familiar changes associated with economic and political development that modify and frequently diminish the role of locality in rural development—usually, though not necessarily, in the course of delivering improved well-being for the rural population. The processes underlying these changes operate jointly, but from different starting points and at different paces. Where a particular society is positioned on those various axes, however, constrains the set of environmental policies that would be workable for it. A Korean-type "New Community

Movement," which has the potential for a stringent conservation component, would seem to require an authoritarian local administration, close settlement with perhaps constraints on migration, and a moderately egalitarian distribution of wealth. Such conditions clearly are not those of South Asia, for example, so transplanting the model there would likely be fruitless. Conversely, more loosely set up nongovernment organizations, like India's Chipko Movement (see below) or various cooperative organizations in Bangladesh, would have been ineffectual in the "hard" states of East Asia. Environmental policies, no less than others, need tailoring to existing institutional arrangements.

But those arrangements are in flux, both at the national level and locally. In many poor countries rural development policy, after several decades of dirigisme, has veered toward promotion of a private property/free-market regime. This retreat of state control, welcome as it is economically and politically, is not necessarily any improvement ecologically (and may be significantly worse) in the absence of an alternative regulatory order at the local level. In any formal sense, such a substitute local order—a framework of economic incentives and legal sanctions—is a long way off. Finding a structure of governance in the meantime that can support the major productivity gains offered by this shift while also delivering ecological stability is an extraordinarily challenging policy task.

In his study referred to earlier, Wade (1988) argues for a middle way of managing depletable, degradable resources at the village level between the extremes of privatization and state control, namely through self-organized corporate action. Wade has much of value to say about the conditions under which corporate responses emerge and what is needed to sustain them. It is clear, however, that those conditions are frequently hard to attain—for the kinds of reasons dealt with in the pages above.

To the extent that rural society in many countries is moving beyond the stage of territorial social organization, an important locus of policy action has been lost. Governments still thinking in terms of village government as the principal agent of reform would be, like generals, equipping themselves to fight the last war, drawing on an armamentarium that implies a local reality already disappearing: one of stable, cohesive villages set at the base of a neat hierarchy of administrative units. But in parts of the Third World (how much of South Asia is a critical question) there seems still to be scope, albeit waning, for village-based organizational initiatives directed at environmental management.

A possible alternative route to ecologically sustainable economic growth is to be found in the loosely structured activist movements that have coalesced around ecological issues in a number of poor countries, India most notably. Best known among them is the Chipko Movement, originating in the early 1970s in the Indian Himalayas as a spontaneous protest by village women against the disruption of their traditional livelihoods by commercial forestry. The movement subsequently spread throughout India (see Centre for Science and Environment, 1985, and Shiva and Bandyopadhyay, 1986). The strength of the Chipko

Movement derives both from the commitment it can draw on in activism on the ground and its sustained voice on forestry issues in national political debate. It has been increasingly influential. More broad-spectrum environmental organizations, rapidly expanding in many parts of the Third World, may be less helpful. They tend to mirror the concerns of activists in the rich countries—wilderness areas, air pollution, climate change—or to adopt radical anticapitalist and communitarian postures. (The private Centre for Science and Environment in New Delhi, compiler of the important "Citizens' Reports" on the state of India's environment, takes the latter stance: corporate greed, both domestic and foreign, is held to underlie India's ecological degradation; population growth and the design of local-level incentives not at all.)

The subject of this essay is considerably narrower in scope than the agenda of most environmental movements. It is not purist in its ecological goals, nor does it seek to gauge the esthetics of environmental change or the economic threats of large-scale (greenhouse-type) ecological trends. It is concerned with preserving and raising the sustainable economic productivity of local ecosystems. To return to the example I started with, the transformation of a landscape from first-growth forest to settled agriculture is a drastic ecological change by any measure, whether it takes place over a few years or many generations. If the long-run stability of the physical base of production is maintained in the process, a permanent gain in economic productivity may be achieved. Notwithstanding the radical goals of today's green movement, under the massive rural demographic expansion that has taken place over recent decades and that will continue for at least several more, those productivity achievements may be the best mark of success we can aim for—and offer a partial defense in ecologically uncertain times.

References

Allan, Nigel J. R. 1987. "Ecotechnology and modernisation in Pakistan mountain agriculture," in Y. P. S. Pangtey and S. C. Joshi (eds.), *Western Himalaya*: Volume II: *Problems and Development*. Nainital, U.P., India: Gyanodaya Prakashan.

Ashish, Sri Madhava. 1979. "Agricultural economy of Kumaon Hills," *Economic and Political Weekly* (New Delhi) 14 (23 June): 1058–1064.

Bajracharya, Deepak. 1983. "Deforestation in the food/fuel context: Historical and political perspectives from Nepal," *Mountain Research and Development* 3: 227–240.

Becker, Gary. 1988. "The family and the state," *Journal of Law and Economics* 31: 1–18.

Blaikie, Piers, John Cameron, and David Seddon. 1980. *Nepal in Crisis: Growth and Stagnation at the Periphery*. Oxford: Clarendon Press.

Boserup, Ester. 1965. *The Conditions of Agricultural Growth*. London: Allen and Unwin.

Boulding, Kenneth E. 1968. "The relations of economic, political, and social systems," in his *Beyond Economics: Essays on Society, Religion, and Ethics*. Ann Arbor: University of Michigan Press.

Centre for Science and Environment. 1985. *The State of India's Environment 1984–85: The Second Citizens' Report*. New Delhi: Centre for Science and Environment.

Eckholm, Erik P. 1975. "The deterioration of mountain environments," *Science* 189: 764–770.
Esman, Milton J., and Norman T. Uphoff. 1984. *Local Organizations: Intermediaries in Rural Development.* Ithaca: Cornell University Press.
Gadgil, Madhav. 1985. "Towards an ecological history of India," *Economic and Political Weekly* 20: 1909–1918.
Gilmour, D. A. 1988. "Not seeing the trees for the forest: A re-appraisal of the deforestation crisis in two hill districts of Nepal," *Mountain Research and Development* 8: 343–350.
Hardin, Garrett. 1968. "The tragedy of the commons," *Science* 162: 1243–1248.
Ives, Jack D. 1987. "The theory of Himalayan environmental degradation: Its validity and application challenged by recent research," *Mountain Research and Development* 7: 189–199.
———, Bruno Messerli, and Michael Thompson. 1987. "Research strategy for the Himalayan region: Conference conclusions and overview," *Mountain Research and Development* 7: 332–334.
Mahat, T. B. S., D. M. Griffin, and K. R. Shepherd. 1987. "Human impact on some forests of the middle hills of Nepal, Part 4," *Mountain Research and Development* 7: 111–134.
McEvoy, Arthur F. 1988. "Toward an interactive theory of nature and culture: Ecology, production, and cognition in the California fishing industry," in Donald Worster (ed.), *The Ends of the Earth: Perspectives on Modern Environmental History.* Cambridge: Cambridge University Press.
McNicoll, Geoffrey. 1975. "Community-level population policy: An exploration," *Population and Development Review* 1: 1–21.
Messerschmidt, Donald A. 1987. "Conservation and society in Nepal: Traditional forest management and innovative development," in Peter D. Little and Michael M. Horowitz (eds.), *Lands at Risk in the Third World: Local-Level Perspectives.* Boulder, Colo.: Westview Press.
Mortimore, Michael. 1989. *Adapting to Drought: Farmers, Famines and Desertification in West Africa.* Cambridge: Cambridge University Press.
Netting, Robert McC. 1974. "Agrarian ecology," in *Annual Review of Anthropology* 3: 21–56.
Olson, Mancur. 1971. *The Logic of Collective Action.* rev. ed. New York: Schocken.
Ooi, Jin Bee. 1988. *Depletion of the Forest Resources in the Philippines.* Singapore: Institute of Southeast Asian Studies.
Pelzer, Karl J. 1968. "Man's role in changing the landscape of Southeast Asia," *Journal of Asian Studies* 27: 269–279.
Pfister, Christian. 1983. "Changes in stability and carrying capacity of lowland and highland agro-systems in Switzerland in the historical past," *Mountain Research and Development* 3: 291–297.
Poffenberger, Mark. 1980. *Patterns of Change in the Nepal Himalaya.* Delhi: Macmillan.
Rhoades, Robert E., and Stephen I. Thompson. 1975. "Adaptive strategies in alpine environments: Beyond ecological particularism," *American Ethnologist* 2: 535–551.
Ryder, Norman B. 1975. "Reproductive behaviour and the family life cycle," in *The Population Debate: Dimensions and Perspectives,* volume II. New York: United Nations.
Shiva, Vandana, and J. Bandyopadhyay. 1986. "The evolution, structure, and impact of the Chipko Movement," *Mountain Research and Development* 6: 133–142.
Wade, Robert. 1988. *Village Republics: Economic Conditions for Collective Action in South India.* Cambridge: Cambridge University Press.

Demographic Responses to Food Shortages in the Sahel

ALLAN G. HILL

TO MALTHUS AND HIS CONTEMPORARIES, it seemed self-evident that crop failures and food shortages would inevitably produce starvation, sickness, and excess mortality among the poor in the stricken populations. Certainly, the African historical record is full of periods of crisis mortality, many of them linked to famine and some of them occurring well into the twentieth century (see Watts, 1983 for a chronology from northern Nigeria, and Carreira, 1982 for a well-documented account of famine-induced mortality in the Cape Verde Islands). Although the data on famine mortality for most of sub-Saharan Africa are very poor, close examination of some of the better documented crises suggests that war and epidemics may be much more important killers than rank starvation. As Figure 1 shows, there is some geographical coincidence between wars and famine in contemporary Africa (Griffiths, 1988). Even the highest estimates of mortality in the Sahelian droughts of the 1970s and 1980s pale into insignificance in comparison with the numbers of excess deaths associated with the great 1918–19 African influenza epidemic, with the Biafran war of secession of 1967–69, or with the wars in Uganda.[1] There may well be some circumstances in contemporary Africa where a direct relationship between food production and mortality can be observed, but generally, the world has changed in several important ways that make it less and less likely that mass starvation will be the inevitable result of either natural calamity or excessive population growth. We cannot of course exclude the possibility of future high mortality levels resulting from civil or international war.

In this essay, my main interest is in demographic changes associated with food shortages arising not from war but from other dislocations or breakdowns of the economy—some provoked by natural calamities but most if not all having roots in the structure and organization of production and its social and cultural context.

FIGURE 1 Africa: Famine areas and location of major wars

SOURCE: Griffiths and Binns (1988): 49.

The state and local social organization

Mass starvation in Africa due to natural disasters or economic failure, perhaps surprisingly, is becoming less common. Some of the consequences of recent droughts in the region, so vividly displayed on Western television screens, are no more than the usually high levels of mortality and destitution prevailing in the more remote rural areas among the least privileged sections of society. In "normal" times, the outside world rarely has the chance to discover these high levels of rural poverty, but when the destitute collect on roadsides, in refugee camps, or on the outskirts of towns and cities, the depths of rural poverty are made visible to foreign audiences, perhaps for the first time. Many of the mortality figures reported for such dislocated populations are of course exceptional simply because of the abnormal age and sex structure of the destitute groups and because their death rates are being compared with figures for the general population with a quite different demographic composition. In addition, there is a widespread tendency to regard the current food crises in sub-Saharan Africa in simple Malthusian terms—as the most recent in a long chain of largely inevitable disasters, triggered by drought, economic decline, and disease. Misunderstanding of the situation is made greater by the mass of enthusiastic but wrong-headed journalistic articles on the root causes of African food shortages, blaming, variously, the climate; the population (animal and human) for growing too fast; the bureaucratic state; and neo-colonialism on the part of the major grain-exporting countries (see Borton and Clay, 1988 and Allison and Green, 1986 for guides to some of this now huge literature). Only recently have comprehensive case histories of crises in the early 1970s started to appear in print (Garcia and Spitz, 1981).

I shall argue that there is a marked historical break between the past and the post-independence era in Africa, specifically in the Sahel, that affects the way in which food crises are related to demographic trends. The sharpness of this break has been underestimated, since the focus in seeking explanations has wrongly been on the climate and ecology rather than on the more important processes of social change and adjustment. These processes operate on at least three levels, international, national, and local, each of which has a bearing on food production and availability.

The sudden discovery of Sahelian poverty by outsiders produced a massive international response of a size quite unthinkable in earlier times. In the rich countries, public concern for the Sahelian poor reached sufficient heights for the major bilateral, multilateral, and private donors to stage an enormous rescue program. In the late 1970s, the value of food aid alone amounted to 10 percent of the gross domestic product of some individual countries in which total international development assistance was already as much as 20 percent of gross national product (World Bank, 1981: Table 22).

A large part of the explanation of this heightened concern for the African poor in the richer countries is undoubtedly the development of modern transport and communications, providing simultaneously the means and the motivation for a large-scale international assistance program. Within Africa itself, the steady growth of transport and communication facilities has been important both in facilitating responses to food crises and in lessening the chances of a major food crisis developing. In nineteenth century northern Nigeria, for example, lack of transport and a poorly developed price and crop production information system contributed to the several famines (Watts, 1983: 135). Despite the long historical connection of the Sokoto Caliphate and the Emirates of northern Nigeria with trans-Sahara trade, local transport facilities were poorly developed. Even during the 1926–27 famine, three-quarters of the food relief brought from the railhead at Kano to Katsina came by camel and donkey (Watts, 1983: 310). Compare this with the sacks of grain being dropped from Hercules aircraft to Ethiopian peasants in the crisis of the early 1980s.

At the national level, probably the most important change bearing on demographic responses to food shortages has been in the ideological domain: the growth of the view—in sub-Saharan Africa as elsewhere—that the state is responsible for the welfare of its citizens, and should in some measure regulate the production and distribution of essential commodities, not least, agricultural commodities. In many states, the post-independence tendency, based on socialist ideology, was to take into public ownership or to create anew the modern means of production by the establishment of state corporations with monopolistic rights. Much of the production system thus became subject to bureaucratic control, creating a dualism between "modern" and "traditional" parts of the economy that persists despite recent reforms (Hart, 1982; Richards, 1985).

In the Sahel, the growth of the role of the state in the production and marketing of agricultural commodities brought to light the existence and function of many traditional arrangements for controlling access to and use of the region's natural resources. Legislation nationalizing land for use by major state-owned commercial farming enterprises cut across long-established patterns of land tenure and use and may have destabilized complex systems based on reciprocity and sharing of limited land and water resources. The situation in the Niger flood zone of central Mali has been particularly well studied by Gallais (1967 and 1984) and Cissé (1985), and further details are contained in a large five-volume survey by ILCA (1983). But other examples of new legislation disrupting older systems of tenure and rights, thus contributing to rural instability and local pressure of population on resources, are widespread. Government measures promoting the commercialization of the rural economy have had similarly disruptive effects. Examples are the enforcement of head taxes paid in cash and state purchase of agricultural commodities by local marketing boards at prices generally below the regional average (see Harriss, 1982 for a full account). Such changes clearly have both positive and negative aspects, but one effect of the acceptance of the principle of state intervention is that major food shortages are now unlikely to occur without eliciting some kind of alleviating response by one or other agency of the bureaucratic state.

The particular effects of expanded state roles that are of most interest in the present essay are those on the functioning of local drought-management systems. Perhaps this link has been overlooked because the institutional changes coincided with a series of low rainfall years in the Sahel and neighboring regions (see Figure 2). The reasons for this dry cycle are not entirely clear, but the work of

FIGURE 2 Annual rainfall departures as a percentage of long-run averages in four West African climatic zones

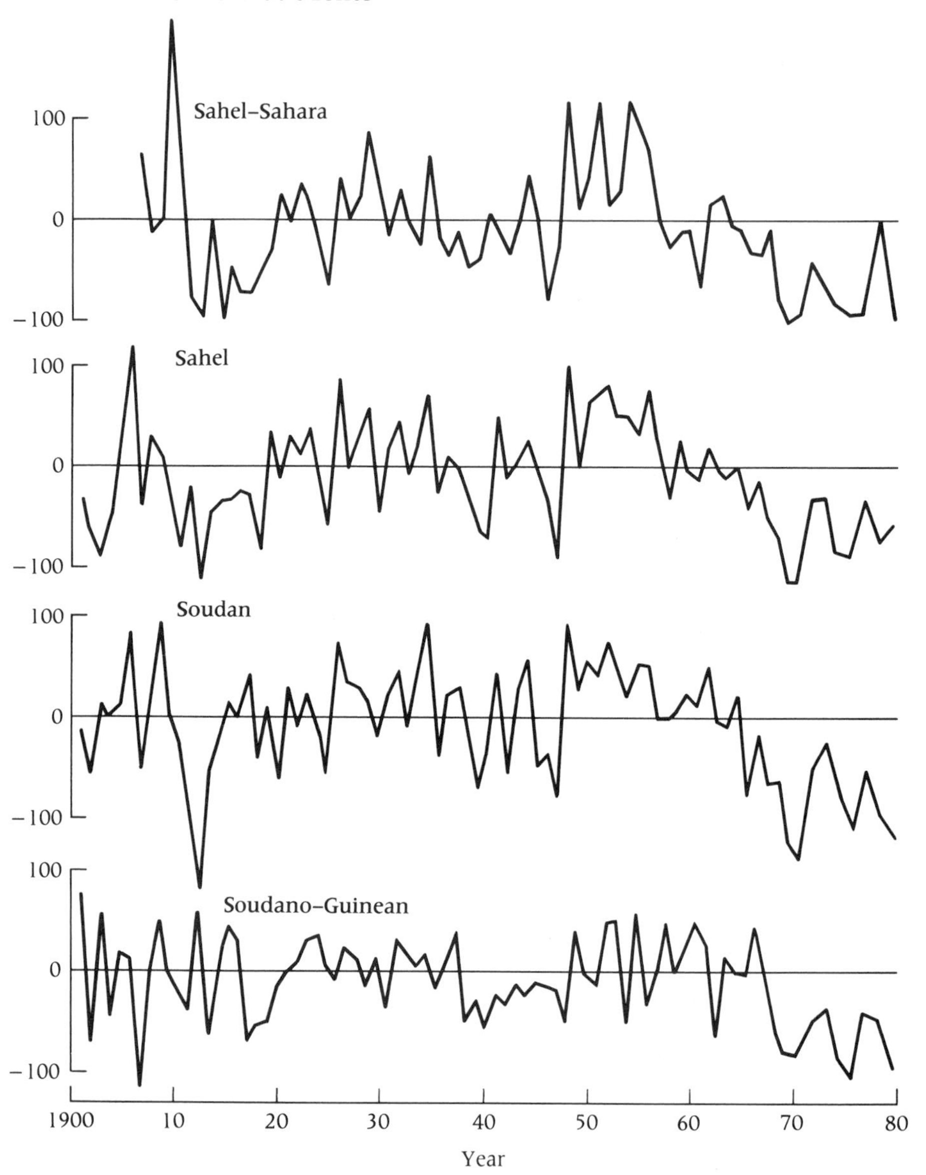

SOURCE: Farmer and Wigley (1985): Figure 3.15.

Farmer and Wigley (1985) provides evidence that they are probably external to the Sahel. The recent droughts therefore cannot be blamed on the contemporary rural populations of the Sahel—the people who, in some accounts, have been held culpable because of their allegedly profligate use of the region's natural vegetation and water resources and for allowing irresponsible growth in the human and animal populations of the zone (see Talbot, 1986 and Garcia and

Spitz, 1981: vol. 1). Whatever their origins, however, the disruptive effects of the droughts on local communities have clearly been severe. Households are split in the course of long-distance moves in search of sustenance. Survival as a household owes much to chance, dependent on the irregular and unpredictable distribution of small amounts of rainfall or the mortality of cattle and of key family members.

Beyond these immediate effects are others, perhaps in the long run more far-reaching. The migration creates situations in which achievement is related to such individual characteristics as commercial acumen, initiative, and education rather than to social status. As a result, traditional power structures in the community are undermined. Returning migrants who have acquired a degree of wealth in the towns may invest it in the rural economy. Others bring back new skills—such as in vehicle maintenance and repair. The employment of pastoral specialists as contract herders is another example of how new skills are altering economic and political relationships between rural social groups (see Swift, 1979; Bonfiglioli, 1985; Oba and Lusigi, 1987). Rural communities have thus become much more "open" to outside influence; competition is growing between traditional power structures and the new urban-oriented groups; and increasingly, personal connections and information networks extend from urban labor markets to the most isolated village or camp.

Summarizing, in attempts to go beyond a crude Malthusian explanation for Africa's recent food crises, we have to consider the changes occurring at various levels of social organization. New national institutional structures are developing in the place of the older community-based institutions, as a result of political nation-building forces and economic change—the latter influenced also by the major trading nations and international organizations. In Marx's characterization of production and social organization, all human beings "enter into definite connections and relations with one another and only within these connections and relations does their action on nature, does production, take place" (Marx, 1849: 80). In the Sahel, these connections are becoming more explicit. We are witnessing a breakdown of what Watts (1983) called "the moral economy"[2] and its replacement by a set of contractual relationships among individuals and between individuals and the state. Community-level organizations have only a limited scope for action in this wider, more open setting. To what extent can traditional institutions still buffer high-risk households from food crises? More generally, what effects do these changes have on household vulnerability to food deficits?

The distinctiveness of Sahelian demographic regimes

The distinctiveness of many of the pastoral and agro-pastoral populations living in a broad belt across sub-Saharan Africa from Senegal to Somalia has been recognized for a long time. The ethnographic literature contains descriptions and analyses of most of the major ethnic groups but very little comparative work on, for instance, links between the groups' ecological circumstances and their social organization. With the accumulation of descriptions based on census or survey inquiries, it is now possible to map regional variation in marriage systems and fertility regimes and sketch the outlines of mortality patterns by ethnic group for most of Africa (Lesthaeghe, Kaufmann, and Meekers, 1986; Lesthaeghe, 1989; Blacker, 1990), and thus to give quantitative content to the concept of a Sahelian demographic regime.

Data on the rural fertility levels of selected Sahelian communities in central Mali and of some comparison countries are presented in Table 1, together with estimates of the effects of the marriage pattern and other proximate fertility determinants on the total fertility rate. The Sahelian communities were the objects of demographic surveys directed by the author during 1981–85. Although not necessarily representative of experience everywhere in the Sahel, these groups cover both pastoral and agro-pastoral communities and the main ethnic groupings of the western Sahel. Findings from these surveys are used frequently in what follows.

From the detailed information available on the proximate determinants of fertility in other rural populations in Africa, it appears that the marriage pattern, including temporary separation of married couples, is responsible for the relatively modest levels of fertility measured in some Sahelian societies (Hill, 1985b; Hill and Thiam, 1987). Substantial proportions of women of reproductive age in some pastoral communities in the Sahel are either single or not living in a sexual

TABLE 1 Fertility and its proximate determinants in selected rural African populations

Population	Total fertility rate	Bongaarts's indexes				"Potential fertility"
		C_m	C_i	C_c	C_s	
Mali (1981–83)						
Bambara	8.1	0.92	0.63	1.0	0.90	15.5
Delta Tamasheq	6.6	0.67	0.65	1.0	0.92	16.5
Gourma Tamasheq	5.2	0.68	0.61	1.0	0.86	14.6
Seno-Fulbe	6.0	0.89	0.61	1.0	0.88	12.6
Seno-Rimaibe	6.3	0.89	0.65	1.0	0.85	12.8
Masina-Fulbe	7.5	0.88	0.68	1.0	0.83	15.1
Masina-Rimaibe	6.6	0.86	0.69	1.0	0.89	12.5
Ghana (1980)	6.5	0.84	0.64	0.94	NA	12.9
Kenya (1978)	7.7	0.81	0.64	0.94	NA	15.8
Senegal (1978)	7.3	0.92	0.65	0.98	NA	12.5
N. Sudan (1978)	6.4	0.79	0.69	0.98	NA	12.0

NA = not available.

NOTES: C_m is the index of nonmarriage; $1 - C_m$ is the proportional reduction in potential fertility attributable to this variable alone. The other indexes—C_i (postpartum amenorrhea), C_c (contraception), and C_s (spousal separation)—are interpreted in the same way. "Potential fertility" (sometimes called total fecundity) = Total fertility rate ÷ ($C_m \times C_i \times C_c \times C_s$).

SOURCES: Casterline et al. (1984: Appendix C); Hill (1985a: Table 2), and survey data.

union. Contraceptive use and induced abortion are nowhere significant brakes on reproduction in rural sub-Saharan Africa. Other factors do have greater effects in lowering fertility from its potential maximum level, but show little variation among populations. Lengthy breastfeeding is still practiced almost universally in rural African populations, with its consequence of long duration of postpartum infecundability and thus of long birth intervals. Abnormal levels of sterility are also important in restricting fertility locally. It is worth noting that an increasing weight of evidence indicates that the theoretical capacity to reproduce—fecundity—is constrained hardly at all by levels of undernutrition short of rank starvation.[3]

For populations experiencing "natural fertility" (no deliberate control of conception), the Sahelian total fertility rates shown in Table 1 are relatively low. What makes the demographic regime of the Sahel so distinctive, however, is the combination of this relatively low level of fertility with high mortality. UN estimates of mortality levels (United Nations, 1987), in conjunction with the fertility levels indicated in Table 1, yield natural increase rates that are quite low, although positive in most years. Excess mortality in a particular year due to an epidemic (measles, influenza, or the like) may result in negative natural increase in that year. Crop failures or other circumstances creating widespread food deficits may have a similar result, both because of the direct effects of undernutrition on resistance to disease and because of indirect effects on fertility and mortality stemming from the forced breakup of family groups. Historically, the relatively slow population growth of sub-Saharan Africa may have been due more to relatively low fertility than to high mortality.

The principal individual-level strategy for coping with drought in the Sahel is to move, a response that has major consequences for the demographic development of the community. Herders are more mobile than farmers, but the organization of herding and of associated pasture rights introduces certain constraints. Usually one part of a drought-stricken pastoral community has to remain with the milch cows and with the very old and the very young. Depending on the circumstances and the social structure of the community, it may be the poor or the privileged who must migrate. Migration often separates husbands and wives, directly reducing fertility. The concentration of births that usually follows soon after families are reunited is explained by the relatively high numbers of females who are in the fecund state following a period when they are neither pregnant nor breastfeeding. The mortality effects of migration are harder to discover, although there is clear demographic evidence from studies in central Mali of excess child mortality among small, single-parent households following the departure of the able-bodied family members (results from the Mopti-Sevaré survey of 1985).

Demographic effects of food shortages

There are as many estimates as authors on the numbers of deaths in the Sahel attributable to food shortages since the 1970s. The central problem is the absence of good measurements of adult and child mortality before as well as during the crises. Given this situation, it seems that many writers have simply assumed that large fractions of the populations of the Sahelian states must have been lost to famine-related mortality. They have apparently believed that mortality in recent famines has been of the same order and extent as in the great famines of China, India, or nineteenth century Ireland (Peng, 1987; Bhatia, 1967; Woodham-Smith, 1964). Recent general reviews of famine and its demographic effects also appear to make this assumption (see Mellor and Gavian, 1987 and Dando, 1980 as illustrations).

There are undoubtedly recent localized cases where a significant fraction of the pre-crisis population died: total deaths in Karamoja, Uganda, in 1980, for example, were estimated to be 100,000—a quarter of the total pre-crisis population (Coquery-Vidrovitch, 1985: 63). Accumulating evidence suggests, however, that the excess mortality of the 1970s drought in the Sahel was on the order of 100,000 in the western Sahel and perhaps 200,000 in Ethiopia (see Sen, 1981 and Caldwell, 1977), far below the 20–30 million excess deaths estimated for China during 1959–61 (Ashton et al., 1984; Coale, 1984) or even the 1.5 to 2 million excess deaths in Africa thought to have resulted from the 1918–19 influenza pandemic (Patterson, 1981: 413). For the 1980s, despite guesses of 30 million at risk of starvation in Africa in 1984–85 and one million excess deaths (Timberlake, 1985: 7; Allison and Green, 1985: 7), the true figures are probably much smaller—as Hugo (1984) suggested in his important review of the demographic effects of famines. After all, the total population of all the states of the western Sahel (Mauritania to Chad) was no more than about 38 million in 1983 (World Bank, 1986: Table 1). In addition, "normal" mortality rates are extremely high with expectancies of life at birth in the "deep rural" populations probably no more than 30–35 years (Hill, 1985a: chapter 2 and van den Eerenbeemt, in Hill, 1985a: 82). In such populations, concentrations of high-risk individuals in camps and similar locations are bound to give the impression of high mortality. (Groupings of the destitute in refugee camps can be fatal for young children when few are immunized and epidemics break out—see Aaby et al., 1983, on this mechanism.) With half the children dying before their fifth birthday in normal periods, the figure found for the Masina Fulani of central Mali (Hill, 1985a), in a population of 10,000 people there would be around 150 infant deaths annually and about the same number of deaths of children aged 1–4 years without any allowance for crisis mortality.

In an extensive search of the literature on demographic changes linked to food shortages in Africa, I have found remarkably few references that quoted numbers or rates based on censuses, surveys, or systematic attempts to register births and deaths.[4] Writers have not hesitated to cite figures from tiny studies and generalize the results to whole regions, and to treat results from surveys in emergency camps or distressed communities as typical of the rural population at large. The attempts at larger reviews by Caldwell (1977) and Hugo (1984) are exceptions. The Malian demographic surveys thus are important in providing a statistical base for investigating the demographic effects of drought in at least one part of the Sahel.

Although it may be possible to identify mortality crises in the past from African censuses (see Hollingsworth, 1981 for one attempt) or from retrospective reports on children ever born and surviving (Pearson, 1985), a more reliable method when vital registration data are unavailable is to calculate life table probabilities of survival from full maternity histories. The data from the surveys in rural Mali in 1981–82 have been analyzed in this way, and period-cohort probabilities of dying by age five are shown in Figure 3. Although there are undoubtedly some irregularities in the data due to omissions and misreporting, there are no signs of a peak in child mortality during or just after the 1973–74 drought. Child mortality levels have held steady since the 1950s, as comparison with 1956–58 and 1961 surveys reveals (Hill, Randall, and Sullivan, 1982: Figure 4). (There are similar findings on the fertility side. When the maternity history data are tabulated by age of mother and period of birth, no clear signs of fluctuations in fertility can be identified in the period-cohort rates from the mid-1970s onward.)

Since the reports on child survival stem from mothers, it might be argued that the results are biased because any excess mortality linked with the drought

FIGURE 3 Child mortality trends in central Mali, 1960–85

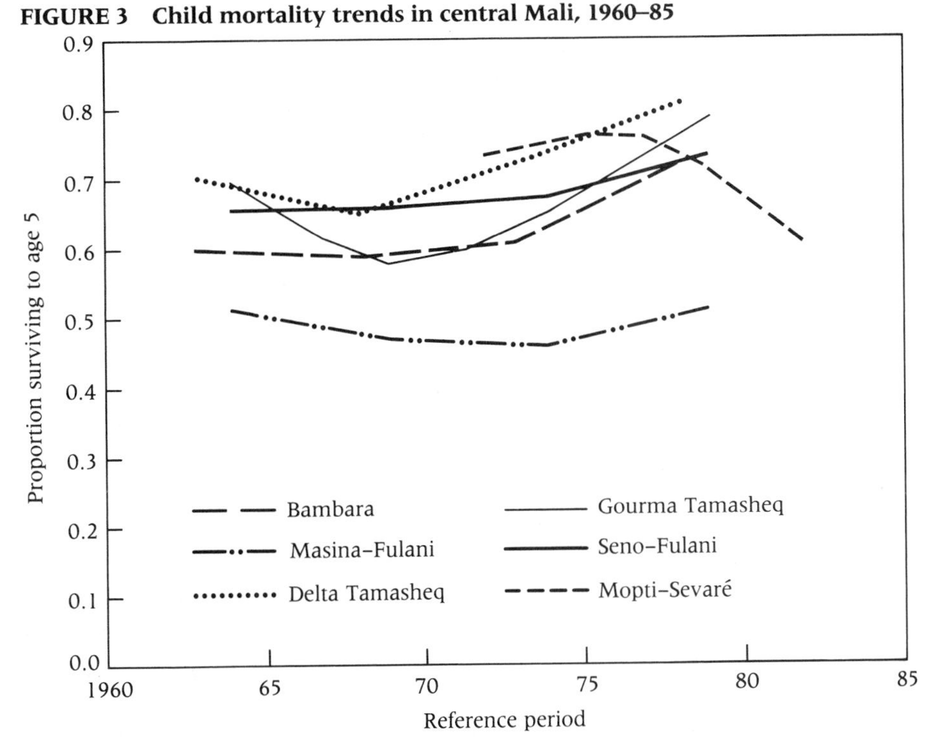

NOTE: Estimates for the Fulani are for the Fulbe and Rimaibe combined.
SOURCE: Birth histories collected by author (Hill, 1985a).

may have removed high-risk mothers as well as their children from the population, either by migration or by mortality. As a test of the importance of migration, a sample of households in two neighboring towns in central Mali was interviewed in 1985 as part of a baseline survey before beginning a new health program. The two towns, Mopti and Sevaré, are popular centers for many of the seasonal and longer term migrants from the rural areas surveyed in 1981–82. The disaster-struck population ("la population flottante" or "les sinistres") was oversampled in order to provide enough cases for a separate analysis. The probabilities of dying were calculated in the same way as for the rural populations but using the additional information on duration of residence in Mopti-Sevaré as a control.

Child mortality was found to have risen recently in Mopti-Sevaré as a result of the inmigration of destitute groups. The inmigrants arriving in the year before the 1985 survey, the group we would expect to be most affected by the series of difficult years in the mid-1980s, do indeed have much higher childhood mortality than the long-term residents of the town. About 40 percent of the children of the recent inmigrants had died before their fifth birthday, compared with about 30 percent among the settled population. When the probabilities of dying by age five were calculated for the new arrivals for earlier time periods (i.e., reaching back over the 10–15-year period before their arrival in Mopti-Sevaré), however, no substantial change in childhood mortality was found among this migrant group over the 1945–85 period. Previous dry periods, including "La Grande Sécheresse" of 1973–74, appear to have left little mark on the history of childhood mortality even among the groups who left the rural areas in the most recent crisis.

The data from Mopti-Sevaré suggest that the low-income, low-status groups bring their already high mortality patterns with them rather than suffering from short-term crisis mortality rates. Some of the large increases in mortality reported by other researchers from sites where the destitute congregate may be simply due to the importation of the previously unrecognized very high mortality of the rural poor rather than due to any short-term worsening of mortality. The mobility of such elements of the Sahelian population poses real problems of measurement and assessment since, with such a high degree of heterogeneity, large changes in mortality rates can occur as a result of compositional changes brought about by migration. Large short-term swings in the numbers of deaths recorded for particular places may be consistent with steady population-wide mortality averages.

Adult mortality is much more difficult than child mortality to measure directly, but indirectly we can produce workable estimates by asking in surveys or censuses about the survival of mothers and fathers. The Mopti-Sevaré data illustrate the important connection between duration of residence in the town and the proportion with mothers or fathers dead at the time of the survey. Again, errors in age reporting are responsible for some irregularities in the data from older respondents, but generally the pattern is clear: the new arrivals in the town do indeed have higher mortality than the established residents. By converting these proportions orphaned to estimates of adult mortality for different time periods before the survey, we can decide whether there are any obvious fluctuations in mortality level for each of the groups. No clear evidence of the 1973–74 drought can be detected in the data either for the inmigrants, the group we would expect to be most strongly affected, or for the usual residents. Again the data suggest that two mortality regimes have coexisted for some time, and neither shows any marked decline or short-term fluctuations.

Although the mortality and fertility effects of drought were found to be comparatively slight, it would probably be anticipated that the migration effects would be very substantial. Limited space precludes a full examination of the manifold effects of food shortages on temporary and permanent migration patterns. These are in any case well described in Currey and Hugo (1984) and in Oba and Lusigi (1987). The potential size of the effects is large, but the measured effects are in fact rather modest, especially in the most recent (mid-1980s) period.

Buffer institutions

External factors, such as international development assistance and food aid, as indicated earlier, are relatively new factors that are responsible for some of the dampening of the effects of drought on the demography of sub-Saharan Africa. In precolonial times a whole set of reciprocal patron–client relations—a moral economy—served to limit the effects of periodic drought and famine on all levels of society. Are similar institutional arrangements working to the same end today? It is impossible to provide an unqualified answer to such a broad question: the differences among societies are too great. I will briefly sketch some aspects of social organization relative to the management of food shortages for the three main ethnic groups—the Bambara, Fulani, and Tamasheq of central Mali—that were the subject of the more detailed studies described earlier.

The Bambara

The Bambara are among the most populous of the ethnic groups of the western Sahel.[5] They are culturally part of the even larger Mande community extending from Senegal through Mali to Burkina Faso. Resistant to the spread of Islam, first at the end of the sixteenth century and again during the nineteenth century, they formed powerful warlike kingdoms able to undertake major conquests and finally to offer strong resistance to the French advance early in this century. Traditionally, Bambara society consisted of three main social classes—nobles who fought and farmed, slaves or captives, and blacksmiths-storytellers ("griots"). In the present day, despite some residual resistance to intermarriage with ex-slaves, more significant social groupings are the male and female age sets and the village youth associations for males aged about 14 to 30 (Monteil, 1924; Toulmin, 1986a). Each village usually consists of several major exogamous lineages or clans governed by a village council of elders. The residential unit is a large extended household, containing in a single compound but in several separate houses the household head, his wives and children (the Bambara are polygynous and practice the levirate), as well as other married men and their families who are normally related to the household head.

The core of the village economy is the cultivation of bulrush millet of different varieties and cycle lengths, intersown with other crops such as cowpeas. Average rainfall in the northern Sahel ranges from about 500mm to 1,000mm, falling in a series of sudden downpours between June and September. The variability of rainfall is very high and its geographic spread is uneven. Families with only a small amount of land in one location face high risks of reaping a small crop or indeed no crop at all.

The production process is highly labor-intensive, demanding long hours of manual labor, involving the preparation of the fields by ploughing with oxen and digging; weeding in the main growing season; and harvesting. Bambara society is generally "labor hungry." Several strategies serve to keep the residential units extremely large. These include early and universal marriage for women (in marked contrast with pastoral communities), retention of control over sons and their offspring by the household heads, and numerous customary systems of rights and obligations. In the 1981 demographic survey of 51 villages in the Kaarta region north of Segou in Mali, the average household size was 14 persons; some compounds contained more than 100 people (Randall, 1984: 12). Among women aged 20–49, 97 percent were currently married at the time of the survey; widows and divorcees are quickly remarried, both for their labor and for their accumulated wealth in fields and possibly livestock.

Mortality is high. About a quarter of children die before their first birthday and only two-thirds survive to age five. Sons are of key importance since only they can perpetuate the lineage name. Agricultural tasks and social activities are strongly segregated by gender (e.g., only women winnow) so that it would be socially impossible for a single man or a single woman to cultivate alone.

As in many societies, the potentially conflicting interests of the individual household and the wider community are a source of tension. This is seen most clearly in the organization of production. Two types of fields are used to cultivate millet: village fields that encircle the village in a continuous belt, and bush fields that are scattered in discontinuous blocks further from the village. Work in the village fields is termed *forobaforo* ("big field") activity and is undertaken jointly by complex households, which usually hold their land there in several separate parcels. These fields are manured and hence permanently cultivated and involve cooperation between households for tasks such as ploughing with oxen, weeding, and harvesting. Throughout the main agricultural season (June to January) household members of working age are expected to be available for cultivation activities in the village fields six days a week from dawn to dusk. Social pressure, coercion, and occasionally force are employed to ensure participation. Individuals can also undertake cultivation in their private fields after the joint work is finished. These activities are termed *senoforo* ("night field"). Sons often cultivate for their mother, who retains the grain obtained for herself, her children, and her husband. Retired people over about age 55 are also allowed to cultivate for themselves.

A complex system of checks and balances regulates the responsibilities and rights of individuals within households and of households within the community. In this essay, it is the recent adjustments to these balances that are of interest, particularly as they concern responses to drought. In the 1960s, the commercial success of groundnut cultivation meant that individuals were able to acquire wealth by growing peanuts in their private fields at the expense of millet cultivation. Some of this wealth was then invested in well-digging and livestock, which were essentially private investments and offered a way by which private fields could be manured and improved without the owner's having to share the benefits with the larger community. This growth of private initiatives seems to have been considered socially divisive, but the problem disappeared with the collapse of the market for peanuts and the series of dry years that ensued in the 1970s and 1980s. Men, formerly free in the late afternoons and on Fridays, have now abandoned almost all private cultivation activities and are required to work full time on behalf of the household. Women, however, retain their own fields. Cultivation also requires heavy investments in ploughs, oxen, donkeys, and carts, the outlays for which require joint efforts by all productive members of the household, including some who migrate in the dry season to work elsewhere.

The Bambara community usually consists of several large lineages with many marriage links between them. There is still a strongly sanctioned commitment to community survival, and weaker households are helped over the annual late dry season crisis when food stocks are low and labor demands are heavy. Help is often available for work in the fields, and grain-deficit households can obtain millet by their women undertaking harvesting and winnowing tasks for others, including farmers in neighboring villages. The village youth associations act as a redistributive labor pool providing labor to the poorest and least able households. Other less formal reciprocal work groups are also common.

According to Toulmin (1986a), larger households are more successful than small ones in the production of food in the semi-arid regions cultivated by the Bambara.[6] Her work, buttressed by a series of detailed measurements made over several years, demonstrates that there is a strong economic rationale behind the Bambara strategy of cooperative effort based on very large residential units. This strategy inevitably implies strong social support for high fertility. Practically, this means ensuring that girls are married early and that virtually all women of reproductive age are in the pool of potential mothers. The emphasis on male descent and perpetuation of the lineage probably means some overall reduction in the fertility of individual women, but in the polygynous Bambara society, male fertility is socially more important. It seems from the studies by Toulmin (1986a and b) and by Martin (1985) that dry periods and periods of economic recession tighten rather than loosen communal bonds. The Bambara survival strategy stresses the responsibilities of individuals to the household and to the community and uses the full force of customary law to enforce these obligations. All this is possible only with an open land frontier; in the Kaarta region of Mali studied by Toulmin and Martin, densities were as low as 10 persons per km^2. The ability of communal institutions to adapt to the closing of the land frontier is so far untested in most areas settled by the Bambara.

The Fulani

The large literature on the Fulani (see the classic studies by Dupire, 1970 and Stenning, 1959) is difficult to summarize since the community of Fulfulde speakers extends over a wide variety of ecological and economic conditions from Senegal to Chad. One of the hallmarks of Fulani society, apart from its strong cultural attachment to livestock herding, is a complex system of castes and classes. This system is undergoing rapid mutation as rights to agricultural and grazing land are altered in the post-independence era (see Gallais, 1967 and 1984 and Cissé, 1985 for details). Although the traditional tenure arrangements in the nineteenth century Masina empire were not typical of other large areas in West Africa, several recent accounts draw attention to the continuation of institutional arrangements in the inner Niger flood zones of central Mali that make it possible for common property resources to be jointly exploited by several different communities (see ILCA, 1983 for evidence on this from a large survey). Outside the Masina, several authors draw attention to the existence of traditional lending associations ("Laawol fulfulde") in Fulani society (see Riesman, 1984: 186–187). Fieldwork in the Niger Range and Livestock Project area among the Wodaabe, a Fulfulde-speaking pastoral group in central Niger, has illustrated how the modern state has adversely affected the traditional capacity of herders to cope with a series of dry years by making pastoral production a marginal activity through the erosion of common property rights and encouragement of outright land ownership by farmers (White, 1986 and Swift, 1984).

For central Mali outside the Masina, much information on household production and consumption among the Seno Fulani has been assembled recently by joint teams from the London School of Hygiene and Tropical Medicine and the International Livestock Centre for Africa. (Summaries appear in Hesse et al., 1985 and Hilderbrand et al., 1985b.) Again, differential access to land and water rights by the cattle-owning Fulbe nobility and the freed agricultural population of Rimaibe, partially constrained by traditional relationships, is a central feature of this Fulani group. Migration and a resort to "famine foods" are just two among many common responses to food deficits. The most notable changes taking place in the production system are the gradual decline of ethnic specialization and the introduction of a more mixed economic base by both Fulbe and Rimaibe. To some extent, economic transformations of this kind, occurring partly in response to the general marginalization of pastoralists, are tending to undermine traditional exchange and dependency relationships.

In terms of survival strategies both during the annual dry season and in exceptionally dry series of years, the Fulani reaction is quite different from that of the Bambara. Rather than combining into large households and herding units to minimize risks for individuals, the Fulani reaction is to subdivide into small family units, some parts of which may never reunite. Cattle herds are split by the age and sex of the livestock, and sheep and goats may be herded separately, enabling the dispersed family unit to better exploit scarce grazing resources. Detailed work by White (1986, 1987), Swift (1984), and Loutan (1985) among the Wodaabe has yielded important insights into such management strategies. Rather than livestock-rich households being the largest, the reverse is true. The explanation offered by White (1987) is that the possession of a larger herd means that new households can be formed more quickly and split off, since the underlying principle of the pre-inheritance system practiced by the Wodaabe, as by many Fulani groups, is that the heifer given to a child at birth should form the core of the offspring's own herd by the time he or she attains the age of majority. In the Wodaabe households with the most animals, cash income was estimated to be twice as high per person as for households with the smallest number of animals (White, 1986: Table 3). The largest households also had less need to dispatch members in search of jobs outside the pastoral economy or to rely on borrowed animals. There is some shame in having poor families represented in the administrative "fraction" (division), so that systems of gifts, loans of milch cows, and *habba nai* loans (involving the temporary gift of a cow until she delivers three calves) are all established ways of redistributing livestock.

Given that the Wodaabe herders are organized in small, highly mobile family units, their collective voice is weak compared with the more highly organized, socially stratified communities (such as the Tamasheq herders discussed below). In addition, the Wodaabe system of pre-inheritance, in which the herd is held in a kind of collective trust for the next generation, means that weaker members cannot be forced out of the pastoral economy. The severity of recent livestock reductions by both death and forced sales appears to indicate that traditional coping mechanisms are no longer adequate (White, 1987: 4–5). Despite the splitting of already small herds and labor migration, many Wodaabe have had to become contract herders, which restricts their mobility (since owners want to maintain contact with their stock) and reduces the chance of their receiving loans of animals under the traditional *habba nai* sharing system (lenders fear infection from animals of outsiders) (White, 1987: 10–11).

In Niger in 1980, the national herd was back to pre-1973 drought levels, but ownership of the livestock had been redistributed, leaving only about 10 percent of Wodaabe families with enough animals even for bare subsistence (White, 1987: 482). Pressures that are bringing about such changes in the pastoral economy are not all related to drought—indeed White maintains that most are not. The joint effect of taxation in cash, unfavorable terms of trade between herders and farmers, encroachment of the agricultural area onto pastoral land, and the growth of the political and economic power of urban-based groups, all have forced the Wodaabe into a dependent relationship with others. Contract herding has grown as nonpastoralists invest in livestock. Out-migration is an established need because of small herd sizes. While customary sharing of limited resources undoubtedly alleviates the poverty of a few households, the welfare of the whole herding group has been hard hit by adverse economic and political developments that will be difficult to reverse. Further splitting of herds and households as well as long migrations for work will have a negative effect on human fertility and may indeed adversely affect the growth and survival of small children. The indirect and longer term demographic consequences of the drought on groups such as the Wodaabe are therefore probably much more important than any direct effects linked to temporary increases in mortality or reductions in fertility.

The Tamasheq

Tamasheq or Tourareg society provides an even clearer illustration of the importance of class or status and prestige in affecting drought survival strategies and their importance in the prevailing demographic regime (see Randall and Winter, 1985; Winter, 1984; Hilderbrand et al., 1985a for details). Continued survival of the upper-class cattle owners during periods of stress is assured by exporting surplus labor, usually the dependent Iklan (Bella) groups, to agricultural areas or nearby towns and by calling in previous loans to kinsmen often spread over a wide area. The contrast with the social solidarity of the Bambara could not be sharper, since, in periods of hardship in Tamasheq society, cattle (wealth) will usually move upward in the social hierarchy, leading to increased concentration in the hands of the owners of large herds. The main difference between the Fulani and Tamasheq pastoral societies is that in Fulani society, the Rimaibe agricultural groups have always had rights to fields and have acquired more extensive rights to land and water in the post-independence era. The Tamasheq lower classes, the Iklan, have not enjoyed this advantage since they have never been able to acquire substantial numbers of cattle except when they have formed independent camps.

Selective migration, forced or voluntary, is the most important strategy open to pastoralists for coping with regular dry season stress or exceptional drought. The numbers moving can be substantial: in the 1973 drought, some 20,000 displaced persons, largely pastoralists, were found around the town of Gao in eastern Mali (Davies and Thiam, 1987: 9). These were people who extended their usual transhumance movements to include the town of Gao. Milk production is at its lowest in the dry season, so animals have to be sold in order to buy cereals. Livestock prices always fall in the dry season as supply outruns demand, but in 1973 and again in 1984 herds were stripped to such an extent that reconstitution of the former herd became much more difficult. In a survey conducted in 1984, only 13 percent of households in the displaced-persons camps of Gao felt they had the resources necessary to rebuild their herds (Mali, 1986). This contrasts sharply with the situation after 1973 when the *Injiten* system of institutionalized loans of animals within a fraction was able to ensure that even the most impoverished were able to restart, albeit as shepherds rather than as herd owners in the first instance (Davies and Thiam, 1987: 12–13). In the 1980s, many destitute households have become near-permanent members of the Sahel's refugee camps as wealth has become concentrated in fewer hands. Traditional redistributive systems are unable to function as the continuing drought pushes both farmers and pastoralists into the displaced-persons camps.

In Mali, perhaps half the 7th Region's goats and two-thirds of its cattle and sheep died in the 1984–85 period, raising the proportion of the nomadic population without any livestock to over one-third, the figure reached in the 1973 drought (Davies and Thiam, 1987: 23). Rebuilding a goat herd consisting of 35 animals will cost some 140,000 FCFA (about US$500 at 1987 prices), well beyond the earning or even saving power of the most energetic migrant family. With the influx of migrants, daily wage rates in the towns have fallen to levels ranging from 300 FCFA ($1) for construction workers to only 50 FCFA (18 cents) for artisans, cleaners, and the like—at a time when a modest meal of millet and a simple sauce costs some 150–400 FCFA ($0.52–$1.40) per family (Davies and Thiam, 1987: 26 and 37).

Clearly, when the dry season crisis becomes severe, producing wholesale economic collapse, large proportions of the pastoral population lose their entitlements to food and their means to earn a livelihood. No redistributive system will alleviate the suffering for the entire community. The terms of trade overall are against the herders, allowing those with access to land and water (in the Malian case, mostly Sonrai farmers in the Niger valley) to benefit from cheap agricultural labor (Bella and other destitute herders) and to obtain a stake in the pastoral economy by the employment of former herd owners as contract shepherds.

There are no data on the demographic effects of structural changes such as those occurring among the pastoral populations of the Sahel. It is clear that the distinctiveness of Sahelian marriage systems may disappear if the connection between herd growth and the formation of new households is lost. Distinctive inheritance systems, including pre-inheritance (Fulani), matrilineal descent, and bridewealth-based marriage arrangements, all seem to delay marriage and to increase the social and economic independence of women, whether married, widowed, or divorced. Growth of state power in the area of land ownership and rights to water and wood appears to threaten many traditional community-sharing institutions in the pastoral zone since the majority of development efforts view pastoralism from the perspective of settled peoples (Riesman, 1984).

This brief summary merely hints at the contrasting social and institutional arrangements in just three Sahelian populations that condition the nature and capacity of each to respond to seasonal and more prolonged periods of stress. The arrangements are far from constant and are heavily influenced by policies and legislation enacted by the modern state. Perhaps the principal conclusion to emerge from the mass of recent fieldwork, including that in Mali, is the extraordinary flexibility and diversity of possible responses to food scarcity. Since many agro-pastoral communities of the Sahel are so "open" and dependent on links with other groups, it is very unlikely that any single policy on food production or distribution will have similar effects across the whole region. In part because of this capacity to respond flexibly to new ecological and economic conditions, and in part because of international development assistance in a variety of forms, fertility and mortality changes will not show any simple and direct link with food shortages. One established demographic response to stress, migration, is likely to remain an important factor, both in itself and in influencing mortality outcomes—in that attempts to constrain migration through national or international controls can result in major mortality crises.

In practice, the most valuable intervention to preserve the welfare of pastoral communities such as the Tamasheq would be measures to offset seasonal falls in their purchasing power. The importance of off-farm dry season work is gradually being appreciated by development planners. The willingness of individuals and families to migrate and change jobs in the dry season is an important aspect of traditional behavior that can compensate for some of the seasonality of the food production cycle; it can also contribute to the more even spread of services, including food aid, provided by the state.

Conclusion

In the vast literature concerned with the drought of the early 1970s in Africa, only a few authors queried the claims in the newspaper headlines that several million Africans were on the brink of starvation and that thousands were dying already. Although many of the sample populations they cited were small, several health workers noted the poor but not disastrous nutritional status of young children (e.g., Seaman et al., 1973). At the same time, Caldwell and scientists from specialist organizations such as the US Centers for Disease Control produced much smaller estimates for the probable numbers of "excess" deaths. Only a few projects that had started before the drought were able to provide more or less continuous observations throughout the crisis (see Faulkingham and Thorbahn, 1975; Faulkingham, 1977). The scattered evidence nonetheless strongly suggested that the direct effects of the drought were less than imagined, although no one wanted to speculate about longer term changes. This presents us with a puzzle, since there is little doubt that the 1970s and the 1980s have been an exceptionally dry period in the whole Sahel region (recall Figure 2). In addition, the worldwide oil price rise of 1973 and falls in the price of many primary commodities sent the economies of most developing countries into a decline from which they have yet to recover.

How were major mortality crises averted when previous experience of drought and its consequent economic shocks predicted disaster? In trying to frame an answer to this key question, we have seen that there have been both positive and negative changes in people's exposure to risk, governed particularly by society-specific forms of social organization and by local institutional arrangements. This is not to ignore the effect of direct interventions by government through food relief and health services. (These interventions are not without drawbacks. Wider vaccination coverage, for example, has undoubtedly reduced the chances of major epidemics; on the other hand, clustering people together in feeding centers has probably had very negative effects on their health because of the introduction of unsuitable foods and the increased risk of disease transmission.) The more important role of government actions in influencing the demographic outcomes of food shortages has probably been an indirect one: working through the sometimes deliberate but usually inadvertent changes those actions have wrought in the social order and in patterns of social life. Such changes, affecting a society's capacity to buffer the impact of stress, are quite distinctive among different ethnic groups, as the three cases discussed above illustrate. Serious problems for the future may well arise where traditional modes of dealing with food crises at the community level have in effect been destroyed.

The most systematic attempt to put African poverty and destitution in its historical context is by Illiffe (1987). In his book, he introduces the important distinction between "structural poverty," which is always with us, and "conjunctural poverty," whose immediate causes are generally natural disasters and political instability (pp. 4–5). In the transition from "epidemic starvation" to "endemic undernutrition of the very poor" (p. 6), the closing of the open land frontier in tropical Africa occupies a central place. Extending Illiffe's analysis, we might surmise that current difficulties in the Sahel are attributable to changes in the extent of "structural poverty," offset in some instances by alleviation of the intensity of "conjunctural poverty" crises. Studies of the Sahelian pastoral and agro-pastoral communities and the "down and out" members of rural society strongly suggest that some traditional institutional sources of support are being eroded as economic and political changes favor private initiative in areas such as settled, capital-intensive agriculture. A surprising number of social institutional bases for support continue to function, albeit in some novel ways. The long experience of most Sahelian societies in coping with the annual period of "soudure" (making ends meet) may be their best long-term protection against deeper crises.

Notes

1 Statistics on the excess mortality associated with major mortality crises are notoriously unreliable except in countries with good census data, but Patterson (1981: 413) provides good evidence that the 1918–19 influenza pandemic in Africa was responsible for some 1.5 to 2 million excess deaths. The claim that the Biafran civil war "claimed a million lives" (King, 1986: 67) and that "the cumulative death toll over the past decade [has passed] the 600,000 mark" in Uganda (King, 1986: 12) are largely unsupported by documentary evidence. Similarly, "famine deaths" in the Ethiopian provinces of Tigray, Wolo, and northeast Shewa during 1971–73 are reported by King (1986: 36) to total 200,000 people. Using data from a network of small studies in northern Ethiopia in 1984–85, Seaman (1987) has estimated excess mortality to be at least 500,000 but possibly double this figure (p. 24). The excess mortality in northern Sudan over the same period, based on extrapolation of the experience in Dafur, is estimated at some 250,000 people, mostly children (Seaman, 1987: 26–28). Seaman also points out that the main causes of this excess mortality were infectious diseases, whose effects were rendered more deadly by the concentration of ill-fed people with reduced resistance to disease in crowded refugee camps (Seaman, 1987 and personal communication).

2 The term originates with E. P. Thompson, *The Making of the English Working Class* (New York: Vintage Books, 1966), p. 203.

3 In the series of World Fertility Survey studies in some 40 developing countries with widely varying dietary compositions and levels of food intake, "potential fertility," a theoretical measure of the biological capacity to reproduce, varied only within narrow limits (Casterline et al., 1984). Many nutritionists now interpret the range of body sizes found in different countries and social classes as an adaptive response to the combined effects of such factors as food scarcity, communicable disease levels, and work burdens. This adjustment would mean that the reproductive potential of Sahelian communities is essentially the same as that of most other human groups, including those in areas of Africa with higher food production.

The literature on nutritional status and reproduction, potential or realized, is large. Good general summaries of the biological factors affecting reproduction are available in Butz and Habicht (1976), Gray (1979), and Mosley (1978 and 1979). The particular effects of famine are dealt with by Bongaarts and Delgado (1979) and by Stein and Susser (1978). Broader demographic repercussions of famine are succinctly dealt with by Hugo (1984).

4 In a systematic search of several major machine-readable databases, none of the references examined contained figures on recent birth and death rates that can be accepted as representing true levels of fertility and mortality in any sizable population. As a typical example, the study of Biellik and Henderson (1981) cites an infant mortality rate for Karamoja in Uganda of 607 per thousand live births, but this is based on the number of infant deaths reported in the year before the survey among just 150 households.

5 This section on the Bambara is based on the following original sources: Fulton and Toulmin (1982), which provides a general description of the Bambara socioeconomic system; Toulmin (1986a and b), which contain the important data on the relative success of larger households as well as much rich ethnographic material stemming from several years' residence in two communities north of Segou in Mali; Randall (1984) on Bambara demography; and Martin (1985) on food intake and seasonal food shortages.

6 The merits of large household units are quite widely recognized in different agricultural societies. Richards (1987: 117) points out that in the rice farming system of central Sierra Leone, smaller than average households suffered more frequently from preharvest hunger. He also cites some specific examples where chance mishaps have adversely affected the welfare of whole households, often smaller ones (pp. 116–117). The importance of the Bambara system is that, perhaps because of the extra hazards of rain-fed agri-

culture in the semi-arid zone, the society has developed a system of social control that prevents large households from breaking up due to disputes of one kind or another. This is not true of other Sahelian societies: in Richards's studies the mean household size was close to 5.5 people, and in Tamasheq society the usual response to any sort of disagreement is for the individual or family in dispute to join another camp, usually in the same "fraction."

References

Aaby, Peter, J. Bukh, I. M. Lisse, and A. J. Smits. 1983. "Measles mortality, state of nutrition and family structure: A community study from Guinea-Bissau," *Journal of Infectious Diseases* 147: 693–701.

Allison, Caroline, and Reginald Green. 1985. "Sub-Saharan Africa: Getting the facts straight," *Institute for Development Studies Bulletin* 16, no. 3: 1–8.

Ashton, Basil, et al. 1984. "Famine in China, 1958–61," *Population and Development Review* 10, no. 4: 613–645.

Biellik, R. J., and P. L. Henderson. 1981. "Mortality, nutritional status and diet during the famine on Karamoja, Uganda 1980," *The Lancet*, 12 December (8259): 1330–1333.

Bhatia, B. V. 1967. *Famines in India 1860–1965*. London: Asia Publishing House.

Blacker, John C. G. 1990. "Infant and child mortality: Development, environment and custom," in *Disease and Mortality in Sub-Saharan Africa*, ed. Richard G. Feacham and Dean T. Jamison. Oxford University Press for the World Bank.

Bonfiglioli, Angelo M. 1985. "Evolution de la propriété animale chez les Wodaabe du Niger," *Journal des Africanistes* 55, no. 1–2: 29–37.

Bongaarts, John, and Hernan Delgado. 1979. "Effects of nutritional status on fertility in rural Guatemala," in *Natural Fertility*, ed. Henri Leridon and Jane Menken. Liège: Ordina.

Borton, J., and Edward Clay. 1988. "The African food crisis of 1982–86," in *Rural Transformation in Tropical Africa*, ed. D. Rimmer. London: Frances Pinter.

Butz, W. P. and J. P. Habicht. 1976. "The effects of nutrition and health on fertility: Hypotheses, evidence, and interventions," in *Population and Development: The Search for Selective Interventions*, ed. Ronald Ridker. Baltimore: Johns Hopkins University Press for Resources for the Future.

Caldwell, John C. 1977. "Demographic aspects of the drought: An examination of the African drought of 1970–74," in *Drought in Africa*, ed. D. Dalby, Harrison Church, and F. Bezzaz. London: International African Institute.

Carreira, Antonio. 1982. *The People of the Cape Verde Islands*. London: Hurst.

Casterline, John B., Susheela Singh, John Cleland, and Hazel Ashurst. 1984. "The proximate determinants of fertility," *World Fertility Survey Comparative Studies* No. 39. Voorburg: International Statistical Institute.

Cissé, Salmana. 1985. "Land tenure practice and development problems in Mali: The case of the Niger delta," Chapter 7 in Hill (1985a).

Coale, Ansley J. 1984. *Rapid Population Change in China 1952–1982*. Washington, D.C.: National Academy Press.

Coquery-Vidrovitch, Cathérine. 1985. *Afrique noire: permanences et ruptures*. Paris: Payot.

Currey, B., and Graeme Hugo. 1984. *Famine as a Geographical Phenomenon*. Amsterdam: Reidel.

Dando, W. 1980. *The Geography of Famine*. London: Edward Arnold.

Davies, Susanna, and Adam Thiam. 1987. "The slow onset of famine: Early warning, migration and post-drought recovery: The case of Gao ville," Save the Children Fund and Food Emergency Research Unit, London, Report no. 1.

Dupire, Marguerite. 1970. *Organisation sociale des Peul*. Paris: Librairie Plon.

Dyson, Tim, and Mike Murphy. 1985. "The onset of fertility transition," *Population and Development Review* 11, no. 3: 399–440.

Farmer, G., and T. M. L. Wigley. 1985. "Climatic trends for tropical Africa," Research Report for the UK Overseas Development Administration, University of East Anglia.

Faulkingham, Ralph H. 1977. "Ecological constraints and subsistence strategies: The impact of drought in a Hausa village," in *Drought in Africa*, ed. D. Dalby, Harrison Church, and F. Bezzaz. London: International African Institute.

———, and P. F. Thorbahn. 1975. "Population dynamics and drought: A village in Niger," *Population Studies* 29, no. 3: 463–477.

Fulton, Duncan, and Camilla Toulmin. 1982. "A socioeconomic study of an agro-pastoral system in central Mali." Bamako and Addis Ababa: International Livestock Centre for Africa.

Gallais, Jean. 1967. *Le delta intérieur du Niger: étude de géographie régionale*. Memoires de l'IFAN, no. 79, 2 vols. Dakar.

———. 1984. *Hommes du Sahel*. Paris: Flammarion.

Garcia, R. V., and P. Spitz. 1981. *Drought and Man in the 1972 Famine: A Case History*, 3 vols. Oxford: Pergamon Press.

Gray, Ronald H. 1979. "Biological factors other than nutrition and lactation which may influence natural fertility: A review," Chapter 10 in *Natural Fertility*, ed. Henri Leridon and Jane Menken. Liège: Ordina.

Griffiths, Ieuan. 1988. "Famine and war in Africa," *Geography* 73, no. 1: 59–61.

———, and J. A. Binns. 1988. "Hunger, help, and hypocrisy: Crisis and response to crisis in Africa," *Geography* 73, no. 1: 48–54.

Hardy, N. 1983. *Agriculture in China's Modern Economic Development*. Cambridge: Cambridge University Press.

Harriss, Barbara. 1982. "Agricultural marketing in the semi-arid tropics of West Africa," *Development Studies Occasional Paper* no. 14, University of East Anglia.

Hart, Keith. 1982. *The Political Economy of West African Agriculture*. Cambridge: Cambridge University Press.

Hesse, Cedric, Adam Thiam, Chris Fowler, and Jeremy J. Swift. 1985. "A Fulani agro-pastoral system in the Malian Gourma." Bamako and Addis Ababa: International Livestock Centre for Africa.

Hilderbrand, Katherine, Allan G. Hill, Sara Randall, and Marie-Louise van den Eerenbeemt. 1985a. "Child mortality and care of children in central Mali," in Hill (1985a).

———, Adam Thiam, Andrew Tomkins, and Elizabeth Dowler. 1985b. "Food, work, health, and nutrition: A comparative study of the seasonal effects of two agro-pastoral populations from the Malian Gourma." Report to the Overseas Development Administration from the London School of Hygiene and Tropical Medicine and the International Livestock Centre for Africa, Bamako and Addis Ababa.

Hill, Allan G. (ed.). 1985a. *Population, Health, and Nutrition in the Sahel*. London: Kegan Paul International.

———. 1985b. "The fertility of farmers and pastoralists of the West African Sahel," *Fertility Determinants Research Notes* No. 6. New York: The Population Council.

———, Sara C. Randall, and Oriel Sullivan. 1982. "The mortality and fertility of farmers and pastoralists in Central Mali 1950–81," *Centre for Population Studies Research Paper* 82–4. London School of Hygiene and Tropical Medicine.

———, and Adam Thiam. 1987. "Marriage, inheritance and fertility amongst the Malian Fulani," in *The Cultural Roots of African Fertility Regimes*, Proceedings of a conference at the University of Ile-Ife, Nigeria, 25 February–1 March 1987, edited by Etienne van de Walle. Obafemi University and University of Pennsylvania.

Hollingsworth, T. H. 1981. "The use of two census populations given by single years of age to determine past calamities: An experiment with Kenyan data," in *Proceedings of the Conference on African Historical Demography*. University of Edinburgh: Centre of African Studies.

Hugo, Graeme. 1984. "The demographic impact of famine," pp. 7–31 in Currey and Hugo (1984).

ILCA. 1983. *Recherche d'une solution aux problémes de l'élevage dans le delta intérieur du Niger au Mali*, 5 vols. Bamako and Addis Ababa: International Livestock Centre for Africa.

Illiffe, John. 1987. *The African Poor*. African Studies Series 58, Cambridge University Press.

King, Preston. 1986. *An African Winter*. London: Penguin Special.

Lesthaeghe, Ron (ed.) 1989. *Reproduction and Social Organization in sub-Saharan Africa*. Los Angeles: University of California Press.

———, G. Kaufmann, and D. Meekers. 1986. "The nuptiality regimes in sub-Saharan Africa," Interuniversity Programme in Demography, Working Paper 86–3, Vrije Universiteit, Brussel.

Longhurst, Richard (ed.). 1986. "Seasonality and poverty," Papers from the Institute for Development Studies Conference, February 1985. *Institute for Development Studies Bulletin* 17, no. 3: 1–7.

Loutan, Louis. 1985. "Nutrition amongst a group of Wodaabe (Fulani Bororo) pastoralists in Niger," Chapter 10 in Hill (1985a).

Mali, Direction Nationale des Affaires Sociales. 1986. *Enquête sociale des populations deplacées au fait de la sécheresse*. Bamako: Minist ère de la Santé Publique and des Affaires Sociales.

Martin, Mary. 1985. "Design of a food intake study in two Bambara villages of the Segou region of Mali with preliminary findings," Chapter 13 in Hill (1985a).

Marx, Karl. 1849. "Wage, labour and capital," reprinted in *Selected Works*. Moscow: Progress Publishers.

Mellor, J. W., and S. Gavian. 1987. "Famine: Causes, prevention and relief," *Science* 235: 539–545.

Monteil, Charles. 1924. *Les Bambara de Ségou et du Kaarta*. Republished in 1977 by Maisonneuve et Larose, Paris.

Mosley, W. Henry (ed.). 1978. *Nutrition and Human Reproduction*. New York: Plenum Press.

———. 1979. "The effects of nutrition on natural fertility," in *Natural Fertility*, ed. Henri Leridon and Jane Menken. Liège: Ordina.

Oba, G., and W. J. Lusigi. 1987. "An overview of drought strategies and land use in African pastoral systems," *Pastoral Development Network Paper* No. 23a. London: Overseas Development Institute.

Patterson, K. D. 1981. "The demographic impact of the 1918–19 influenza pandemic in sub-Saharan Africa," in *African Historical Demography*, pp. 403–430. Volume 2 of the proceedings of a seminar at the Centre of African Studies, Edinburgh, 24–25 April.

Pearson, Roger. 1985. "A simple and economical method for accurate measurement of mortality in populations which have suffered a disaster," unpublished M.Sc. thesis, London School of Hygiene and Tropical Medicine.

Peng, Xizhe. 1987. "Demographic consequences of the Great Leap Forward in China's provinces," *Population and Development Review* 13, no. 4: 639–670.

Randall, Sara C. 1984. "A comparative demographic study of three Sahelian populations," unpublished Ph.D. thesis, University of London.

———, and Michael Winter. 1985. "The reluctant spouse and the illegitimate slave," Chapter 8 in Hill (1985a).

Richards, Paul. 1985. *Indigenous Agricultural Revolution*. London: Hutchinson.

———. 1987. *Coping with Hunger*. London Research Series in Geography 11. London: Allen and Unwin.

Riesman, Paul. 1984. "The Fulani in a development context," in *Life before the Drought*, ed. E. Scott. Boston and London: Allen and Unwin.

Seaman, John. 1987. "Famine mortality in Ethiopia and Sudan," paper presented at the IUSSP seminar on mortality and society in sub-Saharan Africa, Yaoundé, Cameroon, 19–23 October.

———, Julius Holt, John Rivers, and J. Murlis. 1973. "An enquiry into the drought situation in Upper Volta," *The Lancet*, 6 October: 774–778.

Sen, Amartya. 1981. *Poverty and Famines: An Essay on Entitlement and Deprivation*. Oxford: Clarendon Press.

Stein, Z., and M. Susser. 1978. "Famine and fertility," in Mosley (1978).
Stenning, D. J. 1959. *Savannah Nomads*. Oxford University Press.
Swift, Jeremy J. 1979. "The development of livestock trading in the nomadic pastoral economy: The Somali case," in *Pastoral Production and Society*, ed. L'Equipe écologie et anthropologie des sociétés pastorales. Cambridge: Cambridge University Press.
———— (ed.). 1984. "Pastoral development in central Niger," MDR/USAID, Niamey, Niger.
Talbot, Lee M. 1986. "Demographic factors in resource depletion and environmental degradation in East African rangeland," *Population and Development Review* 12, no. 3: 441–451.
Timberlake, Lloyd. 1985. *Africa in Crisis*. London: Earthscan Paperback.
Toulmin, Camilla. 1986a. "Changing patterns of investment in a Sahelian community," unpublished D.Phil. thesis, University of Oxford.
————. 1986b. "Access to food, dry season strategies and household size amongst the Bambara of central Mali," *Institute for Development Studies Bulletin* 17, no. 3: 58–66.
Ujo, Eric. 1985. "Levels and differentials in fertility and mortality among some Kanuri of NE Nigeria," unpublished Ph.D. thesis, University of London.
————. 1987. "A determinant of fertility and its cultural context among Nigerian Kanuri," in *The Cultural Roots of African Fertility Regimes*, Proceedings of a conference at the University of Ile-Ife, Nigeria, 25 February–1 March, 1987, edited by Etienne van de Walle. Obafemi University and University of Pennsylvania.
United Nations. 1987. *Mortality of Children under Age 5: World Estimates and Projections 1950–2025*. New York: Department of International Economic and Social Affairs (ST/ESA/Ser.A/105).
Watts, Michael. 1983. *Silent Violence*. Berkeley: University of California Press.
White, Cynthia. 1986. "Food shortages and seasonality in Wodaabe communities in Niger," *Institute for Development Studies Bulletin* 17, no. 3: 19–26.
————. 1987. "Changing animal ownership and access to land among the Wodaabe (Fulani) of central Niger," paper presented at the workshop on changing rights in property and pastoral development, University of Manchester, 23–25 April.
Winter, Michael. 1984. "A study of family and kinship relations in a pastoral Touareg group in north Mali," unpublished Ph.D. thesis, University of Cambridge.
Woodham-Smith, Cecil. 1964. *The Great Hunger: Ireland 1845–9*. London: Four Square Editions.
World Bank. 1981. *Accelerated Development in sub-Saharan Africa*. Washington, D.C.
————. 1986. *Population Growth and Policies in sub-Saharan Africa*. Washington, D.C.

Local Governance and Economic-Demographic Transition in Rural Africa

GORAN HYDEN

THE PREVAILING PERSPECTIVE ON DEMOGRAPHIC ISSUES in Africa has been that the continent is a latecomer to modernization and thus to the demographic transition. In the literature on the subject, one discerns two distinct stages. The first, lasting through the 1970s, was essentially optimistic in its assumption that sub-Saharan Africa, following in the footsteps of other world regions, would gradually undergo demographic transition without too much social trauma. Although scholars were not unanimous in their assessment of the relevance of transition theory to Africa (cf. Teitelbaum, 1975), that theory became the sole guide to policymakers, African as well as expatriate. Its prompt adoption in these circles was enhanced by the notion prevailing in those days that Africa held more promise for development than Asia (in particular South Asia).

Guy Hunter, an experienced and well-respected development analyst, summarized the prevailing outlook when he argued that Africa was the easy continent to develop because it was seen to be free from the social misery that characterized both Asia and Latin America and was not burdened by the legacy of ancient civilizations standing in the way of modernization (Hunter, 1969). The resolutions adopted by the World Population Conference in Bucharest in 1974 echoed this widespread optimism. Reflecting the message of Bucharest that "development is the best contraceptive," international agencies, in collaboration with African governments, were actively engaged in pursuing rural development activities on the assumption that they would accelerate the move toward a "natural" fertility regulation. The integrated development programs, so popular in the 1970s, became the principal instrument through which this objective was carried on.

Governments were regarded as the prime movers of progress in sub-Saharan Africa, where the absence of indigenous private entrepreneurs and voluntary agencies was used as an excuse for leaving central government with a virtual monopoly in development. The idea that government-initiated socioeconomic changes ultimately have important effects on fertility, and on the resultant demographic transition, still has strong supporters both among academics (e.g. Kocher, 1984) and in policymaking circles (e.g. Bulatao, 1984).

The second stage in the literature coincides with the deteriorating economic circumstances in Africa in the 1980s. The tone is either alarmist or cautioning. The former has been particularly true in donor circles. In spite of generous donor funding and massive technical assistance to "population and development" programs, there is no systematic evidence that countries in sub-Saharan Africa have reached the threshold where spontaneous and voluntary fertility regulation begins to have society-wide repercussions. Nor is there much evidence that African governments have the political will and the capacity for implementation to significantly influence demographic processes. As a result, the international community has adopted a "doomsday" perspective on Africa. Some influential figures (e.g. McNamara, 1984) have revived and invoked the Malthusian scenario. Rapid population growth causes economic stagnation and political instability, the argument goes. Such growth is perceived to be at the root of the crisis facing Africa today.

Caution, or cautioning, best describes the message in the more recent research on demographic issues in Africa. The fact that birth rates in Asia and Latin America have declined in an accelerating fashion in the past 30 years, while no such trend is evident in Africa, can no longer be simply explained by saying the continent is a laggard in the demographic transition. Odile Frank and Geoffrey McNicoll (1987), for instance, argue that contemporary Africa is approaching the demographic transition with very different family structures from those that prevailed when Eurasian countries experienced this process. More specifically they demonstrate that African societies are characteristically different from other regions of the world with regard to the allocation of responsibility within the nuclear family unit for such key social and economic functions as ownership of resources, breadwinning, and providing for children. For instance, because subsistence is a responsibility that largely falls on women's shoulders, they tend to be more attuned than African men to the costs of raising children and to the economic burden of large families. The authors contend that the economic-demographic transition in Africa may not be associated with a crystallization of the nuclear family, as suggested and tested by earlier researchers (Caldwell and Caldwell, 1978; Dow and Werner, 1983) but by visiting unions, a social pattern that may also have the effect of lowering fertility, as Caribbean experience indicates.

This is an important insight at the micro level that no doubt will influence both research and policy in the years to come. It parallels conclusions drawn from studies of other sectors, for example, agriculture, where researchers (e.g. Richards, 1985) highlight the policy implications of the peculiarity of African farming systems as compared with those of other Third World regions. The bold "blueprint" designs of development programs that dominated in the 1960s and

1970s simply do not work in Africa because of the great variability in ecological circumstances and the presence of an alternative logic of social action that has little in common with the one from which such designs are derived.

A major outcome of the growing recognition of these points is a greater inclination within the international community to experiment with small-scale operations using local resources and involving local actors in both design and implementation. Nongovernmental organizations (NGOs), both African and international, have proved to be particularly suitable mechanisms for moving in this direction. NGOs have themselves emerged as principal advocates of the new "wisdom" that development will come about only as a result of a "bottom-up" approach.

While the degree of enthusiasm that these organizations and others have mobilized in support of such an approach is admirable, it must be recognized that we know very little about the constraints and opportunities associated with development "from within." This is particularly true when such development is expected to touch on issues as "close to home" as family size and structure. What incentive do people in local communities have to consider adoption of measures to reduce fertility? In what circumstances does such action begin to make sense to them?

This article attempts to explore the role that local institutions may play in this process in rural Africa. "Local" here refers to institutions above the level of the family and includes all those that have a direct bearing on the livelihood and welfare of the rural population, be they local administration officers, local government officials, cooperatives, self-help groups, other member organizations, or private business. "Governance" is used instead of "government" to emphasize that many matters that bear on the economic-demographic transition in rural Africa are handled by institutions other than the official government, whether central or local.

We first examine how the conditions of governance in Africa are different from those of other regions of the world, then discuss the implications of these conditions for governance, and finally analyze the extent to which local governance in rural Africa may be a factor in reducing fertility. Kenya and Tanzania will be used as principal empirical cases.

The conditions of governance in Africa

The conditions of governance in Africa are determined by three fundamental factors: (1) a legacy of land abundance, (2) the absence of a ruling class, and (3) the predominance of social over territorial types of organization.

The legacy of land abundance in Africa manifests itself in several ways. The tradition of land use has been extensive. Until recently in most parts of the continent, increases in agricultural production have been the result of opening up new land rather than intensifying the use of existing plots. Although expansion of the cultivated area is already declining as a major source of agricultural growth in some parts of the continent—notably the East African highlands, the Sahel, and parts of the tropical zone in West Africa—attitudes toward land use are changing only slowly. At least three reasons can be advanced for this inertia. One is the absence of an "intermediate technology" tradition. As Jack Goody (1971) argues, Africa never acquired the technological means, that is, the plow and wheel-based equipment, that were instrumental in improving productivity on the land in Europe and Asia. As a result, there is little by way of an indigenous reservoir of knowledge that is directly applicable to more intensive land use. Africa is only now beginning to develop such technologies, which are still far from having a marked impact on the continent's agriculture.

The second reason is the absence of crop specialization and the preference among African producers for multi-cropping systems. This preference may be part of a wider "risk-aversion" strategy. It certainly has been a contributory factor to effective use of nutrient-poor soils in many parts of the continent. But it has also lowered the interest among agricultural producers in shifting to "modern" technology. Today there is even a school of thought (e.g. Richards, 1985) which argues that sustainable agriculture in Africa is possible only if the rationale of indigenous farming systems is adequately recognized.

The third reason is the absence of an indigenous tradition of land alienation and concentration and the tendency to reverse it in recent years in those few places, e.g. Kenya and Ethiopia, where it had evolved. As a result individuated land ownership never emerged in Africa as a predominant form of tenure. To be sure, there has been a tendency toward privatizing and commoditizing land in several parts of Africa in recent years, but the implications of such a move are not clear. Ronald Cohen (1988), for instance, shows that in northern Nigeria between the 1950s and 1980s a land market developed and customary communal ownership patterns were eroded. Following Hart (1982), Cohen argues not only that commercialization of agriculture is taking place but that it is the only way Africa can increase farm production. Hastings Okoth-Ogendo (1986), by contrast, argues that the emergence of a land market is not necessarily an unambiguous move toward commercialization of agriculture. In areas where the land frontier is closed, purchase of land serves the same purpose as land clearance did in the past: securing access to land for one's offspring. Thus, the outlook associated with land abundance is perpetuated: the land market provides a new frontier of opportunities to secure rights and benefits in accordance with customary rules. Land improvement is usually not pursued if it interferes with accommodating all members of a family, or any other relevant corporate group, on the land to which the group has usufructory rights.

This takes us to the second factor conditioning governance in Africa: the absence of a ruling class. Africa never developed truly feudal societies or produced anything like the highly regimented forms of small-scale agriculture that permitted the rise of the great Asian civilizations. To be sure, there was an ongoing process of social differentiation in both precolonial and colonial days, leading to apparent differences in wealth in most societies. This process, however, did not lead to the emergence of landlord and tenant classes. Various forms of reciprocal arrangements acted to conceal any tendency toward marginalization of a certain class. Thus, typically people in rural Africa to this day have been differentiated primarily in terms of some owning more of the same thing (land, cattle, etc.) than others. Access to the "means of production" has become increasingly skewed but not to the extent of producing a proletariat. A system of relatively autonomous peasant producers still prevails in much of Africa; decisions about ownership and use of land are typically handled by local elders rather than by the state (although government officials are increasingly being called upon to formalize such land deals).

In this kind of economy, relations between those who rule and those who till the land are not firmly rooted in the production systems as such. Appropriations of the surplus product are made by the state through taxes, representing simple deductions from an already produced stock of value. The relationship between peasant and state becomes tributary rather than productive—unlike the situation under feudalism or capitalism where such appropriations are made in the immediate context of production, either on the landed estate or in the factory. In these latter systems, the state is functionally and structurally linked to the productive demands of the economy and can be used by the rulers to steer and control society. The submerged classes have no choice but to respond to the dictates of the system at large.

The third factor conditioning governance in Africa derives from the fact that social differentiation is still contained within communities where affective bonds prevail. The political relationship between village and state is mediated by kinship and other particularized groups. To this day, kinship rather than territoriality tends to define community systems in rural Africa. In this respect, the continent differs from the predominant pattern of historical development in Europe, Latin America, and East Asia, where the state emerged organically in conjunction with a shift in community organization from kinship to territoriality (Befu, 1967; Cain and McNicoll, 1986). The colonial authorities tried to bring about a similar change in Africa but were unable to complete the task. Wherever kinship prevails, rule tends to be "minimalist." Political activities of the rulers, such as taxation, conscription, public works, and adjudication, are usually ad hoc and perceived as quite light by the populace. The colonial authorities regularized all these activities and made them much more onerous by developing a state machinery that had to sustain itself independently of the prevailing kinship organization of society. The new postcolonial governments were faced with a dilemma. On the one hand, their leaders had committed themselves in the struggle for independence to doing away with the burdens imposed by the colonial rulers. On the other, they needed some of these mechanisms, notably the tax system, to ensure rapid progress.

Experience since independence suggests that throughout Africa kinship rather than territoriality has prevailed. In Kenya, for instance, where land consolidation and adjudication in the colonial days were taken over by government, the trend in the past three decades has been toward diluting these powers. Thus, the formal land consolidation program has been accompanied by a parallel "informal" process of subdividing land based on kinship criteria. Similarly, President Moi has in the past ten years given clan elders a greater say on land adjudication matters.

Similar evidence of kinship undermining territoriality is available from Tanzania. The government's ambitious villagization program in the 1970s was possible only because of the absence of strong corporate-territorial community systems, but as the authorities have learned, such systems are not created and institutionalized "over night." One study of village–state relations in Tanzania (Abrahams, 1985) demonstrates that government in the new villages was only as effective as local interests would allow. The officials appointed to run these villages for the ruling party and the state simply could not acquire the legitimacy or the organizational means to effectively implement key policies introduced by the government to strengthen its power over society. Even where bureaucratic interests appeared to prevail, villagers could mobilize particularistic relationships to appeal at higher levels in the government or party in order to be spared certain measures that were deemed adverse or unfair to the village population.

What I am saying here is that the growing social differentiation that has accompanied changes in Africa in the past century is still contained within what I have called elsewhere (Hyden, 1980) an "economy of affection." Occupational groups and social classes are emerging in Africa but they are still encapsuled in affective relations. As land becomes increasingly scarce, rural Africa may move in the direction of "corporate" peasant communities in which territoriality rather than kinship distinguishes village organization, but such a transition, even where it is already beginning, is unlikely to produce dramatic changes in patterns of governance in the immediate future. Although black Africa is not the only part of the world where kin groups and factions rather than territorial communities play prominent local-level roles—Bangladesh (Demeny, 1975; Arthur and McNicoll, 1978) is another—their overall influence on governance at large tends to be particularly pervasive there. It is to the implications of these conditions of governance that I now turn.

Implications for governance

Three implications for governance are of special interest here: (1) the logic of individual action, (2) the role of the state in society, and (3) patterns of local governance.

It is important to note that the debate in recent years as to whether peasants are "moral" (Scott, 1976) or "rational" in the utilitarian sense (Popkin, 1979) has drawn its material from Southeast Asia, a region where corporate peasant communities have long been in existence. In these communities private and collective property rights are clearly differentiated. People are acutely aware of the value of whatever they own and whatever additional resources they can gain access to. Their well-being depends directly on access to a scarce commodity—land. The question on which Scott and Popkin differ is how to explain peasant behavior and action: is it motivated by self-interest or a "collective" rationality shared by the poorer peasants because their subsistence is otherwise at risk?

This is not the place to resolve the dispute between the two authors, only to register that both arguments are a priori valid in a context where people genuinely experience resource scarcity, there is a land market, private and communal property rights have been differentiated and are recognized as such, and the "little tradition" of the local village community is giving way to a more uniform set of urban-based values. These empirical conditions give rise to a polarization of society in which both an assertive utilitarianism and a defensive "moral" economy may coexist and compete with each other.

These conditions do not prevail in rural Africa today. Hence neither the "rational" nor the "moral" peasant is conceptually very helpful in trying to understand the logic underlying individual action there. The conditions of governance described above point toward a different model of man: one in which (1) individuality and communality are not dichotomized; (2) the notion of "limited good" (resource scarcity) is foreign; (3) progress is equivalent to investment in human capital; and (4) development is a matter of exploring an open frontier rather than carefully managing and improving the productivity of an existing resource base.

The African social order is neither purely communalistic nor purely individualistic. It is, as one African philosopher notes, "amphibious" for it manifests features of both communality and individuality (Gyekye, 1987: 154). The success and meaning of an individual's life depend on identifying him- or herself with a group such as a clan. His or her personal sense of responsibility is measured in terms of responsiveness and sensitivity to the needs and demands of the group. Since this responsibility is enjoined equally upon each member of the group, communalism tends to maximize the interests of all the individual members of the society. This does not mean, as many non-Africans assume, that communalism absorbs the individual to the extent that individuality, personality, initiative, and responsibility disappear. As the following Akan proverb, cited by Kwame Gyekye (1987: 158), suggests, individuals have characters and wills of their own: "The clan is like a cluster of trees which, when seen from afar, appear huddled together, but which would be seen to stand individually when closely approached."

The individual is not an atomized entity. He is separately rooted without being completely absorbed by the cluster. As a result, there is not distinct "private" property, nor a notion of a village "commons." Resources are simultaneously owned by a collectivity and an individual. What matters is membership in a community. Through such membership the individual has an automatic claim to a common resource. What is not being claimed at any given time, however, is not the subject of a special agreement for use as a "commons." Rather, it is held in trust for eventual use by individual members of the community.

In this situation the individual is constantly shielded from exposure to the notion of a limited good. Group membership gives the individual access to additional resources by claims to either subdivide existing ones or acquire new ones. For instance, as land has become scarcer in many parts of Africa, the main response has been subdivision of plots rather than enhancement of land productivity. Similarly, as the economy has diversified, new forms of sharing wealth within communal groups have developed. Access to employment through relatives with the "right connections" is a case in point. It is important to point out that these groups are not always strictly based on kinship. Many are, in a Durkheimian sense, modern "clans" where membership is often flexible (Cruise O'Brien, 1975). Whatever the nature of these groups, however, they have the effect of making individuals generally insensitive to limits and scarcities that would be apparent under either a straightforward system of private property or one of territorially defined communities.

Progress in the African context, therefore, tends to take on a different connotation from that understood in other parts of the world. Instead of being tied to improvement of the material base—for instance, enhanced productivity of land—progress is first and foremost associated with investment in human capital. Technology is still largely manual. The necessary know-how is in the producer's head rather than being independently systematized. "Development," Tanzania's former President Nyerere has said, "is the development of people." This viewpoint manifests itself in a clear preference for investment in education and other "social development" programs that enhance the human resource base. The *harambee* self-help activities in Kenya are a good illustration. When given a choice of how to spend money on "development," rural Kenyans overwhelmingly prefer education, health, and domestic water supply over investments in agriculture and physical infrastructure, such as road construction (Holmquist, 1970; Thomas, 1985). It is also significant that family planning does not feature among local self-help activities in Kenya. A human being is not viewed in "cost–benefit" terms. Everyone is an asset, and an expandable asset at that. Regulating human reproduction, therefore, goes against the prevailing social logic.

Development to by far the majority of Africans is still a matter of exploiting an open frontier. Household strategies have concentrated on diversifying options of livelihood rather than depending exclusively on one source. Thus, some members of a household will remain on the land while others seek sources of income elsewhere, whether in rural or urban areas. To be sure, some households are more successful than others in benefiting from diversification. Yet, there is no evidence that the strategy of the less successful households is changing. For instance, women in female-headed households, which tend to be economically poorer, adopt the same strategy of having many children as do those in better-off households, so as to maximize the opportunities for eventually diversifying sources of income. This outlook survives particularly in situations of economic growth, where people experience tangible evidence of success in the pursuit of new income-earning opportunities. Kenya is a case in point. Although the land frontier in Kenya has been virtually closed, new economic opportunities have continued to open up, particularly in the informal sector.

While this strategy is rational for individual households, it is costly to the economies of African states. It diverts attention from improvement in agriculture. Getting producer prices "right" is an important measure to promote efforts to raise agricultural production, but it is far from being sufficient. As long as there is no sense of a limited good and the belief prevails that there is always a chance for one's offspring to secure a livelihood through the "right connections," diversification will be favored over regulation of human reproduction. Education to boost the chances of one's children will be a much higher priority than family planning.

Thus, the problems of the agricultural sector in Africa are likely to continue in the foreseeable future. Especially threatening is the effect of educational expansion—the standard recipe for an urban-based development path. Today's generation of African children is largely removed from the apprenticeship on the land that in the past secured adequate agricultural production. Instead, they are acquiring "book knowledge" in schools that bears little or no relationship to local production technologies on the land. Even where agriculture is being taught, children learn "modern" or "scientific" methods, the wherewithals and inputs of which are largely absent in African countries. If African governments and the international donor community wish to improve African agriculture, they have to deal with the question of how to preserve and develop indigenous production technologies (Richards, 1985) that in the past were deemed backward and inefficient.

This takes us to the role of the state in Africa. It is often pointed out that the African state is an artificial creation of the colonizing powers. More importantly, the state is not the product of an organic process whereby a ruling class emerges by virtue of acquiring control of land for its own private interests. The African state, then, is typically not tied to prevailing production systems. It is not a repository of norms and codes declaring land to be a scarce resource that must be protected or developed. It is in this sense that the African state sits "suspended in mid-air" (Hyden, 1980).

The fact that the postcolonial state in Africa is not structurally or functionally grounded in agricultural production systems explains two of its characteristics. The first is the strong inclination among African government leaders (regardless of ideological persuasion) to regulate agricultural production and marketing with little or no concern about what is "rational" for the central economy at large. Such regulatory policies do not even appear to reflect what, in Marxist terminology, may be described as "naked class interest." The African state cannot be understood merely as an instrument for the exercise of power by a small group of people already in charge of an economic system of their own making. It is better viewed as an instrument for acquiring power by persons who are often acting in a shortsighted fashion because they lack the vision and confidence that follow from being in effective control of the means of production, whether land or factories.

The second characteristic of the postcolonial state is its tendency to be "over-developed," that is, to have expanded its claim on resources beyond what the society can afford. Like the first characteristic, this reflects the absence of the principle of territoriality in constituting the African state. The state continues to be a conglomeration of communities held together by affective ties, its legitimacy highly dependent on its ability to service these communities. The "over-developed" nature of the state emerges as these communities press exclusive demands on their leaders and the leaders, in an effort to secure their own positions and meet those demands, lay claim to public resources. Donald Rothchild (1986) refers to this system of governance as "hegemonic exchange." Such exchange is based upon mutual adjustment of conflicting community interests. It regulates conflict by strengthening the control that authoritative institutions at the center exercise over these communities. As such, it is a form of state-facilitated coordination, usually within the parameters of a single- or no-party arrangement, in which individual political leaders, serving as patrons of these communities, engage in a process of mutual accommodation on the basis of informally understood norms and rules. It allows for tacit exchanges of interest, frequently applying the proportionality principle to resolving such contentious issues as political coalition formation, political recruitment, and resource allocation. The consequence of this form of governance has been a "soft" state in which authoritative and effective action tends to be blocked by a combination of "centrifugal" community demands and efforts by the political leadership to manage these in such a way that stability of the regime is secured.

The old proverb that "a chief rules over people, not over land," which can be found in various parts of Africa, summarizes quite well the contemporary patterns of governance on the continent. These patterns have the effect of making African governments structurally much less well-placed to influence development than governments in either Asia or Latin America. It is no coincidence that the literature on the subject increasingly refers to the failure of the African state (Rothchild and Chazan, 1988; Wunsch and Olowu, 1990).

The prevalence of "hegemonic exchange" has led to a strengthening of central over local government institutions. Responsibility and resources available to local government authorities in Africa are generally lower than in other parts of the world. Even in states where local governments are little more than arms of the national government, the percentage of total budgetary resources allocated by them is far greater than in Africa. For example, from France at the lower range (17 percent) to Sweden at the higher (66 percent), local-level institutions are major partners in the provision of public services. Equivalent figures for Africa are as low as 2 percent. This pattern is also evident in the distribution of government personnel. African countries as a whole have the lowest percentage of their modern sector employees at the local level (2.1 percent). This figure is low compared with Asia (8.0 percent), Latin America (4.2 percent), and the OECD states (12 percent) (Wunsch and Olowu, 1990).

These figures reflect a persistent trend throughout Africa since independence whereby local self-governing institutions have been supplemented with or replaced by field offices of central administrative agencies. This is true for both Kenya and Tanzania, where in the late 1960s the statutory powers of local governments were abrogated and responsibility for such key rural development functions as primary school education and primary health care was taken over by central government ministries. There have been measures to decentralize authority to field agencies in both countries, but these have only become implementing agencies of increasingly centralized development strategies (Moris, 1981). Only in countries such as Botswana, where indigenous institutions have been "upgraded" to serve as local authorities, has some form of local government been retained (Wynne, 1986).

The typical pattern is that local communities have organized themselves informally on a self-help basis to meet development needs. These groups or associations build on traditional forms of cooperation. Some may be officially registered with the central government but most remain unofficial. Many of the latter are project-specific. Local pride often drives these efforts: there is considerable community pride in a new school or a new water project. Starting a local self-help project may also be a way of seeking additional funds from government. Such at least has been the strategy used by self-help projects in Kenya. One consequence is that many projects that were started as a "political gamble" are never completed because of failure to secure government assistance. Even projects whose physical construction is completed often fail to meet community needs because they are pursued without regard for how to meet recurrent costs and maintenance (Thomas, 1985).

Local governance in Africa, then, is typically characterized by an uneasy rivalry between the lower levels of a centrally controlled bureaucracy, on the one hand, and, on the other, the leadership of a myriad of local groups and associations struggling to secure outside resources for their own, often parochial, objectives. The lower levels of government administration are deliberately kept inert so as to enable higher-level political patrons to take credit for decisions made to satisfy local interests. This type of patronage politics, though by no means exclusive to Africa, becomes particularly important to regime stability in societies where the state sits suspended over multiple, relatively self-contained, communities. But in addition to leaving the state "soft," it has the tendency to erode any move toward genuine local governance. Instead, it encourages the twin notions that the state is a "widow's cruse," that is, resources can always be found through the right connections, and that development means favors dispensed from above.

What are the implications of this institutional arrangement for changes in demographic behavior? That is the subject of the final section of this essay.

Local governance and demographic change

The prevailing view of the prospects for fertility decline in Africa has continued to be pinned to two principal assumptions: (1) that modernization will facilitate decline and (2) that a government policy on population is a prerequisite.

No African country made Bernard Berelson's list, prepared in 1978, of developing nations whose chances were "certain," "likely," or "possible" to reach a crude birth rate of 20 per thousand population by the year 2000 (Berelson, 1978). In socioeconomic terms these countries were still regarded as being far from attaining development levels conducive to the kind of demographic transition now under way in other Third World regions. As a result, writers on population in Africa have tended to stress the need to accelerate rural development and other efforts to modernize society. For instance, the underlying theme of a major collection of essays on rural development and human fertility (Schutjer and Stokes, 1984) is that rising income and higher levels of education do have a significant influence on fertility. In discussing the relationship between income distribution and fertility, one of the contributors, James Kocher, argues that genuine options for government policies and programs to make a difference in Africa are limited but significant. The following three categories of policy can, in his view, over a period of two or three decades, have a noticeable impact on fertility rates in sub-Saharan Africa: (a) major commitment of public resources to schooling up to secondary level, especially for girls; technical training; and the provision of health care and family planning services; (b) adoption of development strategies that emphasize policies and investments stimulating labor-intensive sectors; and (c) strategies of technology development, promotion, and adoption that emphasize technologies that complement rather than displace labor (Kocher, 1984: 229–230). A recent World Bank study essentially comes to the same conclusion, although it argues that improvements in income are less central to fertility decline than improvements in education—especially for women—and better health care. It is not the ability to afford new goods but aspirations for them that influence parents to have fewer children (Bulatao, 1984).

Although most African governments have pursued education and health policies along the lines suggested above, the interests of demographers have centered particularly on countries that have also adopted an explicit population policy. Thus, Ghana and Kenya, both of which adopted such a policy in the 1960s, have attracted special attention in scholarly and donor circles. These countries have been treated as forerunners. As a result, investments in family planning facilities (usually combined with primary health care) and demographic training and research have been particularly extensive there. Twenty years later, however, there is little evidence that their population policies have made a difference. Analysts are therefore faced with the question: is this because the government effort has not been strong enough or is it because the fundamental variables operate differently in the African context? Put another way, is the answer "more of the same" or a readiness to consider and adopt alternative approaches?

These questions take on special significance at this juncture when much of Africa faces economic hardship and governments feel growing pressure to trim their budgets. Will the current economic crisis in Africa necessarily delay the demographic transition? Drawing on the preceding analysis of the conditions of governance, I will examine in a tentative fashion the prospects for such a transition in Kenya and Tanzania.

Kenya Kenya poses a special challenge to demographic analysis. The total fertility rate of 8.0 children per woman, first recorded in 1979, has shown little sign of decline and is one of the highest yet documented for any country. Those aged 14 or below now constitute half the population. Few if any societies have ever been faced with the task of educating, inculcating moral values, and providing gainful employment to their members with a dependency ratio of this size.

Although arable land is almost exhausted and marginal lands can be opened up for agriculture only at great capital outlay, fertility control is still an issue that Kenyans, by and large, consider irrelevant or of only low priority. Ambivalence prevails in government circles. For a long time after the government first adopted a population policy (1967), its implementation remained of greater concern to international donors than to Kenyan officials. Only in the 1980s did a more marked commitment to reducing the population growth rate emerge in government circles. This new commitment, however, remains to be translated into effective action. A National Council on Population and Development has been established but appears to lack the political authority that is necessary to influence the government machinery at large. In other words, quite apart from the issue of strategy, questions must still be raised about the political will and executive capacity of Kenya's government to deal effectively with rapid population growth in the short to medium term.

Family planning in Kenya is also pursued by private and voluntary sector agencies. Two examples are of special interest here. One is the recently started Family Planning/Private Sector program, which targets workers in urban industries and rural plantations. Although it is too early to predict what kind of impact this program will have in the longer run, initial response in terms of contraceptive prevalence rates has been encouraging. The other example is the experimental program carried out by the Presbyterian Church in Chogoria, Meru District, a relatively prosperous area on the slopes of Mount Kenya growing coffee and tea. Labeled a "major success" in a recent evaluation (Goldberg et al., 1986), it is the result of a concentrated and labor-intensive effort to encourage smaller family size among the rural population. The significance of these two programs lies in the fact that they are highly targeted—in one case on wage-earners for whom the "cost" of children may appear real, in the other on a local community—and that they are locally managed with assistance from nongovernmental sources. They differ from the government's general family planning program, which offers services throughout the country in the context of a broader primary health care effort.

These examples also demonstrate that reduction in fertility rates is possible, although it is still far from clear that a real change in demographic behavior is on its way even among people covered by these special programs. There is certainly reason to be cautious about the prospect that Kenya will reach even the modest target set by the World Bank of a total fertility rate of 6.1 by the year 2000 (World Bank, 1986), let alone the goal of a 2.0 percent annual population growth rate by the end of the century that the government of Kenya has had under active consideration (USAID, 1987). The fact that Kenya's economy has held up quite well in comparison to many others in the region, that it has continued to receive foreign aid in large quantities, and that its politics has remained clientelist, i.e. driven by community demands, has in turn sustained a strong "economy of affection" and encouraged a continued pronatalist outlook among the population at large. Kenya has demographically been a captive of its relative economic success, and there is little evidence today that Kenyans seriously consider the "frontier" closed. To be sure, an increasing number of households are experiencing real hardship, but to deal with it family members prefer the strategy of diversification over fertility control. What two Kenyan social scientists wrote about the demographic behavior of their fellow-countrymen some years ago still appears valid:

> [T]he greater the number of children a person has, the greater the amount of wealth he is considered to have and consequently the greater the amount of respect, honour and prestige he will be accorded by members of his community. Hence in order to earn such honour, prestige and respect, both the husband and his wife/wives put the necessary effort to get as many children as they possibly can. (Muinde and Mukras, 1979: 77)

The conclusion one can draw from a survey of the Kenyan situation is that neither the central government nor local institutions appear for the time being to be poised for a lead role in changing demographic behavior. Thus, it seems unlikely at this point that Kenya would replicate the relative success of either community-based family planning programs (for example, Thailand) or strong government programs (for example, Indonesia). Of course, Kenya's political leadership may one day adopt a much more aggressive population policy, but such a shift in outlook will probably only come with a more fundamental change in style of governance. The latter would imply a readiness to break out of its current "populist" mold, to sacrifice the affective networks currently shielding people from such phenomena as high dependency ratios, and to make themselves role models of a strategy that is contrary to the predominant views of the Kenyan population. In the meantime, incremental progress may be registered by private or voluntary agencies working in conjunction with local institutions. It is unlikely, however, that these will produce a movement in support of "spontaneous" fertility control. A more likely scenario is that they will continue to achieve local successes thanks to the trust that such agencies may be able to build in local communities or among select groups of people for whom the notion of fertility control may not be totally alien.

Tanzania Tanzania's demographic features are much less well known than those of Kenya. Data from the 1978 census have only been marginally analyzed. (Data from the 1988 census are not yet available.) No series of survey-based demographic estimates of the kind that has helped to confirm the principal features of the population scene in Kenya is available in Tanzania. Based on a preliminary analysis of the 1978 census data, Israel Sembajwe (1978) estimates the population growth rate at 3.2–3.5 percent and the total fertility rate at 6.3. As in many other African countries, fertility appears to have been rising in the past two decades.

Unlike Kenya, Tanzania has never adopted a distinct population policy. Instead, its development strategy has consistently been based on the assumption that "development is the best contraceptive." Thus, it has placed special emphasis on greater equality in access to social services, compulsory primary education, free medical services, provision of modern farm inputs, and a more even income distribution. In the first ten years after the Arusha Declaration (1967)—the blueprint in which this policy was originally outlined—Tanzania served as a model for many other Third World countries.

As documented by an increasing number of sources, however, Tanzania's policies have come to very little except creating an oversized state apparatus that is unable to execute policy and deliver services. The combination of bold but often ill-conceived policies, on the one hand, and an inefficient bureaucracy, on the other, threw the national economy out of gear. With an overvalued local currency, peasant export crops earned little. With a state-controlled distribution system, allocation of consumer goods became subject to political favor. In this situation, it is not surprising that the country's peasants began to withdraw from the formal economy and lost confidence in government policy (Hyden, 1980).

Unlike the markets in West Africa, however, the parallel, unofficial markets are not as vibrant in Tanzania. The economic crisis has severely affected everyone—both rich and poor. For a long time, consumer goods, even basic commodities, were simply not available. Although they are now on the shelves in shops across the country, they are priced at a level that makes them at best a dream for the vast majority of the country's urban and rural populations. With the International Monetary Fund and other donors putting pressure on the government to trim its budget and improve its financial control system, there is much less scope for political largesse. There are definite limits to the pursuit of affective obligations. Thus paradoxically, although Tanzania has no land shortage like Kenya, the frontier appears closed to the Tanzanians in a way that it does not yet seem to Kenyans. In societies where there is no sense of a limited good but progress is tied to the dynamics of expansive social networks, operating on the principle of reciprocal obligations among members, the cost of large families only begins to be apparent to people when everybody is confronted with a real barrier to sharing and diversifying. This is what has happened in Tanzania (and is beginning to be seen in other African countries). The economy of affection has had to contract drastically. Urban and rural households, for the first time, have been faced with the fact that large families may be a burden rather than an asset.

How common this notion is remains to be further studied, but preliminary reports from family planning workers in Tanzania suggest that there is today a greater interest in regulating family size than was the case in the 1970s when the economy was still performing satisfactorily. An unpublished study (Kamuzora, 1988) also confirms that rural women in northwestern Tanzania now desire fewer children (four) than in the economically expanding 1970s. It is no longer a social taboo to discuss the subject at village meetings. Although the relationship between economic decline and a growing interest in fertility regulation requires further investigation, it does indicate the need for new hypotheses for the study of demographic change in Africa.

Local governance may prove to be an even more important factor in influencing policy effectiveness in a country like Tanzania than in Kenya. One reason is that neither the ruling party nor the state has the commitment or the capacity to play a leading role in this field. Another is that Tanzania's population has been brought together into villages. Although the extent to which the local population views decisionmaking structures in these villages as legitimate appears to vary from village to village (Abrahams, 1985), villagization has increased people's sense of vulnerability at times of economic crisis. Thus, ironically, Tanzania's new villages may gain strength not because of party mobilization or government support but in response to the challenges that stem from what the average Tanzanian today experiences as the failure of party policy.

Efforts by villagers to deal with demographic issues in a vacuum are unlikely to amount to much. Linkages to intermediary organizations that can service local community efforts are important, yet, unlike Kenya, Tanzania has very few of these nongovernmental organizations. Most of them are church-based and not all of them would consider family planning a high priority. Thus, Tanzania lacks the institutional infrastructure that may make it possible to take full advantage of what appear to be intriguing opportunities for bringing about changes in demographic behavior at the village level. An effort to strengthen existing nongovernmental organizations and to create new ones could make a difference, but this would imply a considerable change in the national policy framework in a country where the ruling party's supremacy is enshrined in the constitution and the state bureaucracy is accustomed to being the sole implementing body. As a result, a question mark hangs over the prospect of alternative institutional mechanisms, including community-based efforts, evolving to capitalize on what may be a genuine interest in smaller family size in Tanzania.

Conclusions

Issues relating to the prospect for fertility regulation in Africa are still poorly understood. For a long time analysts have looked at these through the lenses of a theory of demographic transition derived from the historical experience of other world regions. Only more recently have various aspects of such a theory come under greater critical scrutiny on the basis of African data. The search for answers has been extended to new and uncharted terrain.

This essay, in looking at the extent to which local governance may make a difference to demographic behavior in rural Africa, is an attempt to broaden the debate about the determinants of fertility regulation. It questions existing assumptions about the "social logic" of African societies and the role of the state in development. It raises issues about the effect of the economic crisis on demographic behavior in Africa. There is reason to examine whether crisis may not prove to be as powerful a variable in support of fertility regulation as any other associated with modernization. In societies where the "economy of affection" still prevails, one can plausibly argue that economic stagnation rather than economic growth may provide the greatest impetus to limiting family size.

Local governance may take on particular significance in countries facing economic hardship, although its impact on demographic issues is likely to be hampered in many such places by the relative absence and weakness of nongovernmental institutions capable of serving and supporting community-based efforts. Thus, its full importance can only be adequately understood in the broader context of governance and the conditions determining the way African countries are being governed.

References

Abrahams, R. G. 1985. "Villagers, villages, and the state in modern Tanzania," Cambridge African Monograph, African Studies Center, Cambridge University.

Arthur, W. B., and G. McNicoll. 1978. "An analytical survey of population and development in Bangladesh," *Population and Development Review* 4, no. 1 (March): 23–80.

Befu, H. 1967. "The political relation of the village to the state," *World Politics* 19, no. 4 (July): 601–620.

Berelson, B. 1978. "Prospects and programs for fertility reduction: What? Where?," *Population and Development Review* 4, no. 4 (December): 579–616.

Bulatao, R. A. 1984. "Reducing fertility in developing countries: A review of determinants and policy levers," *World Bank Staff Working Papers* No. 680, Population and Development Series No. 5, Washington, D.C.

Cain, M., and G. McNicoll. 1988. "Population growth and agrarian outcomes," in *Population, Food, and Rural Development*, ed. Ronald D. Lee et al. Oxford: Clarendon Press.

Caldwell, J. C., and P. Caldwell. 1978. "The achieved small family: Early fertility transition in an African city," *Studies in Family Planning* 9, no. 1 (January): 2–18.

Cohen, R. 1988. "Adversity and transformation: The Nigerian light at the end of the tunnel," in *Satisfying Africa's Food Needs*, ed. R. Cohen. Boulder, Colo.: Lynne Rienner Press.

Cruise O'Brien, D. 1975. *Saints and Politicians: Essays in the Organization of a Senegalese Peasant Society*. Cambridge: Cambridge University Press.

Demeny, P. 1975. "Observations on population policy and population program in Bangladesh," *Population and Development Review* 1, no. 2 (December): 307–321.

Dow, T. E., Jr., and L. H. Werner. 1983. "Prospects for fertility decline in rural Kenya," *Population and Development Review* 9, no. 1 (March): 77–97.

Frank, O., and G. McNicoll. 1987. "An interpretation of fertility and population policy in Kenya," *Population and Development Review* 13, no. 2 (June): 209–243.

Goldberg, H. I., et al. 1986. "Family planning in Chogoria: A major Kenyan success story," paper presented at the Annual Meeting of the American Public Health Association.

Goody, J. 1971. *Technology, Tradition and the State in Africa*. London: Oxford University Press.

Gyekye, K. 1987. *An Essay on African Philosophical Thought: The Akan Conceptual Scheme*. Cambridge: Cambridge University Press.

Hart, K. 1982. *The Political Economy of West African Agriculture*. Cambridge: Cambridge University Press.

Holmquist, F. 1970. "Implementing rural development projects," in *Development Administration: The Kenyan Experience*, ed. G. Hyden et al. Nairobi: Oxford University Press.

Hunter, G. 1969. *Modernizing Peasant Societies*. London: Oxford University Press.

Hyden, G. 1980. *Beyond Ujamaa in Tanzania: Underdevelopment and an Uncaptured Peasantry*. Berkeley: University of California Press.

Kamuzora, C. 1988. "Impending fertility decline at low levels of development in Kigarama Village, Tanzania," unpublished study, Department of Statistics-Demographic Unit, University of Dar es Salaam.

Kocher, J. E. 1984. "Income distribution and fertility," in Schutjer and Stokes (1984), pp. 215–239.

McNamara, R. S. 1984. "Time bomb or myth: The population problem," *Foreign Affairs*: 62, no. 5: 1107–1131.

Moris, J. 1981. *Managing Induced Rural Development*. Bloomington: International Development Institute, Indiana University.

Muinde, J. N., and M. S. Mukras. 1979. "Some aspects of determinants of fertility in Kenya," Working Paper, Population Studies and Research Institute, University of Nairobi.

Okoth-Ogendo, W. H. O. 1986. "Some issues of theory in the study of tenure relations in African agriculture," paper presented at the Annual Meeting of the African Studies Association, Madison, Wisconsin, 30 October–2 November.

Popkin, S. L. 1979. *The Rational Peasant: The Political Economy of Rural Society in Vietnam*. Berkeley: University of California Press.

Richards, P. 1985. *Indigenous Agricultural Revolution*. Boulder, Colo.: Westview Press.

Rothchild, D. 1986. "Interethnic conflict and policy analysis in Africa," *Ethnic and Racial Studies* 9, no. 1 (January): 66–86.

———, and N. Chazan. 1988. *The Precarious Balance: State and Society in Africa*. Boulder, Colo.: Westview Press.

Scott, J. C. 1976. *The Moral Economy of the Peasant: Rebellion and Subsistence in Southeast Asia*. New Haven, Conn.: Yale University Press.

Schutjer, W. A., and S. C. Stokes. 1984. *Rural Development and Human Fertility*. New York: Macmillan.

Sembajwe, I. S. 1978. "The 1978 population census of Tanzania: A preliminary assessment of its implications," Bureau of Resource Assessment and Land Use Planning, University of Dar es Salaam, Tanzania.

Teitelbaum, M. 1975. "Relevance of demographic transition theory for developing countries," *Science* 188: 420–425.

Thomas, B. 1985. *Politics, Participation and Poverty: Development Through Self-Help in Kenya*. Boulder, Colo.: Westview Press.

USAID (United States Agency for International Development). 1987. "USAID analysis and strategy for assistance in family planning and fertility reduction in Kenya," USAID/Kenya, Office of Population and Health, Nairobi.

World Bank. 1986. *Population Growth and Policies in sub-Saharan Africa*. Washington, D.C.

Wunsch, J., and D. Olowu. 1990. *The Failure of the Centralized State: Institutions of Self-Governance in Africa*. Boulder, Colo.: Westview Press.

Wynne, S. 1986. "Institutional development within the conceptual heritage of Kweneng Bakgalagadi: A study of fundamental units of collective action," paper presented at the Annual Meeting of the African Studies Association, Madison, Wisconsin, 30 October–2 November.

PROPERTY RIGHTS

Responses to Rural Population Growth: Malthus and the Moderns

MICHAEL LIPTON

MOST DEVELOPING REGIONS HAVE REACHED, or are rapidly approaching, some sort of land frontier. What are the responses on the part of rural populations to increasing land scarcity, in terms of agricultural technology, employment, and demographic behavior? What are the outcomes in welfare terms, particularly for food? To answer these questions we must understand (a) the relationships between population growth, rural economic response, and food availability, and (b) the institutions affecting the distribution of land, persons, and available food. The complexity and interdependence of (a) and (b) have led to the development of several theories of the relationship between population growth and the agricultural/rural response. Malthus's early and later writings present two somewhat variant models of increased fertility, as a response to more food or employment. Boserup (1965, 1981) has analyzed effects of increasing populations in generating technologies that grow more food; and Hayami, Ruttan, and Binswanger (hereafter HRB) have argued that increasing populations induce rises in labor use per hectare (see Hayami and Ruttan, 1971, 1985; Binswanger and Ruttan, 1978). Distributional outcomes are additionally governed by the impact of a growing labor force on the structure of entitlements, a subject that Sen (1981) has brought into the explanation of famine.

This essay argues that these theories, often seen as opposed in logic and policy implications, are in fact parts of a single approach. The approach, grounded in Malthus's work, can be summarized in six propositions. Figure 1 shows their relationships.

(1) Malthus's early writings held that more food, higher real wages, or better health increases population (and hence labor supply) per efficiency unit of land. This depresses food availability per person (and wage rates), reducing average claims upon food, and pressing population down again. I shall refer to this proposition as BONMAL: better outcomes *normally* mean aggravation later.

FIGURE 1 Malthus, Boserup, HRB, institutions, and Sen

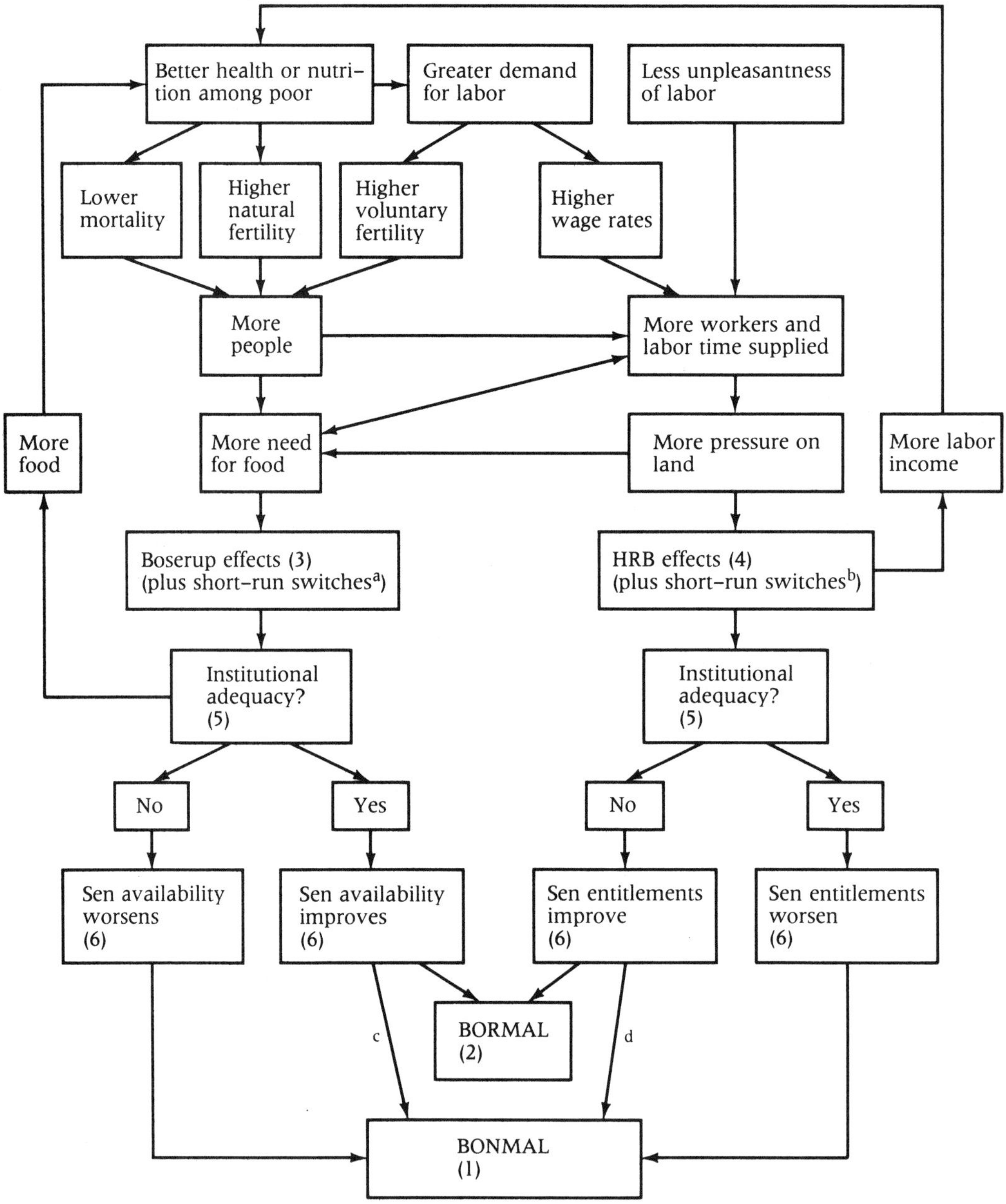

Better health or nutrition among poor

Greater demand for labor

Less unpleasantness of labor

Lower mortality

Higher natural fertility

Higher voluntary fertility

Higher wage rates

More people

More workers and labor time supplied

More food

More need for food

More pressure on land

More labor income

Boserup effects (3) (plus short-run switches[a])

HRB effects (4) (plus short-run switches[b])

Institutional adequacy? (5)

Institutional adequacy? (5)

No

Yes

No

Yes

Sen availability worsens (6)

Sen availability improves (6)

Sen entitlements improve (6)

Sen entitlements worsen (6)

c

BORMAL (2)

d

BONMAL (1)

NOTE: (1) to (6) are the propositions set out in text.

[a]Such as shifts toward foods requiring less land per calorie.

[b]Such as more labor-intensive uses of existing land.

[c]If entitlements do not improve.

[d]If availability does not improve.

(2) In his later writings, Malthus somewhat modified his stance on population response. He had always recognized that "misery" can be avoided, not just by undesirable "vice,"[1] but by "restraint"—especially delayed marriage. Increasingly in his later work, anticipating recent evidence, he maintained that education and declining mortality encourage such restraint. I shall call this modified position BORMAL: better outcomes *remediably* mean aggravation later.

(3) Boserup (1965, 1981) claims that population growth calls forth new technology to produce more food. This is consistent with BONMAL (the view that more food, in turn, normally stimulates damaging population increase), and also with BORMAL (the view that such stimuli are remediable).

(4) Hayami-Ruttan-Binswanger argue that rural population growth calls forth increasingly labor-intensive technologies. Boserup's claim is neither necessary nor sufficient for the HRB proposition.

(5) Institutional factors largely determine the strength of both Boserup and HRB effects. Only if both are powerful can BONMAL reliably be transformed into BORMAL. In particular, the interaction of local institutions with those of international technology transfer has produced too weak Boserup effects in much of Africa, and too weak HRB effects in much of Asia.

(6) Inadequate Boserup effects roughly give rise to a long-run "food availability decline," and inadequate HRB effects to a long-run "failure of entitlements" to food, in the senses used by Sen (1981, 1986) in analyzing the short-term causes of famines.

In using (1) through (6) to briefly explore interactions between food and population, I can only touch upon some key issues. How important, in determining fertility change, is couples' demand for net income from children—and hence for child quantity and quality—as opposed to the "supply" of children from natural fertility, and to the costs of fertility regulation (Easterlin and Crimmins, 1985)? How important are institutions (of family, lineage, community, and state) in transmitting—or resisting—pressures, old or new, to reduce fertility (Cleland and Hobcraft, 1985, esp. p. 5)? What is the evidence on "demographic transition," and on the effects of income distribution on it (Repetto, 1979; Kocher, 1984; Winegarden, 1985)?

Brevity also dictates simple assumptions. Most of the following arguments assume a single, food-staple crop; a rural society rather equal internally, and rather closed to external trade and factor movements; a fixed responsiveness to technical progress; and population growth determined only by health status, family size norms, and access to food.

Malthus: From BONMAL to BORMAL

Malthus confronted a new agro-demographic situation in Europe. The Neolithic and Medieval agricultural revolutions had raised food output very slowly—probably because their adoption by each small rural group normally required drastic change in its political institutions (Lipton with Longhurst, 1989: 319–324)—and mostly by quite easily achievable expansion of the area of food-bearing land. Meanwhile population also grew very slowly, interrupted by major pandemics[2] and many local wars.

During 1740–1850 a new sort of agricultural-cum-demographic change transformed European farming (Deane and Cole, 1967). Much of Europe (and, Malthus foresaw, soon some colonies too) was becoming land-scarce. Better hygiene, nutrition, and social organization had greatly reduced epidemics (Cassen, 1978: 14–17), speeding up population growth to an unprecedented, sustained one percent per year (World Bank, 1984: 56–59). The new technologies, which more than maintained food supply per worker, rested upon piecemeal experiments by farmers in lead districts (Riches, 1967) and required no basic political change. Malthus in 1798 saw these trends accelerating.[3] Was there any predictable, limited path of human numbers or well-being?

Malthus argued that humans would respond to increased food resources (or, by implication, to better survival prospects under given food resources) by faster population growth. Evidence from the United States suggested that "population, when unchecked, [will roughly] double itself every 25 years" (Malthus, 1824/30: 27). In the short run, an induced switch to more intensive food-growing, to more labor use, or to food redistribution might help in maintaining food consumption levels (ibid.: 28–29): respectively, short-run versions of Boserup and HRB, and Sen's "entitlements."[4]

For Malthus, however (ibid.: 35–37), such responses—by providing extra food for the poor—normally must (early BONMAL), or at least remediably may (mature BORMAL), induce earlier marriage, less abstinence, and hence faster population growth. He correctly predicts the existence of only three logically possible outcomes to any such prolonged, nonconvergent excess of human increase over food increase:[5] undesirable "positive checks," that is, rising age-specific death rates ("misery"); undesirable "preventive checks" ("vice"); or desirable "prudential checks" ("restraint"). Moderns accept Malthus's inference:[6] that "human institutions [should exert] all influence . . . to diminish the amount of vice and misery" (ibid.: 38), that is, to encourage restraint.

How is this to be done? The later Malthus emphasizes that "extreme healthiness [greatly increases] the prudential check" on birth rates, marriage age, and marital fertility.[7] So do widespread "civil liberty and education," which render the poor less "ready to acquiesce for the sake of present [sexual] gratification in a very low standard of comfort and respectability" (ibid.: 40–49). He thus anticipates evidence (Singh and Casterline, 1985; Weller, 1984; and especially United Nations, 1987) that substantially reduced child mortality, modernized employment, and prolonged education (especially of women) reduce total fertility. But a good deal of these improvements, especially for the poorest, may be needed to bring fertility down; small improvements beginning from low levels may well raise natural fertility more than they raise the demand for contraception. Therefore, development programs designed for very poor regions or groups require substantial, prolonged complementary efforts on child mortality, or female employment or schooling, if BONMAL is to be avoided.

BORMAL also accords with evidence that parents' and children's schooling induces parents to "substitute quality for quantity" by producing fewer but better-educated offspring (Schultz, 1981). Furthermore, both Malthus and the moderns argue that, by raising psychological subsistence (the expected "standard of comfort and respectability"), development efforts can induce couples to practice prudential restraint so as to avoid being pressed down to a lower physical subsistence (Booth and Sundrum, 1984: 62).

For believers in BORMAL, to blame the poor (as overbreeders) for their poverty "does them great injustice [, being disproved by frequent] prudential restraint upon marriage" (Malthus, 1824/30: 51–52). BORMAL is largely about policy selection to encourage later marriage. New knowledge, changed moral tastes, and extension of the analysis beyond areas of the European delayed-marriage pattern (Hajnal, 1982) induce us to add, respectively, encouragement of breastfeeding, contraception, and reduced family size norms within marriage; but the 1824 prospectus is modern in spirit, worlds away from the prediction of inevitable "positive checks."

But are both Malthus and the moderns battling with a nonproblem? Even with diminishing returns to labor, need rural population pressure threaten the economy's capacity to provide productive work and access to food? Or is the threat vanishing in a Southern sea of cash crops (traded for Western food surpluses), rural industry, and urbanization?

Nonagricultural escape routes?

For urbanization, UN data project that rural populations will experience absolute decline in 1985–90 of over 0.7 percent yearly in temperate South America (and 0.2 percent in tropical South America); rural populations in India and Indonesia will experience decline by 2000, and in China by 2005. Even in Africa, rural population growth is projected to fall from about 2 percent annually in 1985–90 to around one percent by 2010 (UN, 1985: Tables A1–8, especially p. 102). Most development projects being considered today—notably for agricultural research—will bear fruit in 10–20 years. Do the UN projections relieve the pressure on such projects to provide workplaces, food, or food entitlements?

Unfortunately, these projections seriously understate the problem. First, they underestimate initial rural population. Definitions of "rural" exclude places with over 500 inhabitants in Papua and Peru; over 2,000 in Angola, Cuba, Ethiopia, Honduras, Kenya, Liberia, and the Philippines; and so on (ibid.: 63–71). Second, the estimates understate relative growth of rural populations, by steadily excluding communities as natural increase "graduates" them across stationary rural–urban definitional thresholds, such as a threshold of 5,000 persons (Lipton, 1977: 225; and 1982: 16; Stolnitz, 1983: X52–57; Silver and Grossen, 1980: 252–254), or as they expand to capture nearby places (ibid.;

McGee, 1981: 161–162). Third, projections of rural/urban growth ratios are even more understated, because the UN projection model (which has low r^2s and barely significant beta-coefficients) assumes a "world urban/rural ratio norm" tending toward infinity (UN, 1980: 10, 31; UN, 1985: 4–6), and excludes economic or political reactions that limit urbanization. Fourth, faster rural population increase is not clearly correlated with (let alone relieved by) faster urbanization (Mohan, 1982). Finally, urbanization seldom provides net gains for the poorest rural groups (Connell et al., 1976).

Hence—although the urbanization/industrialization complex (apart from relieving the land of the need to support extra workers directly) could, for several reasons,[8] offer major prospects for rapid fertility declines—urbanization will help the African and Asian poor much less than the projections suggest. Nor will industrialization remove "Malthusian stress" from most Third World agriculture. The proportion of workers in industry between 1965 and 1980 rose only from 12 to 13 percent in India, and from 5 to 8 percent in low-income countries of sub-Saharan Africa (World Bank, 1984: 182–185, 238–239). The 1980s have seen falling rates (and employment intensities) of industrial growth in the less developed countries (World Bank, 1986: 238–239).

Can employment-based entitlements be created for growing populations, while overcoming land shortages, by rural industry (and exchange of its outputs for food)? Micro-studies show that rural industry already accounts for surprisingly high proportions of output and income (Chuta and Liedholm, 1979). However, first, it has higher marginal capital/labor ratios—costs per new job—than agriculture. Second, across South Asia, rural-industrial artisanship is slowly giving way to outwork. While this shifts some income from men to women, and may provide forms of employment that induce fertility decline (UN, 1987: chs. 7, 9), it further raises marginal capital/labor ratios. Third, poor women are likeliest to gain from outwork that is compatible with child care; such jobs may not discourage fertility.

In most developing areas, agriculture must provide most extra livelihoods, if growing rural populations are to maintain or improve access to food per person. "The absolute number of workers in agriculture in Africa . . . will double" in 1985–2010 (World Bank, 1986a). Most rural industries produce inputs, process outputs, or absorb consumer incomes of agriculture—and therefore tend to grow to the extent that agriculture does. Hence it is on agriculture's responses, as they interact with fertility change, that prospects of movement from BONMAL to BORMAL largely depend. But what form of agriculture?

If many growers shift from food staples to cash crops, their local terms of exchange of the latter for the former will sharply worsen. When poor producers expand their supply of a food staple, their demand normally expands almost commensurately. Moreover, for most key food-exchangeables (cash crops), price-inelastic world demand (Godfrey, 1985) prevails. The global emphasis, in poor countries whose population growth threatens claims on food, must be on expanding supply of, alongside work-based entitlements to, food staples.[9]

The Boserup model

Boserup (1965, 1981) and a neat presentation by Robinson and Schutjer (1984) suggest the following stripped-down version of the model. It assumes: an entirely food-producing, isolated, subsistence community, that is, one where non-food demands are met free or not at all; one food crop, or a set mixture; not-very-unequal income and output per worker; endogenous technical progress; and exogenous population growth, similar for all households and age groups. In response to it the community strives to maintain subsistence, not by reallocating food or curtailing energy expenditure, but by growing more food—initially along the same production function, for example *NP* (Figure 2). However, whether moves rightward along, say, the "Long fallow" production function indicate land expansion or intensification, they meet diminishing returns to labor, respectively because land quality declines or because land quantity is fixed. This eventually pushes people below acceptable levels of food intake, until, at *P*, they are driven to evolve improved technology, that is, to restore subsistence (or better) by shortening fallow. Later, at *R*, they shift to settled cultivation; later still, to multi-cropping; and so on.

Boserup's mechanism does not require that population growth relaxes a labor constraint, or renders labor-intensive methods more attractive. She em-

FIGURE 2 Boserup thresholds

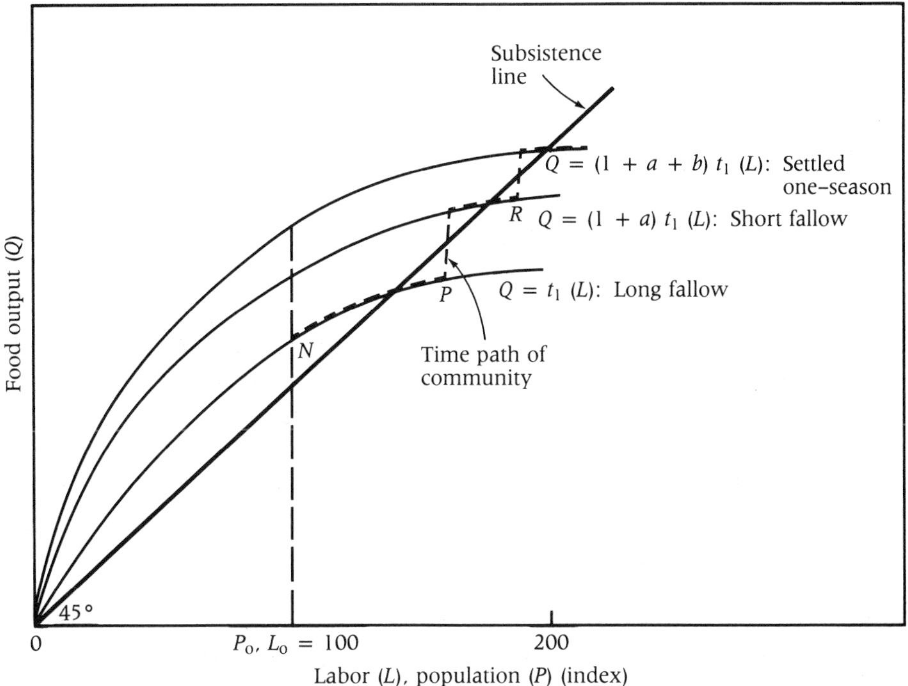

$a > 0$, $b > 0$, $dQ/dL > 0$, $d^2Q/dL^2 < 0$

phasizes (1965: 45–51) that there is much spare labor unused in long fallows, and even, seasonally, in short fallows. Parts of it do get used up in the early stages of exogenous population growth—labor input must, and can, grow faster than population, to maintain subsistence along the diminishing-returns curve of a given technology. At *P* or *R*, when food crisis looms and the community moves to a new technology to feed itself, labor inputs may well rise: probably, initially, by "laboresque investment" (Sen, 1968) to meet the costs of transition to the new technology (Boserup, 1981: 45); possibly, later, to cultivate more land, or the same land more intensively. For Boserup—unlike HRB—the necessity of more food, not the availability of more labor, is the mother of population-induced invention. The new farming need not involve more labor than the old,[10] for example, if "fertilizer, weed [or] water control can be effected . . . by industrial inputs" (ibid.: 25). The extra food can be provided by greater capital, skills, or inputs.[11]

HRB and Boserup accept many of one another's modeled responses to exogenous population growth, but it is clearer to keep them separate. Labor-intensification and food-intensification are logically independent. They correspond to distinct necessary conditions for BORMAL [see propositions (4) and (6) above].[12]

Institutional limits

Boserup is not mainly concerned with the institutional factors underlying her posited food-productivity response. Yet the adequacy of the response to exorcise BONMAL depends crucially on institutions, especially those of agricultural research. The model's *operation*, too, is limited institutionally.

First, a new technology, to increase food output per person, is called forth not by hunger alone, but by hunger backed by effective demand. If poor households, despite population growth, acquire no more claims on food, the Boserup process will be aborted. Will institutions (e.g., of kinship) that once provided reasonable equality of claims on food survive rapid growth in labor supply?

Second, the process in Figure 2 implies a threshold mechanism. Extra people feel steadily growing need, yet move down *NP*, the diminishing-returns curve of a given technology. At *P*, necessity bites hard enough to call forth an invention. Then there is a sudden jump to the superior technology. Do the institutions that can demand or supply technical change respond in this way?

Third, no convincing explanation—certainly not labor shortage—is given as to why the pressure of need did not evoke that superior technology until the extra people arrived.[13] Boserup initially (e.g. 1965: 54) appeared to attribute, to rural people with early technology, a backward-bending supply curve or high leisure preference. She wisely omits such "explanations" in recent work, but nobody has yet suggested satisfactory alternatives.

Fourth, shifts from gathering to agriculture—and thence to settlement—have involved major institutional upheavals. Substantial concentrations of force over labor, or other changes in social forms, were associated with Neolithic settlement, with Europe's medieval agricultural revolution and its links to feudal power, and with macro-irrigation "despotisms." Despite a few remarks on land tenure, Boserup says little about institutional factors easing, or delaying, such dramatic community responses to exogenous population change.

Fifth, to assume that population change is exogenous may be to evade Malthus's question: by inducing population growth, cannot agricultural "betterment" make things worse? Boserup's correct assertion—that exogenous population growth, though making things worse at first, later induces new technology that makes outcomes better—assumes away subsequent endogenous (and perhaps Malthusian) responses of population to extra food (after technical progress).[14] Two syntheses, permitting population growth to be endogenous and technical progress exogenous (as in Malthus) or vice versa (as in Boserup), demonstrate that borderline values for system variables—given an initial exogenous change in population growth rate—determine whether real wages per worker will ultimately rise or fall (Pryor and Maurer, 1982; Lee, 1986). But it is institutions, especially those of labor supply, demand, and organization, that decide on which side of the borderline the system variables lie.

Boserup, agro-technology, and infrastructure

Boserup's account, however, is on the whole convincing. A BONMALthusian, knowing only the 1798 edition of the *Essay* and the world's population history in 1940–60, would expect most of Africa, with its major land resources, to have postponed "positive checks" while its first batch of extra people expanded into (apparently) spare land; and South and East Asia, with few chances to expand land area after 1960, to have run heavily into "positive checks." Yet vital statistics and agricultural growth performance tell the opposite, Boserupian, story. In South and East Asia (where the pressure of population has induced rapidly diminishing marginal returns to labor on given technology), institutions have helped necessity to mother invention. Many areas there have shifted to more "food-producing" technologies, and thus have avoided (or postponed) the vulgar-Malthusian nemesis. In much of Africa (where spare land seemed almost as good as currently used land, so that marginal returns to labor did not fall sharply enough to compel a shift to new technologies as population grew), such a nemesis is a more credible threat.

Malthus alone cannot explain this; Boserup can. She points out (1965: 59–61) that population pressure in many documented instances preceded intensifying shifts (cf. Pingali et al., 1987). However, her argument that "a certain density of population is a precondition for the introduction of given techniques" (ibid.: 65) is more plausible for farming itself than for ancillary services (transport, power), despite the alleged economies of scale for the latter from increased population density. First, as density increases, the diversion of investment resources, from on-farm land and water improvements to much larger scale shared technologies, is often an output-reducing temptation, not a cost-reducing gain: the roads use up the savings and expertise needed for farming. Second, physical grid infrastructure—usually subsidized or even free to users—biases farming away from "local options," toward long-distance trade and exchange along the grid. Third, access to infrastructure is unequal (see Lipton, 1987).

The cost of infrastructure per user is clearly reduced by large-country effects (i.e., there are efficiency gains for a country with a relatively large total population); sometimes by population density; but perhaps not at all by rural population growth (which is neither necessary nor sufficient for increased density). Many "social overhead costs . . . do not vary with the size of the population" (NAS, 1985: 66). In North India, "population density has a significantly positive effect on the intensity of irrigation and on the net cropped area" (not all one-way causation) but a "negative impact on research investment, road expenditure, electrification, and credit provision" (ibid.: 38; Evenson, 1988).

Demand limits on Boserup response in Asia

In agriculture itself, however, most populations sooner or later respond to prolonged increase by some form of Boserup-style technology transition. Will such a response suffice to restore sustainable food output per person? "The food potential of the world has been narrowed down by populations [that] did not know how to match their growing numbers by more intensive land use without spoiling the land for a time or forever" (Boserup, 1965: 22). Rightward movement along a curve in Figure 2—which can continue for some time if population growth starts when the community is at a point on the curve well to the left of the subsistence diagonal, and with spare time available for extra work—can mine out the soil. As (say) *P* is approached, this course will have rendered progress with new methods more difficult, perhaps compelling "the choice between starvation and migration" (ibid.: 41). This risk increases secularly because each leap to a new curve costs more, by way of investment per unit of extra food, than the last one did: diminishing returns apply to steps (to each successive new technology) as well as to output gains with a given technology.

Thus, while most communities with restricted prospects for migration (or for non-food production and trade) eventually try to respond to gradual depression below the subsistence diagonal by shifting to a new food-growing technology, they may succeed—if at all—only after the diminishing-returns path along the old technology has carried them too far below the subsistence line to get back above it *and* repay debts *and* undo environmental harm incurred while below it.

Soil mining with the old technology may have reduced sustainable output with the new technology, as is the case in much of East and Northeast Africa (World Bank, 1986a: 24). In part, this happens because population growth not only increases pressure upon common property resources directly, but also sharpens other pressures on them from unequal access and commercialization (Repetto and Holmes, 1983; Jodha, 1985a, 1985b; Lipton, 1985b).

Thus Boserup responses are consistent with steadily diminishing long-run food intake per person (Figure 3). As labor falls further below the subsistence line, it is weakened and reduced in quality. This is less BORMAL than BONMAL with the threat of "positive checks."

The risks of inadequate "Boserup responses" are clearest in intensive agricultures. In response to population growth, these have already invested doses of successively costlier capital (per unit of extra output) to achieve, over the centuries, successive leaps to new technology curves T_2, T_3, etc. (Figure 3). Such countries may be too poor to meet further population pressure with even costlier leaps. Without booming oil revenues and rapid fertility transition (World Bank, 1984: 65), Geertz's fears (1963: 146) about the future of "agricultural involution" in Java would have been realized, despite optimistic Boserupian interpretations (Booth and Sundrum, 1984: 209–212).

FIGURE 3 Inadequate Boserup process

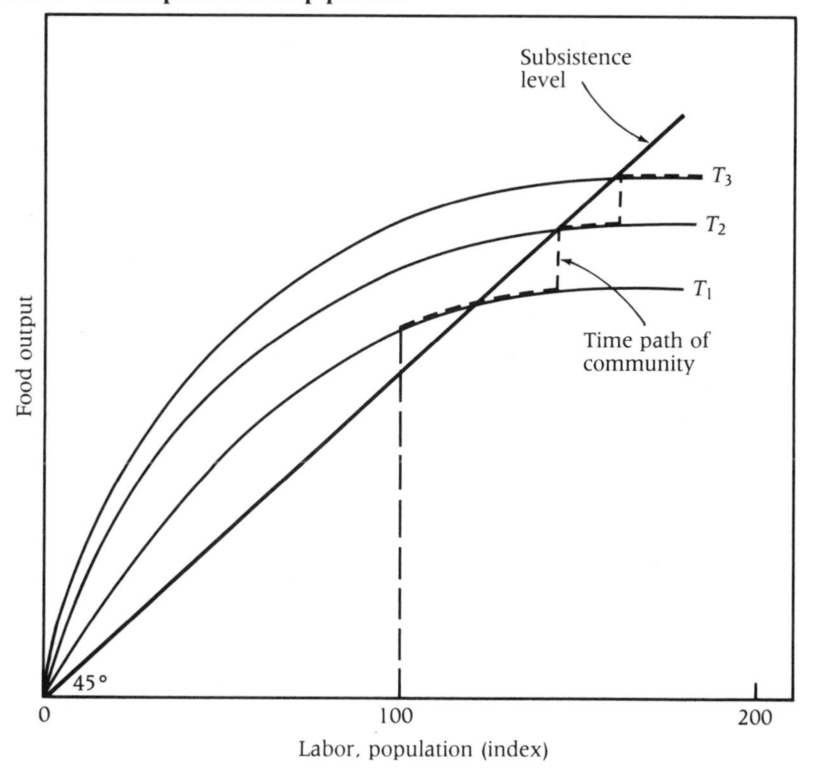

Since 1965, the green revolution might suggest a "bias for hope." Yet the biophysical limits on further yield response, in some increasingly heavily populated South Asian soil–water regimes, were implicitly underestimated by Boserup (1965: 87–89). The biggest yield advances of the green revolution have been in exceptional, water-controlled, moderately heavily populated rice deltas and wheat flatlands—and, lately and promisingly but not in very densely populated places, with semi-arid hybrid sorghum and finger millet. The most "involuted" areas of eastern India and southern Bangladesh do not, so far, offer low-cost technological responses to extra demands for food. They may also feature stubbornly high fertility, both for institutional reasons and because it is indicated for individually optimizing poor couples.

Decades of increasing land scarcity and concentration, in these problem areas, often have brought severe inequality of income. Growing populations of workers receive small shares of extra gross national product, and therefore generate little extra demand for food, or (as they are near-landless) for appropriate new farm technologies. Hence demands on both indigenous and formal research institutions for quick, large outward shifts of "Boserup curves" are weak. The alternatives—modern descendants of Poor Law provisions, alongside incentives to their beneficiaries to reduce fertility—are not often triggered by effective demand, either; so reform of the research process is critical. There are successes—tubewell planning in eastern Uttar Pradesh, modern rice varieties in parts of Java and Bangladesh—but effective demand, by farmers and by consumers, has (unlike need and numbers) grown too slowly to provide sufficient pressure on researchers to shift the Boserup curves in most of India's eastern "poverty square."

Supply limits on Boserup response in Africa

Africa's problem of research-cum-institutional response to population growth is different. Intra-rural inequality is generally less; in most rural places, though land scarcity looms, extra people can still farm extra, albeit worse, land. So rural population growth brings extra effective demands on indigenous and formal research, both for extra food—imported or home-grown—and for effective smallholder techniques to shift the Boserup curves upward. However, in much of Africa, research and other institutions are ill-placed to *supply* such techniques—as against (for example) India's "poverty square," where lack of effective *demand* for Boserup shifts is the main problem.

Africa's problem of institutional responsiveness to demands for Boserup shifts has four aspects. First, does the institutional set-up encourage response via soil-mining, rather than via investment embodied in soil-fertility enhancement? By its effects in increasing the costs of managing common property and raising time-preference, African population growth (alongside the uneasy transition of trust in resource management from local chiefs to government) has increased the pressures to "mine" grazing land, water, and especially (via shorter fallows without fertilizers) soils. This has cut soil fertility, and perhaps accelerated deforestation and overgrazing, leading to climatic instability and desertification (NAS, 1985: 41–46; Lipton, 1985b; World Bank, 1986a: 46). In a situation in which there is apparent "spare" land, and in which impending land scarcity is obscured by high discount rates, institutions for technical change are encouraged to supply not Boserup shifts but short-run solutions that are long-run depletions.[15]

Second, what of Africa's institutions to mobilize savings and to finance investments that "embody" on-farm the results of research to develop new Boserupian technologies? Well over half the gross investment, public and private, of low-income African countries is now financed by net aid. Past increases in aid to African agriculture have in part displaced domestic investment. Private savings show no upward trend in most of the region, and institutions for collective savings enhancement—such as Mexico's *ejidos* or South Korea's rotating credit and savings associations—are rare.

Third, do the institutions of formal research respond any better to Africa's population-led demands for more intensive food-growing technologies? If new, intensifying farm techniques—ploughs, modern cereal and root-crop varieties, hand-pumps—are being constantly developed, discussed with farmers, tested for smallholder profitability and safety, and, if suitable, promoted for adoption, then even sluggish savings and high temptations to soil-mine can be overcome. Conversely, "no country has been able to benefit from science-based technical change in the absence of an agricultural research system capable of doing both basic and applied scientific research" (Binswanger and Pingali, 1985: 85; cf. Booth and Sundrum, 1984: 95). Research investments (per agriculturalist, hectare, or dollar of agricultural net value added) have been much larger in Africa than in other developing regions, but have shown much lower returns. Crop scientists have been more costly; engaged in ways much less congruent with the area composition of local food crops; much more often foreign-funded; and more seldom linked to *national* commitments of agricultural policy, current resources, or personnel (Lipton, 1985).[16]

Fourth, even where Boserup responses have been impressive—notably in Kenya until the mid-1970s—many institutions of community authority, family, and private property transfer (Cain and McNicoll, 1988) in Africa militate against later BORMAL response by population, that is, against rapid fertility decline. In the great majority of African countries, national-level institutions had produced family planning policy that was far weaker in 1981 (and had shown less improvement since 1972) than in comparably poor Latin American or Asian countries, though per capita public spending on African programs was no less (World Bank, 1984: 149, 200–201).

"Sub-Saharan Africa" and "South Asia" are lazy expressions. Most institutional problems of Bihar or Nepal are "African" in flavor; most in Kenya or Zimbabwe, "Asian." The low average density of Africa's four giants (Ethiopia,

Nigeria, Sudan, Zaire), however—and hence their difficulties in funding regional government services and in maintaining civil order—do typify a situation in Africa that has made for weak and cost-ineffective institutional supply responses to population-induced demand for Boserup transitions.[17] Prevailing soft options, from transport to market relaxation (Lipton, 1987, 1987a)—while good in themselves—have, by arousing undue expectations, undermined the hard work of institutional reform, especially in generating better food-growing technology for smallholders.

HRB responses and their limits

Hayami-Ruttan-Binswanger present evidence that extra labor supply due to a growing population and workforce calls forth research into, and adoption of, increasingly labor-using farm methods. This is reinforced by a parallel process outlined by Chayanov (1966): extra children induce adults to raise labor supply, even from a given workforce, so as to use up leisure and feed the new mouths. Yet, even with Chayanovian support, successful Boserup transition may induce little or no HRB labor intensification. This matters; as population increases, a growing proportion of poor rural households comes to depend, for command over food, mainly upon income from employment rather than from farm operation. It is therefore increasingly important—if a Boserupian shift is to enable these households to acquire enough extra food as their size and number grow—that the new technology increases their income-based claims on food by raising the demand for labor, especially hired labor.[18] This labor is in long-run price-elastic supply, especially during rapid population growth; the real wage rate will not rise greatly when demand for labor rises. So the question is: will new farm technology increase employment in hired-labor households faster than their population increases?

The HRB hypothesis suggests a generally favorable answer. HRB can be right—scarcer, dearer farmland per unit of labor can induce a more labor-using, land-saving pattern of inventions, embodied in innovations and gross investment—even if Boserup's hypothesis is irrelevant (because rent/wage ratios rose without prior increases in workforce or population) or wrong (because, although population has grown, the HRB responses raise labor/land ratios but not the community's command over, or production of, food). Similarly, Boserup can be right about long-run response to population growth—per-person output of, or command over, food can be restored or increased through a responsive shift to a new technology—even though HRB is irrelevant (e.g., because no inventions are necessary, as farmers can shift toward new but already known methods) or wrong (e.g., because food output per person is maintained, but by innovations that, in saving land, save even more labor). There are instances of HRB with or without Boserup, and of no HRB with or without Boserup.

Boserup or HRB responses, while likely, are neither logically necessary nor, if achieved, necessarily adequate. The task of policy is to manage incentives, institutions, and technologies so as to obtain adequate Boserup and HRB responses to population change—alongside fertility transitions reducing the strains upon such responsiveness—and not to assume that either response is inevitable, implies the other, or suffices to safeguard the food entitlements of the poor. HRB recognize this (Hayami and Ruttan, 1985: ch. 12), and they give many examples of divergence from "their" path of induced, appropriately factor-saving, innovations (Binswanger and Ruttan, 1978).

For three reasons, profit-maximizing farmers plus farmer-responsive researchers *need* not yield HRB outcomes. First, suppose that rising labor/land ratios, by reducing wage/rent ratios, induce farmers to select more labor-using, land-saving farm practices, crops, and so on (given available techniques) (Salter, 1966). Each farmer then moves to a new optimal farm structure, giving high output-per-hectare, relative to output-per-worker. With this structure, a farmer gains just as much from any invention that provides a cost reduction of *x* percent, whether via unit labor costs or unit land costs. Hence farmers appear to have no motive to pressure researchers for inventions, or themselves to seek innovations, more labor-intensive than before.[19]

Second, there are not just two factors, land and labor, to whose prices farmers respond in choosing techniques. When a third factor, capital, is introduced—and it usually must be, to embody farmers' technology shifts (Boserup, 1981: 45)—farmers may do best to save more of the factor that has become more plentiful (labor) than of the scarcer factor (land), through the use of the extra amounts of the third factor (capital, weedicide) per unit of output. Hicks (1939) saw this as exceptional, but recent work (e.g., Quizon and Binswanger, 1983: 532) has enlarged the scope of such theoretically possible events. This may contribute to explaining labor displacement by tractors, weedicides, and so forth even in areas of rapid population growth.

Third, HRB responses even in a two-factor model—where labor and land can be combined along any given curve in Figure 2 without any investment requirements—need not produce labor-absorbing outcomes in general equilibrium (ibid.: esp. p. 526). If the economy is modeled with more than one region, or with an industrial as well as an agricultural sector, a rising labor/land ratio in one sector or region can readily be associated with falling total employment nationally.[20]

Thus HRB responses may be subverted, despite population growth, by labor-displacing technology. Moreover, such technology usually manifests itself despite population growth. The non-local nature of most invention processes helps to explain their frequent failure to require more labor in an HRB response to locally rising labor/land ratios. In five historical series out of nine, "significant positive [beta-] coefficients were obtained" when regressing land/labor ratios upon a rental/wage index, "clearly inconsistent with the [HRB] hypothesis." In three cases, "such behavior might have been caused by an exogenous labor-saving bias" in the United States, plus "biased borrowing" of US technology by Japan, France, and Great Britain, despite their higher person/land ratios (Booth and Sundrum, 1984: 94–95, in part citing Binswanger and Ruttan, 1978: 63–64). Today also, "money and research skills are overwhelmingly in rich countries, and produce . . . labor-saving not labor-using innovations" (Birdsall, 1985: 49). This raises a crucial institutional issue: how to design indigenous research, or select borrowings, to reach HRB paths of labor-using response to population growth.

It is not only foreign research that induces a non-HRB path of rural innovation. (1) Industrialization, as a policy aim supported by incentives and public allocations, normally shifts inputs and outputs—including rural and agricultural ones—in a labor-displacing way. (2) Dualistic agricultures can mean, as in Argentina (de Janvry, 1978), that big farmers—able to afford (and to risk) rapid innovation, but faced with labor search and supervision costs and/or having monopsony power in labor markets—select capital-intensive innovations, while smallholders lack resources to respond to extra labor supply by purchasing equipment or inputs that shift technologies toward high labor/land ratios. As a result, even if labor/land ratios are increasing, there is likely to be farmer pressure for inventions that lower the relative unit costs of less labor-intensive activities (Grabowski, 1979). (3) Even family farmers, faced with rising demand for food, may adopt near-to-hand technologies that, while they intensify farming, displace labor in the longer term, as in parts of Uttar Pradesh (Ray, 1985).

In the short run, unless labor owns the capital that displaces it, these perverse non-HRB responses harm the poor. In the long run, BONMAL would imply that HRB (and Boserup) responses merely induce higher total fertility; conversely, the labor-displacing effect of some new agricultural techniques (e.g., tractors), and their effect in raising the returns to education, accelerate fertility decline by reducing the "child labor" motive for large families (see, for example, the report of this outcome in the Indian Punjab by Nag and Kak, 1984: 676–677).

It is wildly unlikely, however, that labor-saving is good for the poor. The decline in poverty in the Punjab (Chadha, 1983) is due to labor-using and food-cheapening aspects of the green revolution. It has taken place despite labor-saving tractors, threshers, and weedicides. There have to be less arbitrary, uneconomic, and cruel routes to fertility transition than anti-HRB techniques that create unemployment as workforces grow.

Institutions and HRB

Social institutions, except in the very long run, are unlikely to adapt optimally to HRB pressures from growing farm populations for labor-using investment and invention. If population growth polarizes farm sizes, as in Bangladesh, such pressures are weakened because larger farms are more inclined toward capital-intensity, and very small farms toward family labor, as compared to the more hired-labor-intensive, medium-small farms that they replace. Also, ex-operators of these latter farms swell the supply of labor for hire, as do part-timers from the new micro-farms.

Social obstacles in extensive agricultures, too, delay HRB responses. Shifts from transhumance to ox-ploughing may be impeded by socially embedded patterns of residence and family-labor management (often reinforced by long distances between grazing and crop areas). Response may be delayed until there is a much larger labor force. By then, human and livestock populations may have so overgrazed that the long-deferred adoption of HRB technology brings only a frustrated Boserup response (Figure 3). (This is how I read Delgado and McIntyre, 1982, in the light of my own observations in Botswana.)

Yet, if local institutions (perhaps "adapted" to person/land ratios of a generation ago, but still embodying elite interests and power) delay labor-using HRB responses, then the extra population's pressure for Boserupian food supply enhancements will be addressed by "imported" labor-displacing technology.[21] For this is steadily made less costly (relative to HRB technology) by Western research. Thus the green revolution and associated inputs in South and East Asia in the 1960s showed an employment elasticity of output expansion around 0.4, but in the early 1980s only about 0.1 (Jayasuriya and Shand, 1986). The institutions of research in (and for) less developed countries are not yet geared to the increasingly pressing need to seek out HRB technology choices, especially those that use hired labor.

Fertility effects of Boserup and HRB responses

Could HRB responses induce higher fertility? Couples' fertility decisions respond "rationally," if weakly, to the changing economic costs and benefits of extra children (Schultz, 1981; Birdsall, 1988; Rodgers, 1984: esp. chs. 2–3). Now suppose that an exogenous rise in population is met by an HRB response. The new, labor-using technology may well raise the demand for labor by more than population growth increases workforce supply. That would raise employment per member of the workforce, and perhaps the real wage-rate also. In turn, this would tend to increase endogenous fertility, via the couple's incentive to produce children, who may later earn wages and make transfers to their parents. Usually, the exogenous population rise had resulted from a fall in mortality concentrated among infants and children.[22] HRB responses would then "rationally" include a search for new technologies workable by or with child labor. Success would strengthen the incentive for couples to produce more children.

The Boserup process increases food availability, but thereby may encourage higher fertility via the income effect (allegedly positive) on couples' capacity to "demand" children as consumer goods. As Malthus recognized, this is remediable by linkages from rising income, via better health, expectations, and education, to reduced family size norms. Less readily remedied is the BONMAL worry about the HRB process: it increases poor workers' food entitlements by making agriculture more labor intensive, but may thereby encourage higher fertility via the substitution effect (surely positive) on couples' incentive to demand children as investment goods. The unit value to the couple of the children's later labor is increased more than the cost of child care.

Such fertility-enhancing effects have been identified in a careful study in Bangladesh (Cain and Mozumder, 1980). Conversely, both for the Indian Punjab and for Java (Nag and Kak, 1984), fertility was apparently reduced by capital-intensifying, labor-displacing changes in farming practice during the 1970s. BONMAL seems to be a more intractable enemy to sustainability for HRB than for Boserup processes.

Boserup/HRB as Sen availability/entitlement responses

The sustained mortality declines that have underlain the exogenous population expansion of 1945–90 occurred both in regions (mostly in Africa) where food output and availability per person have fallen, and in regions (mostly in Asia) where—although they have risen—food entitlements per potentially hungry person have not. These problems are largely due to inadequacy both of the Boserup response to growing population, and of the HRB response to a growing workforce. Both responses must be adequate if the food problems analyzed by Sen (1981)—availability and entitlement—are to be avoided. The problem is compounded by the effect of even smallish Boserup and HRB responses in inducing further (endogenous) population growth. Big HRB and Boserup responses—and small or negative fertility reactions to them—are not automatic. They need to be encouraged by policy: above all, perhaps, by measures to support crop research.

Entitlement failures among vulnerable groups have led to famines despite increasing national and even local output (and often availability) of food staples per person (Sen, 1981, 1986). Even in countries that have avoided famine over recent decades, such as India, these failures have prevented clear improvements in the nutritional status of chronically hungry poor groups. In 1965–86, India's green revolution boosted food availability. Up to 30 million tons of public sector food stocks accumulated, yet the poor and the landless barely maintained their former average level of energy intake. Unskilled workforces have grown so rapidly that employment (and real wage rates) per worker have been insufficient to permit substantially increased claims, via labor income, on food per adult equivalent.[23]

The gap-filling effects of extra nonmarket entitlements via demand for labor (e.g., Maharashtra State's Employment Guarantee Scheme), asset distribution for the poor (e.g., *antyodaya* in Rajasthan), and food provision (e.g., the Noon Meals Scheme in Tamil Nadu) have not yet greatly reduced the incidence and severity of poverty. This "South Asian story" features a green revolution that brought successful Boserup adjustments, but HRB adjustments too slow to swamp labor-displacing effects (threshers, weedicides, etc.). Hence entitlements—and thus food demand—grew too slowly to match food supply. South Asia's rural workforces have outpaced, not available food, but in most areas the growth of demand for labor from HRB responses, and therefore have seldom gained extra (labor-market-mediated) claims on food per person.

Where the poor eat what they farm, non-HRB technology responses to population growth do much less to erode demand-based entitlements to the fruits of agricultural progress. Thus, even if the sources of growth of output in finger-millet were capital-intensifying, the poor in the semi-arid areas of Karnataka State in India—being mainly small farmers, rather than mainly landless laborers—would benefit. Nor is the problem so acute where land reform had kept "pre-innovation" farms fairly equal; where landless workers are rare; or where their increase can be absorbed in extra off-farm work. Thus poor people's claims on food increased (despite possibly anti-HRB and capital-intensifying farm growth) with modern rice varieties in Taiwan and South Korea, and with maize hybrids in Zambia.

In general, though, the problem remains. Boserup technology shifts, while inadequate so far in much of Africa, are a likely response to rural population growth. It is desirable that these should be HRB shifts, but for several reasons they may not be. Even if they are, their effect on entitlements may not be enough to outweigh other, labor-displacing, events. Thus Binswanger and Pingali (1985: 98) review "evidence on the labor-saving benefits and the consequent increases in yields per man-hour through a change to animal-drawn ploughs and tractors" in grass-fallowing. This is consistent with rising labor income only if (1) laborers' households have rights in land or capital, or (2) the growth rate of (food) output exceeds the sum of the rates of growth of population and of labor productivity—that is, in practice, if there is plenty of spare land to farm without dramatically falling total factor productivity.

Thus, if HRB responses are weak, Boserup shifts—even if more than adequate to maintain food availability (Figure 2)—may leave poor people's entitlements to food dwindling as the growth of rural workforces increases unemployment, even at a subsistence minimum wage-rate. This can bring not only "positive checks," but also weaker food prices, discouraging additional Boserup shifts.[24]

Failures of exchange entitlements are seldom simply market failures; excellent markets plus poverty mean ineffective demand (NAS, 1985: 141). However, there are ways out. Creation of major nonmarket food entitlements for the poor is the most widespread and promising, but requires "poor power" to press for it; creates bureaucracies; and risks pauperization. A shift toward more grain-demanding animal production (eaten mainly by the nonpoor) can keep up grain demand, but at the cost of both jobs-per-hectare and grain supplies for human food. Intra-rural migration appears to worsen the distribution of entitlements: in many agriculturally fast-growing villages, a labor force swollen by immigration competes for work from fewer farmers, owing to their concentration and emigration (Kikuchi and Hayami, 1983: 251; Leaf, 1983; Breman, 1985). Agro-asset redistribution is politically very difficult, but can resolve the problem by placing land (with which to adopt Boserup inventions) with the households likeliest both to select HRB techniques and to grow and eat the extra staples they need. There is also evidence that institutional change that equalizes asset ownership alleviates the risk of rising fertility during the response to Boserup or HRB processes (Repetto, 1979; Stokes and Schutjer, 1984: 207; Kocher, 1984: 225).

"We'd love to graduate to these problems," many in Africa, mired in frank deficiency of food output, might reply. Yet large parts of Kenya, Rwanda, Zimbabwe, and Malawi have such problems already, in normal years. Elsewhere, too, the pressures to treat the food-availability diseases of population growth with purely Boserupian, but not HRB, technologies—and thereby to worsen entitlement diseases—will mount. Many African countries can import tractors. Fewer have the institutions to generate Boserupian responses sufficient to maintain food availability per person. But suppose the tractors do raise food availability, but throw people out of work. Lacking institutional changes (such as land redistribution) that favor HRB responses also, Boserupian responses will not suffice. Steady increases in labor supply in many African agricultures—while land scarcity (and imported technology) spreads—will bring failures of entitlements, even if enough food is available.

Research–institutional response

BONMAL pessimism and Boserup–HRB response-optimism often share a deterministic flavor. If the "picture" of Figure 1 is roughly correct, however, nondraconian policies, by affecting incentives, institutions, and allocations, can improve nations' population–food paths—as indeed the divergence of such paths since 1945 suggests. "Despair not; life gives still for human effort scope."[25]

The planning era, say 1950–75, emphasized public allocative approaches to the problems of increasing, speeding, or spreading gains from Boserup–HRB responses (more public money for appropriate research), or of minimizing "adverse" BONMAL fertility reactions if such responses succeeded (more money for family planning). The market era, say 1979–87, shifted emphasis toward private incentive approaches: remove state interventions that militate against farmers' pressure for, or use of, appropriate inventions (especially via farm price repression), or against small family size norms (e.g., by unduly subsidizing education). The essays in this collection are directed to a third option: not a glib synthesis of plans and markets, but analysis of group institutional roles, processes, and policy options, as they affect both Boserup–HRB responses and their impact on the fertility transition.

In closing, I address a narrower, but crucial, issue. Can policy, by altering the institutional framework of agricultural research, improve Boserup–HRB responses and reduce their risk of inducing fertility increases?

The output of agricultural research is not positively related to the employment needs of growing populations of rural poor, as reflected in agricultural or rural population density (Evenson, 1988). Conversely, however, weak indigenous formal agricultural research is linked to the fact that many African countries, in the face of population growth, can adopt only imported technical change, thereby often actually decreasing labor-intensities (NAS, 1985: 40).

Reliance on informal research—farmers' own experiments—has often worked well, despite weak formal national agricultural research and strong labor-saving impacts from abroad, with rural population growth at 0.5–1 percent yearly; but it is insufficient at 2 percent or above (Rounce, 1949; Binswanger and Pingali, 1985). Boserup (1965: 69) shows that in the 1950s many Indian and African farmers had rationally rejected labor-intensive techniques. Faced with slow population growth, such farmers could intensify on their own volition, either along gently diminishing-returns curves or by modest technical progress. Faced with rapid population growth, farmers need help from local formal agricultural research, adapting new, Boserup-and-HRB, methods to suit local profitability and safety.

In such a context, the absence of linkage between agricultural researchers and demographic policy analysts is strange. The financial and agricultural research institutions that were behind the thrust in the early 1960s toward research into modern varieties of Third World food staples had been fired by "Malthusian optimism" (Sen, 1986). As more people required more food output,[26] analysts saw the main danger, both to the survival of the poor and to the stability of the polity, in the swift approach to the extensive frontier of usable farmland (Ford Foundation, 1959). The problem was specified as follows: growing rural populations could no longer feasibly expand onto new farm areas (unless far inferior to those already farmed); population would therefore outstrip national food supplies, given the technology; this would increase poverty and hunger; the chain could be broken only by a research-based green revolution.

The problems of national hunger and rural poverty were perceived in "Malthus versus Boserup" terms. International research goals—and national research institutions—were financed to improve prospects for a move to a more food-generating technology, presumably for hungry farmers and townsfolk. Neglected was the need for HRB employment-generating technology, to improve the food entitlements of hungry laborers (though modern varieties did somewhat raise labor demand).

All the same, the strategy of varietal research was implicitly demographic. It sought, by raising food output, to push the outcome of exogenous population growth from a Malthusian to a Boserupian "victory," via exogenous intervention in research. New technology in agriculture was to create a breathing-space for education, rising welfare, and contraception to bring down family size. Plainly, fertility motives and responses were critical.

Yet there is (and was) almost no demographic content in modern varietal research. The right way to fill that void, especially if we seek to help the poor, is much changed by new knowledge, partly reviewed above, since Boserup (1965). The appropriate question is not, as it might have seemed 20 years ago, "Which pattern of staples research and intervention will create the most favorable balance between aggregate food supply and exogenous population size?", but instead, "Which pattern will best help poorer families (and the society that surrounds them) to improve their per-person entitlements to food, and to achieve their goals at a lower family size?"

Agricultural research needs to seek ways to place reliable extra resources, especially command over food, with households most threatened with nutritional danger by loss of entitlements during population change: big families and/or landless and near-landless families (Lipton, 1983, 1983a). In appropriate policy contexts, and provided that the resources are not offered in ways rewarding higher fertility, this will retard population growth.[27]

Varietal research, apart from *breeding* robust and high-yielding seeds, can then *screen* alternatives for release, selecting varieties that help farmers to find alternatives to large family size, in several ways. For example, a variety with timed labor requirements that do not overlap with women's seasonal work peaks[28]—or a variety with nutrients readily absorbable by sick or weaning children—will reduce the number of infant deaths and subsequent replacement births.

The most worrying threat posed by population growth, however, is to the food entitlements of the increasing proportion of poor people dependent mainly on hired farm (or farm-related) labor, rather than on farm ownership. The bargaining position of rural laborers is weakening. What can agricultural research do to help the "new poor"—not created by population growth, but made increasingly dependent by it on labor income?

Two points should be noted. First, the varieties, crops, and farming systems best suited for the rural poor may differ depending on whether those people obtain, say, 15–20 percent of income from own-account farming or are 100 percent landless laborers. Second, in Africa, many past outcomes of agricultural research were not accepted owing to labor shortage; but with person/land ratios doubling every 20–30 years (and with a 10–15-year lag between research design and widespread adoption of a released modern crop variety), "labor saving" is seldom a sensible research goal.

There are several research directions that seem able to reach the new poor majority, farm laborers. Agricultural research might aim at farm systems, crops, and varieties that, while safe and profitable to farmers, greatly raise the demand for labor (in ways not readily avoidable via threshers, weedicides, etc.). Relevant here are emphases on mixed stands, on fertilizer placement in the root zone, and on organic fertilizers. Where governments undertake large-scale food-for-work schemes, research can screen varieties, crop-mixes, and farm systems for their capacity to provide cheap, readily stored foods well adapted to such schemes (and in particular to the needs of small children, and mothers of large families). Formal research could concentrate on, and interact with, informal on-farm experiments, notably in varietal selection and water management (Richards, 1985); these latter are likely to respond more quickly to changing factor scarcities than to formal research. Or research could seek varieties, crop-mixes, and the like for the areas where, even with substantial population growth, the poorest will long rely on own-account farming, not on labor incomes; however, as a sole strategy for agricultural research, this would leave many of the world's poorest unassisted.

Suppose that research goals respond to the short/medium-term need for "food *from* work" of the newly emerging landless majority among the rural poor. How are BONMAL outcomes to be avoided? That is, how can such agricultural research be selected—in scale, region, or type—so as to improve laborers' long-term food entitlements by, at least, not retarding fertility transitions? This issue is little researched (exceptions are Basu et al., 1978; Stoeckel and Jain, 1986). The fertility effects of any option derived from agricultural research are probably very long run; hard to separate from the fertility effects of many other factors; and perhaps impossible to assess in one place because of interactions with migration (usually sex-selective) to and from areas with different types of research orientations and hence of rural prospects for farmers and laborers.[29] Nevertheless, the importance of the issue is such as to demand more research. At present, agricultural research design is based on a tacit—and long-refuted—demographic premise: crude neo-Malthusian optimism. Our new understanding of fertility decisions following the World Fertility Survey (e.g., UN, 1987) needs to be reflected in research design to improve food availability and entitlements.

Although the organization of agricultural research is a huge, much-analyzed topic, interactions with actual and desired population growth and structure are hardly discussed. Yet the proper size, topic-mix, incentive structure, and goal-setting procedures of any research system—if it is likely to produce really important innovations—depend, probably critically, on how farmer and laborer households adjust family size norms to those innovations. Population policy is largely implemented (if at all) via health ministries; yet population outcomes (and most health outcomes) depend heavily, perhaps mainly, on the performance of agriculture. The linkage of these two areas is almost everywhere very weak.

Notes

1 Explicitly including contraception (Malthus, 1824/30: 38).

2 In 1347–49, the Black Death apparently killed about one European in three (Genicot, 1966: 647–651).

3 For example, Jenner's demonstration of cowpox-smallpox vaccination in 1796 revealed new prospects for scientific disease prevention.

4 Even in the first *Essay* (First Edition, 1798, pp. 29–31, cited in Eltis, 1980: 26) Malthus views growing population as a threat to well-being mainly via entitlements—falling wage rates as growing workforces compete for jobs; rising food prices as they compete to buy cereals.

5 This conclusion does not depend on Malthus's famous (and dubious) claim that food resources cannot long grow geometrically (Cassen, 1978: 9).

6 Less acceptable to most moderns—and unnecessary for Malthus's argument—is his confinement of desirable "restraint" to abstinence from, or within, marriage, and his inclusion of contraception in "vice."

7 Malthus cites data on high life expectancy accompanying slow population increase in several "Alpine parishes."

8 Because Asian and African townward migration is selectively male, reducing birth rates among newly arrived urban residents; second, because female "modern" employment tends to reduce total fertility (UN, 1987: ch. 9); third, because, in cities, contraception is less costly to supply and to obtain.

9 Although local shifts, from producing staples toward cash crops, normally bring some improvement in nutrition, even to poor at-risk groups.

10 Per unit of food grown, the leap from *P* to *R* (Figure 2) may leave labor input per unit of food output lower than at *N*, before population growth impels society rightward along the production function. If so, labor input might even fall per hectare and/or per adult-year.

11 Somewhere between Boserup and HRB lies Chayanov's account (1966). Here, extra people induce a new technology to produce more food (as in Boserup), not to switch to the more labor-intensive activity (as in HRB). However, because of the increased marginal utility of food as against leisure, the Chayanovian household in fact always reacts by selecting more labor-intensive methods (after diminishing returns have carried it some way along a given production function), as in HRB, and not by capital intensification that actually displaces labor, as is an option for Boserup.

12 Also, they derive from a different approach and "problematic." The HRB approach is the grandchild of Hicksian "induced *investment*" theory, where a farmer (or other businessman) chooses new investments that save on relatively scarce factors of production, for example, that replace land by labor in response to lower wage/rental ratios (due, e.g., to growth in population and therefore workforce). The child of such theory says that scarce-factor-saving, plentiful-factor-using *innovations* (i.e., new selections, from an existing menu of techniques, when replacing or expanding equipment) are also chosen. The grandchild claims that more abundant labor and scarcer land call forth *inventions* (by farmers, communities, or government researchers) to add new, more scarce-factor-saving techniques to the menu, both for farm technology (Hayami and Ruttan, 1971, 1985) and for rural organizations (Binswanger and Ruttan, 1978)—inventions responsive, after the workforce increases, to the felt need to replace costlier land by less costly labor.

13 This is especially odd in a very poor country such as Bangladesh [NAS, 1985, p. 140 and (citing work by Ghatak and Ingersent) pp. 29–30].

14 If we drop the assumption of uniform population growth, and assume growth is concentrated in households initially poor and with few prospects to expand effective demand for food, the begged BONMAL question is sharper still.

15 Will secure private ownership of land prevent the "soil-mining" response to population growth (NAS, 1985: 41–43)? Not if the extra demands created by population

growth—together with the rising cost of successive shifts in technology—render investments in raising soil fertility increasingly costly, pull up real interest rates, and raise the attractiveness of farming for maximum returns now, rather than for sustainable returns later.

16 Some Boserup responses—for example, migration to lower lying soils that require more intense HRB farming methods but offer higher yields (Binswanger and Pingali, 1985: 73)—can feasibly be researched by, and spread from, experimentally minded farming groups, like the Mende around Bumpeh, Sierra Leone (Richards, 1985). This may suffice to maintain food output per person while population grows at 0.5–1 percent yearly; at 2–4 percent, more formal scientific research is needed as well.

17 Extreme rural–urban inequality in most of sub-Saharan Africa, moreover, counterbalances its advantage over "Asia" in terms of smaller intra-rural inequality. If extra income goes to not-so-poor townspeople, then extra demands (political as well as economic) generated by extra people may not, after all, even in Africa, make themselves felt sufficiently as effective demands for technologies to grow more food.

18 Would it not suffice for the new methods to raise food output, and hence cut its price, so that more of it can be claimed even by a constant labor-income? It is not very likely: (1) Food prices would need to fall at a rate exceeding population growth. (2) Such prices are often largely determined internationally. (3) If food prices are restrained, employers can often respond by restraining the growth of money wage-rates. (4) If labor income stagnates, then farmers switch land away from poor people's foods, pushing their supply prices back up.

19 Salter's paradox arises because it telescopes the long, slow process of farmers' investment response to population growth. In fact, *during* this process, farmers are more interested in land-saving research (relative to labor-saving research) than before it. But the paradox does limit the time during which such pressures on researchers will be felt.

20 There are also Keynesian effects; such regional or sectoral shifts may mean a decreasingly labor-intensive structure of aggregate demand.

21 Or by direct food imports—normally financed by exports of less labor-intensive products than food.

22 And/or from a rise in fertility; recent evidence suggests that this usually happens in demographic accelerations (Dyson and Murphy, 1985).

23 Rising child/adult ratios increased this problem.

24 Such Boserup shifts without adequate labor absorption, and hence with threats to effective demand for food, appear to have happened in Britain in 1740–1830 (World Bank, 1984: 57) and may have been part of the reason why Corn Laws were "needed" to maintain food prices until 1847.

25 However, Arnold continues (in *Empedocles on Etna*): "But since life teems with ill, nurse no extravagant hope."

26 Strictly, more food availability. But extra net food imports were unwelcome to the prevailing consensus, which (1) gave priority to the import requirements of industrialization; (2) viewed less developed countries' export outlook with gloom; and (3) saw food aid as unreliable.

27 Several processes operate. As appropriate modern varieties raise welfare in large, poor families, fears for security in old age—a motive for producing many children—decline. So do expectations of, and new conceptions to replace, child deaths, as better nutrition reduces mortality. Also, as family incomes progress, children move from farm labor to school, so numerous children become a burden rather than a source of quick income. Finally, poor parents, as they acquire income, tend to have more education. Female education particularly is associated with reduced fertility.

28 It is also desirable that inputs purchased for cultivation of modern varieties should require payment—if this cannot be deferred—at times that permit financing from seasonal peak incomes normally accruing to poor people from their hired employment, remittances, or other sources likely to loom larger as person/land ratios increase.

29 This is because agricultural research may then also affect fertility in regions of migrant origin or destination, not just where the research activities are based.

References

Basu, D., R. Roy, and P. Nikhil. 1978. *Impact of Agricultural Development on Demographic Behavior.* Baroda: Operation Research Group.

Binswanger, H., and V. Ruttan. 1978. *Induced Innovation: Technology, Institutions and Development.* Baltimore: Johns Hopkins.

________, and P. Pingali. 1985. "Population growth and technical change in agriculture," in *Proceedings of the Fifth Agricultural Sector Symposium,* ed. E. Davis. Washington, D.C.: World Bank.

Birdsall, N. 1985. "A population perspective on agricultural development," in *Proceedings of the Fifth Agricultural Sector Symposium,* ed. E. Davis. Washington, D.C.: World Bank.

________. 1988. "Economic approaches to population growth," in *Handbook of Development Economics,* ed. H. Chenery and T. Srinivasan. Elsevier Science Publishers.

Booth, A., and R. Sundrum. 1984. *Labour Absorption in Agriculture.* New Delhi: Oxford University Press.

Boserup, E. 1965. *Conditions of Agricultural Growth.* London: Allen and Unwin.

________. 1981. *Population and Technology.* Oxford: Blackwell.

Breman, J. 1985. *Of Peasants, Migrants and Paupers.* New Delhi: Oxford University Press.

Cain, M. 1982. "Perspectives on family and fertility in developing countries," *Population Studies* 36, no. 2.

________, and K. Mozumder. 1980. "Labour market structure, child employment and reproductive behaviour in rural South Asia," ILO, World Food Programme, Population and Labour Policies Programme, Working Paper No. 87. Geneva.

________, and G. McNicoll. 1988. "Population growth and agrarian outcomes," in *Population, Food and Rural Development,* ed. R. Lee et al. Oxford: Clarendon Press.

Cassen, R. 1978. *India: Population, Economy, Society.* London: Macmillan.

Chadha, G. 1983. *Dynamics of Rural Transformation: A Case Study of the Punjab, 1950–80.* New Delhi: Centre for Regional Development, Jawaharlal Nehru University.

Chayanov, A. 1966. *Theory of Peasant Economy,* ed. D. Thorner, tr. B. Kerblay. Homewood: R. D. Irwin.

Chuta, E., and C. Liedholm. 1979. *Rural Non-Farm Employment: A Review of the State of the Art.* East Lansing: Michigan State University.

Cleland, J., and J. Hobcraft (eds.). 1985. *Reproductive Change in Developing Countries: Insights from the World Fertility Survey.* Oxford University Press.

Coleman, D., and R. Schofield (eds.). 1986. *The State of Population Theory: Forward from Malthus.* Oxford: Blackwell.

Connell, J., B. Dasgupta, R. Laishley, and M. Lipton. 1976. *Migration from Rural Areas: The Evidence from Village Studies.* Delhi: Oxford University Press.

Deane, P., and W. Cole. 1967. *British Economic Growth, 1688–1960.* Cambridge University Press.

de Janvry, A. 1978. "Social structures and biased technical change in Argentine agriculture," in Binswanger and Ruttan, 1978.

Delgado, C., and J. McIntyre. 1982. "Constraints on oxen cultivation in the Sahel," *American Journal of Agricultural Economics* 64, no. 2.

Dyson, T., and M. Murphy. 1985. "The onset of fertility transition," *Population and Development Review* 11, no. 3.

Easterlin, R. and E. Crimmins. 1985. *The Fertility Revolution: A Supply–Demand Analysis.* Chicago: University of Chicago Press.

Eltis, W. 1980. "Malthus's theory of effective demand and growth," *Oxford Economic Papers* 32, no. 1.

Evenson, R. 1988. "Population growth, infrastructure and real incomes in North India," in *Population, Food and Rural Development,* ed. R. Lee et al. Oxford: Clarendon Press.
Ford Foundation. 1959. *India's Foodgrain Crisis and Steps to Meet It.* New Delhi: Ministry of Food and Agriculture and Ministry of Community Development.
Geertz, C. 1963. *Agricultural Involution.* Berkeley: University of California Press.
Genicot, L. 1966. "Crisis: From the Middle Ages to modern times," in *The Cambridge History of Modern Europe I: The Agrarian Life of the Middle Ages,* ed. M. Postan. Cambridge University Press.
Godfrey, M. 1985. "Trade and exchange-rate policies in Sub-Saharan Africa," *Institute of Development Studies Bulletin* 16, no. 3.
Grabowski, R. 1979. "The implications of an induced innovation model," *Economic Development and Cultural Change* 27, no. 2.
Hajnal, J. 1982. "Two kinds of preindustrial household formation system," *Population and Development Review* 8, no. 3.
Hayami, Y., and V. Ruttan. 1971 and 1985. *Agricultural Development: An International Perspective.* Baltimore: Johns Hopkins.
Hicks, J. R. 1939. *Value and Capital.* Oxford University Press.
Jayasuriya, S., and R. Shand. 1986. "Technical change and labour allocation in Asian agriculture: Some emerging trends," *World Development* 14, no. 3.
Jodha, N. S. 1985a. "Market forces and erosion of common property resources," in *Agricultural Markets in the Semi-arid Tropics,* ed. M. von Oppen. Patancheru: ICRISAT.
________. 1985b. "Population growth and the decline of common property resources in Rajasthan, India," *Population and Development Review* 11, no. 2.
Kikuchi, M. and Y. Hayami. 1983. "New rice technology, intrarural migration, and institutional innovation in the Philippines," *Population and Development Review* 9, no. 2.
Kocher, J. 1984. "Income distribution and fertility," in Schutjer and Stokes, 1984.
Leaf, M. 1983. "The green revolution and cultural change in a Punjab village," *Economic Development and Cultural Change* 31: 227–270.
Lee, R. D. 1986. "Malthus and Boserup: A dynamic synthesis," in Coleman and Schofield, 1986.
Lipton, M. 1977. *Why Poor People Stay Poor.* Cambridge, Mass.: Harvard University Press.
________. 1982. "Rural development and the retention of the rural population," *Canadian Journal of Development Studies* 3, no. 1.
________. 1983. "Demography and poverty," World Bank Staff Working Paper no. 547.
________. 1983a. *Poverty, Undernutrition and Hunger,* World Bank Staff Working Papers.
________. 1985a. "The place of agricultural research in the development of sub-Saharan Africa," Discussion Paper No. 202. Brighton: Institute of Development Studies.
________. 1985b. "Coase's theorem vs. prisoners' dilemma: A case for democracy in LDCs," in *Economics and Democracy,* ed. R. Matthews. Macmillan.
________. 1987. "Agriculture and central physical grid infrastructure," in *Accelerating Food Production in Sub-Saharan Africa,* ed. J. Mellor et al. Baltimore: Johns Hopkins.
________. 1987a. "The limits of agricultural price policy: Which way at the World Bank?," *Development Policy Review* (June).
________, with R. Longhurst. 1989. *New Seeds and Poor People.* Baltimore: Johns Hopkins.
Malthus, T. R. 1830. "A summary view of the principle of population," abridged and revised from *Encyclopedia Britannica* (1824 supplement), in F. Osborn (ed.), *Three Essays on Population.* Mentor.
McGee, T. 1971. "The role of cities in Asian society," in *Urbanization and National Development,* ed. L. Jakobson and V. Prakash. Sage.
Mohan, R. 1982. "The effects of population growth, of the pattern of demand, and of technology on the process of urbanization," World Bank Staff Working Paper no. 520.
Nag, M., and N. Kak. 1984. "Demographic transition in a Punjab village," *Population and Development Review* 10, no. 4.
National Academy of Sciences, Committee on Population. 1985. *Population Growth and Economic Development: Policy Questions.* Washington, D.C.: National Academy Press.

Pingali, P., Y. Bigot, and H. Binswanger. 1987. *Agricultural Mechanization and the Evolution of Farming Systems in Sub-Saharan Africa.* Baltimore: Johns Hopkins.

Pryor, F., and S. Maurer. 1982. "On induced economic change in pre-capitalist societies," *Journal of Development Economics* 10, no. 3.

Quizon, J., and H. Binswanger. 1983. "Income distribution in agriculture: A unified approach," *American Journal of Agricultural Economics* 65, no. 13.

Ray, S. 1985. "Population pressure and agricultural intensification in Uttar Pradesh," *Indian Journal of Agricultural Economics* 40, no. 2.

Repetto, R. 1979. *Economic Inequality and Fertility in Developing Countries.* Baltimore: Johns Hopkins.

———, and T. Holmes. 1983. "The role of population in resource depletion in developing countries," *Population and Development Review* 9, no. 4.

Richards, P. 1985. *Indigenous Agricultural Revolution.* London: Cass.

Riches, N. 1967 (originally published 1937). *Agricultural Revolution in Norfolk.* London: Cass.

Robinson, W. and W. Schutjer. 1984. "Agricultural development and demographic change: A generalization of the Boserup model," *Economic Development and Cultural Change* 32, no. 2.

Rodgers, G. 1984. *Poverty and Population: Approaches and Evidence.* Geneva: ILO.

Rounce, N. 1949. *The Agriculture of the Cultivation Steppe.* Cape Town: Longmans.

Ruttan, V., et al. 1978. "Factor productivity and growth: A historical interpretation," in Binswanger and Ruttan, 1978.

Salter, W. E. 1966. "Productivity and technical change," Cambridge University, Department of Applied Economics, Monograph 6.

Schultz, T. P. 1981. *Economics of Population.* Reading, Mass.: Addison-Wesley.

Schutjer, W., and C. Stokes (eds.). 1984. *Rural Development and Human Fertility.* New York: Macmillan.

Sen, A. K. 1968. *Choice of Techniques.* 3rd ed. Oxford: Blackwell.

———. 1981. *Poverty and Famines.* Oxford: Clarendon Press.

———. 1986. "Food, economics and entitlements," *Lloyds Bank Review,* April.

Silver, A., and P. Grossen. 1980. "Rural development and urban-bound migration in Mexico," *Research Papers,* No. R-17, Resources for the Future Inc.

Singh, S. and J. Casterline. 1985. "The socio-economic determinants of fertility," in Cleland and Hobcraft, 1985.

Stoeckel, J., and A. K. Jain (eds.). 1986. *Fertility in Asia: Assessing the Impact of Development Projects.* New York: St. Martin's Press.

Stokes, C., and W. Schutjer. 1984. "Access to land and fertility in developing countries," in Schutjer and Stokes, 1984.

Stolnitz, G. 1983. "Urbanization and rural-to-urban migration in relation to LDC fertility." Bloomington: Fertility Determinants Group, Indiana University, mimeographed.

United Nations. 1980. *Patterns of Urban and Rural Population Growth,* Population Studies No. 68. New York.

———. 1985. *Estimates and Projections of Urban, Rural and City Populations, 1950–2025: The 1982 Assessment,* St/ESA/SER-R/58. New York.

———. 1987. *Fertility Behaviour in the Context of Development,* Population Studies No. 100. New York.

Weller, R. 1984. "The gainful employment of females and fertility," in Schutjer and Stokes, 1984.

Winegarden, C. 1985. "Can income redistribution reduce fertility?," in *Fertility in Developing Countries,* ed. G. Farooq and G. Simmons. New York: St. Martin's Press.

World Bank. 1984. *World Development Report 1984.* New York: Oxford University Press.

———. 1986. *World Development Report 1986.* New York: Oxford University Press.

———. 1986a. *Population Growth and Policies in sub-Saharan Africa.* Washington, D.C.

Institutional and Environmental Constraints to Agricultural Intensification

PRABHU L. PINGALI

ESTER BOSERUP HAS ARGUED THAT SOCIETIES across the world have responded to increasing land scarcity in a strikingly similar and predictable manner. These societies have moved from shifting cultivation (land-extensive) systems to sedentary (land-intensive) cultivation systems. This process of agricultural intensification is usually associated with the adoption of technologies for soil fertility maintenance, labor savings, and land investments. During the early stages of intensification these technologies, such as manure use, animal draft power, and the like, are usually farmer generated. Science- and industry-based technical change becomes important only under higher farming intensities and where populations are growing extremely rapidly.

While examples abound of the population and market demand–induced productivity growth through intensification, there are other examples where intensification has led to reduced labor productivity. Declining productivity has sometimes been accompanied by soil depletion, exhaustion, and even abandonment (National Research Council, 1986). The magnitude and severity of the problem depend on the rate of population growth, the extent of institutional rigidities, and the nature of the agroclimatic environment (climate, soil type, and slope characteristics of the region).

This essay reviews and synthesizes case study evidence from Africa and Asia on the institutional and environmental constraints to successful intensification. The following are identified as the major causes of the unsuccessful transition to sustainable permanent cultivation systems: (1) Institutional innovations, especially the evolution of long-term rights to land, although induced by population growth, often tend to occur at a slower pace. (2) Despite secure long-term tenure to land, free rider problems may prevent collective action for making watershed-level investments to prevent land degradation. (3) In marginal environments, the returns to preventive land investments may be low. And (4) government policies may prevent migration from marginal environments despite rapidly declining productivity levels.

The first part of this essay describes the population and market demand–induced process of agricultural intensification from shifting cultivation to permanent cultivation systems. Associated changes in technologies and institutions, especially the evolution of land rights, are also discussed. The second part of the essay provides a typology of the conditions under which soil fertility degradation and soil erosion occur, along the lines mentioned above.

Population density, market infrastructure, and the intensification of agricultural systems

Agricultural intensification is the movement from forest and bush fallow systems of cultivation to annual and multi-crop cultivation systems, whereby plots of land are cultivated one or more times per year. The existence of a positive correlation between population density and the intensity of land use has been demonstrated by Boserup (1965, 1981). She argues from the premise that during the neolithic period, forests covered a much larger part of the land surface than is the case today. The replacement of forests by bush and grassland was caused by (among other things) a reduction in fallow periods due to increasing population densities: "The invasion of forest and bush by grass is more likely to happen when an increasing population of long fallow cultivators cultivate the land with more and more frequent intervals" (Boserup, 1965: 20).

In addition to high population densities, intensive cultivation will also be observed in areas with better access to markets, provided soil conditions are suitable. Intensification occurs for two reasons: (1) higher prices and elastic demand for exportables imply that the marginal utility of effort increases, hence farmers in the region will begin cultivating larger areas; and (2) higher returns to labor encourage migration into areas from neighboring regions with higher transport costs. Carr (1982), Curtin et al. (1978), Iliffe (1979), and Gleave and White (1969) provide examples of intensification induced by improved transport facilities or by proximity to urban centers.

The foregoing discussion leads to the broad generalization that for given agroclimatic conditions, the evolution from shifting cultivation to sedentary/permanent cultivation is driven by population growth and by the higher returns to farming that arise when market infrastructure improves and farmgate prices increase. General support for this proposition has been provided in case studies by Clarke (1966), Gleave and White (1969), Bosehart (1973), Brown and Podolefsky (1976), Lagemann (1977), Ludwig (1968), Bradford (1946), Humphrey (1947), and Seavoy (1973a and b), among others. Cross-country empirical verification of Boserup's proposition was provided by Pingali et al. (1987) and Turner et al. (1977). Causes of population growth and concentration are not discussed in this essay. See Pingali et al. (1987) for a review of the causes of population concentration and consequently the determinants of agricultural intensification.

In using population densities as a basis for predicting the level of agricultural intensification, one ought to standardize arable land by soil quality and climate. Binswanger and Pingali (1988) provide a standardized population density measure: the number of people per million kilocalories of production potential. This measure is called the agroclimatic population density. Country-level estimates of potential calorie production at different technology levels were obtained from an FAO study on Land Resources for Populations of the Future (Higgins et al., 1982). When countries are ranked conventionally by population per square kilometer of agricultural land, Bangladesh comes first, India comes seventh, Kenya falls somewhere in the middle, and Niger is near the bottom. When ranked by agroclimatic population density, the ranking changes dramatically: Niger and Kenya are more densely populated than Bangladesh, and India ranks only twenty-ninth on the list. This explains why one observes intensive cultivation in some sparsely populated countries with poor market infrastructure—Niger and Botswana are examples.

Intensification-induced changes in cultivation techniques and labor use

As discussed by Boserup (1965) and Ruthenberg (1984), the total input per hectare on a given crop is positively correlated with the intensity of farming, holding technology constant. The increase in labor input is due to an increase in intensity with which certain tasks have to be performed (e.g., land preparation and weeding) and due to an increase in the number of operations performed (e.g., manuring, irrigation). A discussion of labor use across intensities of farming is provided below. Table 1 illustrates the increase in operations performed with the intensification of the farming system.

In the forest and bush fallow systems of cultivation, land clearing, planting, and harvesting are the major tasks performed. Fire is the most prevalent technique for land clearance. This form of land clearance, in addition to regenerating the soil, reduces weed growth. Land clearance by fire requires very low levels of labor input: 300 to 400 hours per hectare for forest fallow systems in Liberia and Côte d'Ivoire (Ruthenberg, 1984). The ground, being under tree cover, is soft and hence no further land preparation is required prior to sowing with a digging stick or a hand hoe. Such systems of cultivation require almost no weeding or cultivation, and the period between planting and harvesting is virtually task free.

As the fallow period becomes shorter and the land under fallow becomes grassy, fire can no longer be used for land clearance. Fire cannot destroy grass roots, hence grasses persist through the growing season. The intensive use of a hoe for land preparation becomes essential. Under short fallow systems of

TABLE 1 Comparison of operations and technology across farming systems

Operation	Forest fallow	Bush fallow	Short fallow	Annual cultivation	Multi-cropping
Land clearance	Fire	Fire	NA	NA	NA
Land preparation	No land preparation; digging sticks used to plant seeds	Land is loosened using hoes and digging sticks	Use of plow for preparing land	Animal-drawn plows and tractors	Animal-drawn plows and tractors
Manuring	Ash; household refuse for garden plots	Ash, burnt or unburnt leaves, other vegetable matter, and turf brought from surrounding bushland	Animal and human waste; green manuring; composts; silt from canals	Animal and human waste; composting; cultivation of green manure crops; chemical fertilizer	Animal and human waste; composting; cultivation of green manure crops; chemical fertilizer
Weeding	Minimal	Required as the length of fallow decreases	Weeding required during the growing season	Intensive weeding required during the growing season	Intensive weeding required during the growing season
Use of animals in farming	None	As length of fallow decreases, animal-drawn plows begin to appear	Plowing, transport, interculture	Plowing, transport, interculture, post-harvest tasks, irrigation	Plowing, transport, interculture, post-harvest tasks, irrigation
Seasonality of labor demand	None	None	Land preparation, weeding, and harvesting	Acute seasonal labor demand concentrated around the rainy season and harvest	Acute seasonal labor demand concentrated around land preparation, weeding, harvest, and post-harvest tasks
Fodder supply	None	Emergence of grazing land	Abundant free grazing land	Free grazing during fallow period; crop residues	Intensive fodder management and fodder crop production

NOTE: NA = not applicable.

cultivation, the need for early-season weeding and plant protection becomes pronounced. Also, manuring is required to complement fallow periods for maintaining soil fertility.

Permanent cultivation of land requires labor investments for irrigation, drainage, and leveling or terracing. It also requires the development of more evolved manuring techniques to restore soil fertility. Land preparation and intercropping and weeding become much more important tasks.

Intensification, therefore, leads to both an increase in agricultural employment and an increase in yields per hectare. On the other hand, in the absence of a switch to labor-saving technologies, one could anticipate rapid declines in labor productivity where population densities are increasing rapidly and/or where labor requirements for erosion control, irrigation, and fertility maintenance are high. In other words, labor input per hectare may increase at a faster rate than yield per hectare in the movement to more intensive systems of farming. Pingali and Binswanger (1984) found a significant positive increase in labor use and in yield per hectare with farming intensity, using data from 52 specific locations in Africa, Asia, and Latin America. They also found that labor productivity may not increase with intensification.

The consequences of declining labor productivity, in the absence of other income-earning opportunities, are: declining real wages, increasing land rents, and growing poverty and income inequality (Hayami and Ruttan, 1985; Lee, 1980). Where the land frontier still exists, declining productivity can be overcome by expanding into virgin lands. Where such opportunities are not available, declining trends in labor productivity may be reversed by the rapid adoption of modern seed-fertilizer technologies (Binswanger, 1986). At this stage, farmer-generated technical change is superseded by science-based technical change.

Agricultural intensification and soil preference

The intensification of agricultural systems is constrained by climatic and soil factors. For given agroclimatic conditions, the extent of intensification is conditional on the relative responsiveness of the soils to inputs associated with intensive production such as land improvements, manure, and fertilizers. The responsiveness to intensification is generally higher on soils with higher water- and nutrient-holding capacity. This is primarily because higher water-holding capacity implies lower drought risk. Water-holding capacity is higher the deeper the soils and the higher their clay content. It is low on shallow sandy soils.

For a given toposequence, soils on the upper slopes are relatively light and easy to work by hand, and tillage requirements are minimal. The clay content and hence the heaviness of the soils increases as one goes down the toposequence; consequently, power requirements for land preparation increase. Movement down the slope also reduces yield risks because of increased water retention capacity of the soils. The soils are heaviest in the depressions and marshes at the bottom of the toposequence. These bottomlands are often extremely hard to prepare by hand and are often impossible to cultivate in the absence of investments in water control and drainage. In choosing a place to cultivate, the farmer trades off between lesser labor requirements for land preparation and greater risk of drought. The extremely high labor requirements for capital investments and land preparation make the bottomlands the least preferred for cultivation under low population densities, and they are often found to be under fallow. As population densities increase, however, the bottomlands become intensively cultivated because of the relatively higher returns offered to labor and land investments, especially in rice cultivation. Also, as population densities increase, so does labor supply, making it possible to undertake such labor-intensive investments as irrigation and drainage.

Population pressure leads to a sharp reversal in preference (price) of different types of land. As population densities increase, one observes the cultivation of land that requires substantially higher labor input but that is also more responsive to the extra inputs. Case study evidence on the movement down the toposequence with population growth is provided by Sutton (1979) and Lagemann (1977) for Nigeria; Berg (1981) for Madagascar; Trapnell (1943) and Trapnell and Clothier (1937) for Zambia; Malcolm (1938), Brain (1980), and Collinson (1972) for Tanzania.

Soil-type differences across a toposequence could be micro-variations limited to a few hundred meters or a few kilometers, or they could be macro-variations such that entire regions are part of one toposequence level. For example, large parts of northeastern Thailand can be characterized as upper slopes, although there are, of course, depressions in northeast Thailand within micro toposequences. The central plains of Thailand are largely flood plains and have the characteristics of bottomlands and marshes.

The tradeoff between greater input of labor and risk of a lower yield varies with the agroclimatic zone, becoming less severe as humidity increases. Under arid conditions lower slopes and depressions are the only land that can be cultivated because only here is water retention capacity sufficient to sustain a crop at very low rainfall levels. This is the reason for the intensive cultivation systems of an oasis type that one observes in arid areas even under low population densities. Pockets of arid farming in primarily pastoral areas of Botswana are a good example of this phenomenon.

Under semi-arid conditions the midslopes are the first to be cultivated. As population densities increase, cultivation replaces grazing on the lower slopes and eventually in the depressions. Power sources for tillage are first used in the bottomlands, generally when population pressure makes these lands valuable for cultivation. The reversal of land preferences is quite dramatic. In the semi-arid zones of India, on the other hand, the depressions are intensively cultivated, usually with rice, using elaborate irrigation systems and mechanical power.

Yield risks due to low water availability are not a major problem in the subhumid and humid tropics, hence one finds cultivation starting at the upper slopes and gradually moving downward as population pressure increases. Other soil-related constraints to intensive field crop production in the humid tropics are discussed in the next section. At high population densities the swamps and depressions become the most important land sources for food production, often associated with extremely intensive rice production. One observes such labor-intensive rice production in South and Southeast Asia (Barker et al., 1985) and could expect the same for Africa as population densities increase.

Changes in land rights

It is generally asserted that the most prevalent organization of land rights under low population density is "communal property," where members of particular groups have general cultivation rights; that is, they are assured the right to cultivate a plot, but when they abandon it again to fallow, the cultivation right to that plot reverts to the group. On the other hand, areas of high population density typically have "private property" rights; that is, cultivators have a large array of rights to specific plots. The distinction between communal and private property, however, is a harmful oversimplification, as it focuses on a simplistic legal codification of land rights rather than on the process of "induced institutional innovation" by which societies move from general land rights to specific land rights through the gradual addition of one specific right after another. While there are enormous variations in how this process evolves and while political powers have often interfered with it or speeded it up, there are basic tendencies as described below.

With general land rights, cultivators typically own only the right to cultivate in a particular region. Cultivators acquire the right to use a specific plot and retain this right as long as they actually cultivate the plot; when the current cultivator departs—usually to leave the plot fallow—the use right to the plot reverts to the lineage. In land-abundant environments, outsiders are often welcome and general cultivation rights may thus be available to groups that are broadly defined to include almost everyone. When population densities increase, the groups become more narrowly defined and more people are regarded as outsiders who are excluded from the general cultivation rights. Noronha (1985), Hopkins (1973), and Miracle (1967) review the evidence on the evolution of land rights in precolonial Africa.

With the development of specific land rights, the cultivator can begin to assert certain rights in specific plots, starting with the right to resume cultivation of the plot after a period of fallow. At a later stage he asserts, and will receive, the right to assign the plot to an heir or temporarily to another member of the same group; the use right in the plot no longer reverts to the group.

Rights to graze cattle on fallow plots or on stubble shift from all members of the community to the individual. Eventually individuals acquire the right to sell land to other group members and even to outsiders. Often such changes occur at different times in different regions or subregions of a country. Hopkins (1973) summarizes the evolution of land rights as follows: "[W]here population was dense and the period of fallow short or non-existent, as was the case with permanent cultivation, then claims on individual plots became stronger, and in these circumstances freehold tenure, pledging and sale of land were recognized in customary law" (p. 38).

The transition to specific land rights widens incentives to undertake investments in specific plots, investments that are required for the intensification of production and the preservation of soil fertility (Noronha, 1985). Formal land ownership as characterized by the possession of title also helps the farmer in acquiring credit for making the necessary investments in the land. Evidence on land ownership and investment is provided by Feder and Onchan (1987) and Chalamwong and Feder (1986) for Thailand, while Noronha (1985) reviews the African case study evidence.

Population growth is not the only process that moves societies toward more specific land rights. Any of the other factors that cause intensification have the same effect. We therefore often see large variations in the number of specific rights allocated to individuals within different parts of the same country, depending on the degree of land fertility and market access.

Hayami and Ruttan (1985) review the evolution of property rights to land in Japan, Thailand, and the Philippines. They find population growth to be instrumental in the process, although expanded possibilities for trade also play a role. Binswanger and Pingali (1988) provide field data from ten countries in sub-Saharan Africa on the evolution of land rights. They find a positive association between the degree of privatization of agricultural land and population density. They also find the degree of privatization of land to be positively related to the level of market infrastructure. The degree of privatization of land was determined by using a series of questions on how outsiders could acquire land in a particular society.

When does the evolution in land rights break down? Population and market infrastructure–induced evolution in land rights is sometimes circumvented by government policies, often with detrimental effects as discussed in the next section. In colonial Africa, expropriation of fallow lands for European settler use is an example of such breakdown (Noronha, 1985; Curtin et al., 1978; and Hopkins, 1973). Contemporary Africa and Asia offer several examples of socialist governments that attempted to introduce collectivized agricultural systems and accordingly abolished individual rights to land (e.g. Iliffe, 1979 for Tanzania; and Hung, 1977 for Vietnam). The normal process of evolution can also break down when unusually high population growth rates are experienced, as caused by a sudden increase in migration into the area, and hence an unprecedented increase in the opportunity cost of fallow land.

Property rights and the soil degradation problem

Soil degradation is defined here as soil erosion and/or a decline in soil fertility associated with intensive cultivation, overgrazing, and deforestation. Soil degradation is not a universal problem. First, the threat of soil degradation varies by soil type, temperature, rainfall regimes, and slope. Second, appropriate land use and land investments can prevent the problem, even on high-risk soils. Therefore, degradation problems are mainly restricted to areas where the rate of return to preventive land investments (such as fertility management and terracing) is low. Incentives for corrective land investments depend on the relative land endowments of the region and on whether potentially degradable land is under private or communal control.

Soil fertility depletion

Let us start with farmer incentives for maintaining soil fertility. The preceding section argued that agricultural intensification leads to a substitution of fallow periods with increasingly intensive manuring systems for maintaining soil fertility. The long-term process described above depends on the farmer's incentive for protecting the long-term productivity of the land. These incentives would be high if the farmer had secure long-term rights to currently cultivated and fallow land. Where these rights are not certain, one would observe shortened fallow periods without a commensurate increase in labor for fertility management. The ultimate outcome of this soil mining process is a low yield equilibrium and an outmigration of population from the area. For case study evidence on soil mining and migration see: for Nigeria, Mortimore (1971), Grossman (1972), Morgan (1955), Mabogunje (1959), and Lagemann (1977); for Kenya, Mitchell (1947); for Tanzania, Malcolm (1938) and Rounce (1949); for Uganda, Tothill (1940); for Zambia, Johnson (1956) and Allan et al. (1948); for Botswana, Schapera (1943); for Nepal, Sainju and Ram (1981); for the Philippines, Fujisaka and Capistrano (1985), Capistrano and Fujisaka (1984), and ACIAR (1984); for Indonesia, Conway et al. (1985).

Rights to fallow lands become increasingly uncertain in societies experiencing rapid population growth (through natural increase or through migration) and in societies with a persistent fear of land expropriation (by colonial authorities or socialist governments, for instance).

In the first case, rapid and unprecedented increase in population pressure, perhaps caused by migration from other areas, leads to increased demand for land intensification while farmer-induced technical change in fertility management may lag behind. In the second case, fear of expropriation of fallow land by the government has resulted in the intensification of the agricultural system long before population densities warrant it. The result is continuous cultivation of all land in the society at low yield levels. Noronha (1985) presents a definitive review of this evolutionary pattern for colonial Africa, where the fear of fallow land expropriation was particularly high, especially in areas with European settlers.

In areas where land productivity declines with agricultural intensification and a low yield equilibrium sets in, seasonal or permanent migration is used to sustain farmer livelihood. Morgan (1955), Mabogunje (1959), and Udo (1969) report on farmers in Nigeria who seasonally migrate to cultivate lands in other areas as tenant farmers. Seasonal migrants also work as farm laborers or as urban laborers (see Lagemann, 1977; Sainju and Ram, 1981; and Schapera, 1943, among others).

In areas where uncultivated land still exists, permanent migration to a new area can be observed after the current area has been degraded. The Philippine upland areas provide a striking example of this phenomenon. Upland farmers follow loggers into cleared timberland areas, renting farm parcels from the nominal "owners" (the logging companies) or the "pseudo landlords" from nearby towns. Since the migrants do not have a secure long-term tenure on the land, they tend to maximize short-term returns and follow the logging companies to the next site once degradation has set in. Those remaining behind acquire usufruct rights to the land, primarily by paying land taxes, and a more sustainable land management system evolves (ACIAR, 1984; Fujisaka and Capistrano, 1985). Other examples of landless households in Peru, Thailand, the Philippines, and India are found in Postel (1984).

In cases such as the above, where farmers do not have long-term rights to land, the private rate of discount is higher than the social rate of discount, hence private investments made in sustaining land productivity will be lower than those society would consider optimal. The reason is that those people who contemplate making investments in soil conservation will not reap the full benefits and therefore underinvest (National Research Council, 1986). Dumsday and Flinn (1977) show that, for the semi-arid tropics of West Africa, the level of soil degradation is directly related to the private discount rate.

Soil erosion

Unsuccessful attempts at sedentarization of the agricultural system could also result in an increased incidence of soil erosion on some sloping lands. The previous section addressed the process of agricultural intensification along the toposequence, concluding that the response to intensification is highest on the heavier soils of the lower slopes and valley bottoms. However, the very forces of intensification that lead to land investments lower down the toposequence may cause higher levels of erosion on the upper parts of the toposequence. The mechanics of this process and the conditions under which this problem can be corrected are briefly described below.

On the mid-slopes, the return to land investments is often sufficient to induce, where necessary, private investment in erosion control. Ridges, bunds, and terraces are a few examples of farmer-generated erosion control structures. Successful examples of erosion control investments are generally associated with secure long-term rights to land. See Roche (1988) and Soemarwoto and Soemarwoto (1984) for West Java; Mortimore (1971) for Nigeria; Brain (1980), Collinson (1972), and Rounce (1949) for Tanzania. The probability that erosion control investments will not be made is higher in areas without well-defined property rights and where group action is required for making the investments. Fujisaka and Capistrano (1985) report on the increased incidence of erosion caused by the "get it while you can" attitude of Philippine farmers in Southern Luzon, who cultivated upland areas without a secure long-term tenure to the land. The relationship between land tenure insecurity and the reluctance of farmers to adopt appropriate practices requiring long-term investment is also documented by ACIAR (1984) for the Philippines; Soemarwoto and Soemarwoto (1984) for Indonesia; Schapera (1943) for Botswana; Anthony and Uchendu (1970) for Zambia; and Wilson (1977) for Sudan. Feder and Noronha (1987) review the African case study material on the relationship between land rights and investments.

The importance of tenure security for sustainable land management can be summarized by the following distinctly different examples. In 1947 colonial attempts at terracing upland areas in Kenya that were being subject to high levels of erosion were unsuccessful because the Kenyan farmers feared that terraced fields would be taken over by the European settlers (Mitchell, 1947). In Java the transition from a shifting cultivation system with decreasing fallow periods and ill-defined land rights to an agroforestry system with very clear rights to land resulted in high levels of investments for preventing soil erosion (Soemarwoto and Soemarwoto, 1984). In Vietnam, recent policy reforms that provide farmers long-term leases to land have resulted in an increase in land investments for preventing degradation and sustaining productivity (Pingali and Vo Tong Xuan, 1989). While in the early stages of development de facto possession may not imply a substantial uncertainty regarding the ability to utilize the land in the future, uncertainty tends to increase with population growth, increased commercialization, and the higher income potential brought about by new technology. One would expect the level of land investment to be negatively related to the level of uncertainty regarding tenure (Feder and Noronha, 1987).

Despite secure long-term rights to land one would expect erosion and degradation to persist in areas where collective action is required for watershed-level investments. The free rider problems associated with such collective action are self-evident. Group action for making watershed-level investments in erosion control would be possible only where farmers have an incentive to cooperate or where they are coerced to do so. Incentives for cooperation will be higher in relatively closed communities under severe land pressure. Terraces in West Java are an example of such cooperation (Soemarwoto and Soemarwoto, 1984). The colonial period in Africa provides several examples of collective action through coercion. Johnson (1956) and Mitchell (1947) provide examples from Zambia and Kenya. Both studies concluded that erosion continued despite the attempted collective action.

Erosion problems are usually most visible and persistent on the upper slopes of the toposequence. Several mechanisms are involved. In regions with small endowments of low-lying areas, the cultivated area will expand into the shallow soils of the upper slopes that were previously used for pasture and/or communal forestry. However, yields are low, as are the payoffs to erosion control or intensive inputs. Unlike the case of the mid-slopes, and despite privatization, individual initiative cannot arrest the process of degradation because of the low return to control measures. A low-yield equilibrium will emerge in association with seasonal migration for farm or nonfarm employment in other areas (see Cruz, 1986 for a review of Philippine cases of migration to and the exploitation/degradation of the upper slope lands). Where abundant low-lying or mid-slope land is available, the upper slopes will continue to be used as pastures or forests.

The problem of the upper slopes is often one of controlling access for nonagricultural use, primarily commercial logging. Across Southeast Asia large tracts of upper slope land cleared by loggers are exposed to erosion problems. Where privatization is infeasible or undesirable, community action or government intervention is required both for control of access and for investments in erosion control and forest replanting. Difficult dilemmas frequently arise as traditional users are threatened with a loss of their use rights. Societies are often incapable of solving these ownership or access control problems before serious damage has occurred.

Agroclimatic differentiation in susceptibility to erosion

The extent of the degradation problem varies according to agroclimatic zone. The general principle here is that the rate of return to land investments varies by agroclimatic zones, and the problem is highest in those zones where the returns to such investments are the lowest. Pingali et al. (1987) provide a detailed discussion of the relationship between the agroclimatic zone and the land use and land intensification process. Despite privatization, the returns to protective investments are lowest in the arid fringe areas; where migration to more favorable lands is not possible or is prevented, arid uplands tend to display high levels of degradation. In the semi-arid tropics, because of privatization and moderate returns to land investments in the mid-slopes, soil degradation is reasonably well controlled. The problem areas in the semi-arid zones are the upper slopes, which, depending on the region's land endowments, could be used for grazing, forestry, or low-yield cropping. As discussed earlier, these upper slopes could suffer from degradation either through an inability to control access or through permanent cultivation on extremely marginal land.

Upland areas in the humid tropical zone face special problems for agricultural intensification. Continuous field crop production on most soils of this zone leads to rapid leaching, soil acidification, and/or erosion resulting in declines in soil fertility and yields (Allan, 1965; National Research Council, 1982; Ruthenberg, 1984; FAO, 1984; Kang and Juo, 1981; and Sanchez et al., 1982). Ruddle and Manshard (1980) provide a comprehensive survey of the human impact on the humid tropical forests across the world. The experience of Zaire is illustrative of the problem. In 1960, 150 acres of Zairean tropical forests were cut down, burned, and uprooted. The soil was deep-plowed and cultivated in a manner similar to the temperate zone. Yields dropped dramatically by the second year because deep-plowing impaired soil structure and the exposed soil was severely leached. The entire area was abandoned after only a few years of cultivation, and ten years later the area was only thinly wooded (Ruthenberg, 1984). There are several other examples of humid tropical forests that were cleared for intensive cultivation and subsequently abandoned in a degraded state—for Nigeria and Côte d'Ivoire see FAO (1984); for Ghana see Moorman and Greenland (1980); for the Philippines see Sajise (1986); and for Peru, Sanchez et al. (1982).

What are the economically viable options for increased food crop production in the humid tropics in the face of increasing population densities? In general, a system of farming that closely mimics the dense natural vegetation of the humid forests is what will work in the long run. This is exactly what the shifting cultivators tried to achieve in their multi-storied cultivation of cleared tropical forest lands (Ruddle and Manshard, 1980). For instance, a system of intercropping short- and medium-term crops under a protective cover of trees is feasible, since the ground would be covered for most of the year and therefore the detrimental effects to the soil would be minimized (Okigbo, 1974; Sanchez et al., 1983; Sanchez and Salinas, 1981). (For a detailed discussion of the technological options for the humid tropics see Pingali et al., 1987, chapter 10.)

In the face of rising populations, secure long-term land rights are very important for the sustained development of the humid tropics. Without such rights to land, farmers do not have the incentive to plant trees and perennials for protecting the soil against leaching. As Noronha (1985) points out, in societies where the ownership of trees is tantamount to the ownership of the land the trees are on, farmers temporarily occupying the land are prevented from planting trees or perennial crops. In the humid tropics, where uncertain ownership or tenure prevents the planting of trees, degradation is most likely to occur.

Conclusions

While most societies have responded to population growth and/or increased market demand by intensifying their agricultural systems in remarkably similar and predictable patterns, there are several examples of failures. In the latter cases, increasing land scarcity has been associated with soil fertility depletion, soil erosion, declining agricultural incomes, and migration. This review has found that failures to achieve a sustainable intensive cultivation system are equally similar and predictable. Failures in agricultural intensification can be attributed to one or more of the following reasons: persistence of uncertain long-term rights to land; encroachment of cultivation onto marginal lands; and collective effort required for watershed-level protective investments.

Farmer incentives for regenerating soil fertility and for making erosion-prevention investments are high if secure long-term land rights exist, both to currently cultivated land and to fallow land. Where farmers do not have long-term rights to land, the private rate of discount is higher than the social rate of discount, hence private investments in land productivity will be lower than the societal optimum. In societies with slowly and steadily growing populations, property rights to land evolved systematically, being induced by growing land scarcity. However, the evolution to secure land rights has not been successful in societies experiencing rapid population growth (through natural increase or through migration), nor in societies where this induced institutional change is circumvented by government policies (e.g., land expropriation by colonial authorities or socialist governments).

Even with secure long-term rights to land, one would expect erosion and land degradation to persist in areas where collective action is required for watershed-level investments. The free rider problems associated with such collective action are self-evident.

Given secure rights to land, the rate of return to land investments varies by agroclimatic zone and soil type: degradation problems are most severe where the returns to land investments are lowest. The arid fringe areas, upper slopes in the semi-arid and the humid zones, and shallow sandy soils exhibit the highest levels of erosion, other things being equal. Degradation problems are likely to be most severe in regions with relatively large endowments of marginal lands.

Degradation of marginal lands is also likely to be severe where institutional arrangements are not available for controlling access to grazing and forestry. Where privatization is infeasible or undesirable, community action or government intervention is required both for control of access and for investments in erosion control and forest replanting. Difficult dilemmas frequently arise as traditional users are threatened with a loss of their use rights. Societies are often incapable of solving these problems of ownership or controlled access before serious damage has occurred.

The potential problems of declining labor productivity and environmental degradation are not problems of levels of population densities. Given sufficient time, it is likely that a combination of farmer innovations, savings, and the development of research facilities and institutions for dealing with soil degradation issues will be able to accommodate much more than the current population in most countries, especially in many of the less densely populated ones. However, if all these changes are required quickly and simultaneously because of high population growth rates, they may emerge at too slow a pace to prevent a decline in human welfare.

References

Allan, W. 1965. *The African Husbandman.* Edinburgh: Oliver and Boyd.

________, M. Gluckman, D. U. Peters, and C. G. Trapnell. 1948. *Landholding and Land Usage Among the Plateau Tonga of Mazabuka District.* Cape Town: Oxford University Press.

Anthony, K. R. M., and V. C. Uchendu. 1970. "Agricultural change in Mazabuka District, Zambia," *Food Research Institute Studies* 9: 215–267.

Australian Centre for International Agricultural Research (ACIAR). 1984. *Soil Erosion and Management.* Proceedings of a Workshop held at PCARRD, Los Baños, Philippines, 3–5 December.

Barker, R., R. W. Herdt, and B. Rose. 1985. *The Rice Economy of Asia.* Washington, D.C.: Resources for the Future, in cooperation with the International Rice Research Institute.

Berg, Gerald M. 1981. "Riziculture and the founding of monarchy in Imerina," *Journal of African History* 22: 289–308.

Binswanger, Hans P. 1986. "Evaluating research system performance and targeting research in land-abundant areas of sub-Saharan Africa," *World Development* 14, no. 4.

________, and P. L. Pingali. 1987. "The evolution of farming systems and agricultural technology in sub-Saharan Africa," in *Policy for Agricultural Research,* ed. V. W. Ruttan and C. E. Pray. Boulder: Westview Press.

________, and P. L. Pingali. 1988. "Technological priorities for farming in sub-Saharan Africa," *World Bank Research Observer* 3, no. 1: 81–98.

Bosehart, H. W. 1973. "Cultivation intensity, settlement patterns, and homestead forms among the Matengo of Tanzania," *Ethnology* 12: 57–74.

Boserup, Ester. 1965. *The Conditions of Agricultural Growth.* London: Allen and Unwin.

________. 1981. *Population and Technological Change: A Study of Long-Term Trends.* Chicago: University of Chicago Press.

Bradford, E. L. 1946. "African mixed farming economics as applied to Bukura, Nyanza Province, Kenya," *The East African Agricultural Journal*: 74–83.

Brain, J. L. 1980. "The Uluguru land usage scheme: Success and failure," *Journal of Developing Areas* 14 (January): 175–190.

Brown, P., and A. Podolefsky. 1976. "Population density, agricultural intensity, land tenure, and group size in the New Guinea Highlands," *Ethnology* 15, no. 3: 211–238.

Capistrano, D., and S. Fujisaka. 1984. "Tenure, technology, and productivity of agroforestry schemes," Philippine Institute of Development Studies Paper 84–6.

Carr, Stephen. 1982. "The impact of government intervention on smallholder development in north and east Uganda," Agriculture Development Unit, Occasional Paper No. 5, School of Rural Economics, Wye College.

Chalamwong, Y., and G. Feder. 1986. "Land ownership security and land values in rural Thailand." Washington, DC: World Bank Staff Working Paper No. 790.

Clarke, W. C. 1966. "From extensive to intensive shifting cultivation: A succession from New Guinea," *Ethnology* 5: 347–359.

Collinson, M. P. 1972. "The economic characteristics of the Sukuma farming systems," Economic Research Bureau, Paper No. 72.5, University of Dar Es Salaam.

Conway, G. R., et al. 1985. "The critical uplands of eastern Java: An agroecosystems analysis," Kelompok Penelitian Agro-Ekosistem, Agency for Agricultural Research and Development, Republic of Indonesia.

Cruz, Concepcion J. 1986. "Population pressure and migration in Philippine upland communities," in *Man, Agriculture and the Tropical Forest: Change and Development in the Philippine*

Uplands, ed. J. S. Fujisaka et al. Bangkok, Thailand: Winrock International Institute for Agricultural Development, pp. 87–118.

Curtin, P., S. Feierman, L. Thompson, and J. Vansina. 1978. *African History*. Boston: Little, Brown.

Dumsday, R. G., and J. C. Flinn. 1977. "Evaluating systems of soil conservation through bioeconomic modelling," in *Soil Conservation and Management in the Humid Tropics*, ed. R. Lal and D. J. Greenland. New York: John Wiley and Sons.

Food and Agriculture Organization (FAO). 1984. *Institutional Aspects of Shifting Cultivation in Africa*. Rome: Human Resources, Institutions and Agrarian Reform Division.

Feder G., and R. Noronha. 1987. "Land rights systems and agricultural development in sub-Saharan Africa," Discussion Paper, Report No. ARU 64, World Bank.

———, and T. Onchan. 1987. "Land ownership security and farm investment," *American Journal of Agricultural Economics* 69, no. 1: 311–320.

Fujisaka, S., and D. A. Capistrano. 1985. "Pioneer shifting cultivation in Calminoe, Philippines: Sustainability or degradation from changing human–ecosystem interactions," Working Paper, East-West Center, Honolulu.

Gleave, M. B., and H. P. White. 1969. "Population density and agricultural systems in West Africa," in *Environment and Land Use in West Africa*, ed. M. F. Thomas and G. W. Whittington. London: Methuen.

Grossman, David. 1972. "The roots of the practice of migratory tenant farming: The case of Nikeland in Eastern Nigeria," *The Journal of Developing Areas* 6: 163–184.

Hayami, Y., and V. W. Ruttan. 1985. *Agricultural Development: An International Perspective*. Baltimore: Johns Hopkins University Press.

Higgins, G. M., et al. 1982. *Potential Population Supporting Capacities of Lands in the Developing World*. Technical Report of Project INT/75/P13. Rome: Food and Agriculture Organization, United Nations Fund for Population Activities, and International Institute for Applied Systems Analysis.

Hopkins, A. G. 1973. *An Economic History of West Africa*. New York: Columbia University Press.

Humphrey, Norman. 1947. *The Liguru and the Land*. Nairobi: Colony and Protectorate of Kenya.

Hung, Nguyen Tien. 1977. *Economic Development of Socialist Vietnam, 1955–80*. New York: Praeger Special Studies.

Iliffe, John. 1979. *A Modern History of Tanganyika*. Cambridge University Press.

Johnson, C. E. 1956. *African Farming Improvement in the Plateau Tonga Maize Areas of Northern Rhodesia*. Agricultural Bulletin No. 11. Lusaka: Government of Northern Rhodesia, Department of Agriculture.

Kang, B. T., and S. R. Juo. 1981. "Management of low activity clay soils in Tropical Africa for food crop production," paper presented at the 4th International Soil Classification Workshop, Kigali, Rwanda, 2–12 June.

Lagemann, Johannes. 1977. *Traditional African Farming Systems in Eastern Nigeria*. Munich: Weltforum Verlag.

Lee, R. 1980. "An historical perspective on economic aspects of the population explosion: The case of pre-industrial England," in *Population and Economic Change in Developing Countries*, ed. Richard A. Easterlin. Chicago: University of Chicago Press, pp. 517–556.

Ludwig, H. D. 1968. "Permanent farming on the Ukara," in *Smallholder Farming and Smallholder Development in Tanzania*, ed. Hans Ruthenberg. Munich: Weltforum Verlag.

Mabogunje, M. A. 1959. "Rice cultivation in southern Nigeria," *Nigerian Geographical Journal* 2, no. 2: 59–69.

Malcolm, D. W. 1938. *Report on Land Utilization in Sukuma*. Tanganyika: Department of Agriculture.

Miracle, Mervin. 1967. *Agriculture in the Congo Basin*. Madison: University of Wisconsin Press.

Mitchell, Philipp. 1947. *The Agrarian Problem in Kenya*. Nairobi: Colony and Protectorate of Kenya.

Moorman, F. R., and D. J. Greenland. 1980. "Major production systems related to soil properties in humid Tropical Africa," in *Priorities for Alleviating Soil-Related Constraints to Food*

Production in the Tropics. Los Baños, Laguna, Philippines: International Rice Research Institute.

Morgan, W. B. 1955. "Farming practice, settlement pattern and population density in south-eastern Nigeria," *The Geographical Journal* 121, part 3: 320–333.

Mortimore, Michael J. 1971. "Population densities and systems of agricultural land-use in northern Nigeria," *The Nigerian Geographical Journal* 14 (June): 3–15.

National Research Council. 1982. *Ecological Aspects of Development in the Humid Tropics*. Committee on Selected Biological Problems in the Humid Tropics. Washington, D.C.: National Academy Press.

________. 1986. *Population Growth and Economic Development: Policy Questions*. Washington, D.C.: National Academy Press.

Noronha, Raymond. 1985. "A review of the literature on land tenure systems in sub-Saharan Africa," Discussion Paper, Report No. ARU 43, World Bank.

Okigbo, B. N. 1974. *Fitting Research to Farming Systems*. Ibadan, Nigeria: International Institute of Tropical Agriculture.

Pingali, P. L., and H. P. Binswanger. 1984. *Population Density and Agricultural Intensification: A Study of the Evolution of Technologies in Tropical Agriculture*. Agriculture and Rural Development Department, Research Unit, Report No. ARU 22. Washington, D.C.: World Bank.

________, Y. Bigot, and H. P. Binswanger. 1987. *Agricultural Mechanization and the Evolution of Farming Systems in Sub-Saharan Africa*. Baltimore: John Hopkins Press.

________, and Vo Tong Xuan. 1989. "Vietnam: De-collectivization and rice productivity growth," paper presented at a special symposium on "Priorities for Rejuvenating Vietnam's Agriculture" held at the 1989 American Agricultural Economics Association annual meeting in Baton Rouge, Louisiana, 30 July–2 August. Agricultural Economics Department Paper No. 89–16, International Rice Research Institute, Los Baños, Philippines.

Postel, S. 1984. "Protection forests," in L. R. Brown et al., *State of the World 1984: A Worldwatch Institute Report on Progress Toward a Sustainable Society*. New York: Norton, pp. 74–94.

Roche, Frederick C. 1988. "Java's critical uplands: Is sustainable development possible?" *Food Research Institute Studies* 21, no. 1.

Rounce, N. V. 1949. *The Agriculture of the Cultivation Steppe*. Cape Town: Longmans, Green.

Ruddle, K., and W. Manshard. 1980. *Renewable Natural Resources and the Environment: Pressing Problems in the Developing World*. Published for the United Nations University by Tycooly International Publishing Limited, Dublin.

Ruthenberg, Hans. 1984. *Farming Systems in the Tropics*, 4th edition. Oxford: Clarendon Press.

Sainju, M. M., and B. K. C. Ram. 1981. *Hill Labour Migration: Issues and Problems*. Proceedings of the Seminar on Nepal's Experience in Hill Agricultural Development, 30 March–3 April.

Sajise, Percy. 1986. "The changing upland landscape," in *Man, Agriculture and the Tropical Forest: Change and Development in the Philippine Uplands*, ed. J. S. S. Fujisaka et al. Bangkok, Thailand: Winrock International Institute for Agricultural Development, pp. 13–41.

Sanchez, P. A., and J. G. Salinas. 1981. "Low-input technology for managing oxisols and ultisols in Tropical America," *Advances in Agronomy* 34: 279–406.

________, D. E. Bandy, J. Hugo Villachica, and J. J. Nicholaides. 1982. "Amazon basin soils: Management for continuous crop production," *Science* 216: 821–827.

________, J. H. Villachica, and D. E. Bandy. 1983. "Soil fertility dynamics after clearing a tropical rainforest in Peru," *Soil Science Society of America Journal* 47, no. 6: 1171–1178.

Schapera, I. 1943. *Native Land Tenure in the Bechuanaland Protectorate*. Capetown: Lovedale Press.

Seavoy, R. E. 1973a. "The transition to continuous rice cultivation in Kalimantan," *Annals of the Association of American Geographers* 63: 218–225.

________. 1973b. "The shading cycle in shifting cultivation," *Annals of the Association of American Geographers* 63: 522–528.

Soemarwoto, O., and I. Soemarwoto. 1984. "The Javanese rural ecosystem," in *An Introduction to Human Ecology Research on Agricultural Systems in Southeast Asia*, ed. T. A. Rambo and P. E. Sajise. University of the Philippines, Los Baños, Laguna, pp. 254–287.

Sutton, J. E. G. 1979. "Towards a less orthodox history of Hausaland," *Journal of African History* 20: 179–201.

Tothill, J. D. 1940. *Agriculture in Uganda*. London: Oxford University Press.

Trapnell, C. G. 1943. *Soils, Vegetation and Agriculture of North Eastern Rhodesia*. Lusaka: Government of Northern Rhodesia.

———, and J. N. Clothier. 1937. *The Soils, Vegetation and Agricultural Systems of North Western Rhodesia*. Lusaka: Government of Northern Rhodesia.

Turner, B. L., R. Q. Hanham, and A. V. Portararo. 1977. "Population pressure and agricultural intensity," *Annals of the Association of American Geographers* 67, no. 3: 384–396.

Udo, R. K. 1969. "The changing economy of the lower Niger Valley," *Nigerian Geographical Journal* 12, nos. 1 and 2: 113–124.

Wilson, R. T. 1977. "Temporal changes in livestock numbers and patterns of transhumance in southern Darfur, Sudan," *The Journal of Developing Areas* 11 (July): 493–508.

Depletion of Common Property Resources in India: Micro-Level Evidence

N. S. JODHA

A WELL-KNOWN CONSEQUENCE OF RAPID POPULATION GROWTH in developing countries is increased pressure on land, leading to over-exploitation and degradation of the natural resource base of agriculture. The problem is greater in the low-productivity, high-risk environments such as dry tropical regions. In these regions not only is nature's bio-mass productivity low, but recovery speed of once-damaged land and vegetative resources is slow. Hence, in the face of rising population pressure, these areas run a high risk of permanent ecological degradation (UNCOD, 1977). Furthermore, these regions are among the poorest in the Third World (World Bank, 1975), and most of their inhabitants subsist on traditional agriculture. The social and economic consequences of population growth in these regions are, therefore, primarily manifested through strains and stresses faced by traditional farming systems, including the management patterns of the fragile resource base sustaining them. This essay illustrates the situation with the help of a case study of conditions in India. The focus is on rural common property resources, which have historically been an important source of collective sustenance of the people, especially the rural poor. Using micro-level data from various dry tropical parts of India, the essay narrates the process of shrinkage and degradation of common property resources.

Statement of the problem

Rural common property resources (CPRs) are broadly defined as resources to which all members of an identifiable community have inalienable use rights. In the Indian context CPRs include community pastures, community forests, waste lands, common dumping and threshing grounds, watershed drainages, village ponds, and rivers and rivulets as well as their banks and beds. The first three resources are particularly important because of their large area and their contributions to people's sustenance in dry regions. Historically, the maintenance of these resources reflected people's concern for their collective sustenance and for protection of fragile land resources by keeping them under grass, bush, or trees.[1] CPRs often helped generate multiple earning options and the flexibility required by farmers to adjust their farming systems to variable agroclimatic conditions (Jodha, 1986b). The changing situation of CPRs in the dry tropics must be viewed in the context of traditional farming systems in these regions.

Traditional farming systems in dry tropical regions of the developing world are currently passing through a transitional phase (McCown et al., 1979; Sandford, 1983; Livingstone, 1984; Jodha, 1986b). Having evolved during times of low population pressure and absence of rapid technical and institutional interventions, such systems were land-extensive.[2]

In the face of increased population pressure, increased integration of agriculture into the monetized market economy, and a variety of external interventions, however, traditional resource-extensive (i.e., land-extensive) farming is becoming less feasible. As elsewhere in agriculture (Boserup, 1965), the strong trend is toward increased intensity of use of land resources. This is reflected in ploughing of low-productivity submarginal lands, reduced duration and frequency of fallow, and emphasis on short duration or multiple cropping. In India, community pastures, forests, and waste lands—the major focus of extensive resource use in the past—now bear the burden of land use intensification. It happens in several ways. First, CPRs are privatized and converted into crop lands irrespective of their use capabilities (Jodha, 1986a). Second, the use intensity of CPRs, even when they are not ploughed, is increased. The latter happens because of shrinkage of CPRs and reduced availability of periodically fallowed crop lands. The final consequence is physical degradation of CPRs.

The depletion of common property resources involves the operation of several factors, jointly and individually. They can be grouped under (1) demographic factors, (2) market factors, (3) technological factors, and (4) public interventions. The literature on the subject covers different dimensions of these factors (Hardin, 1968; Ciriacy-Wantrup and Bishop, 1975; Boserup, 1965; Runge, 1981; Livingstone, 1984; Norgaard, 1984; Bromley and Chapagain, 1984; Singh, 1986; Ostrom, 1986). Figure 1 sketches the role of these factors in depletion of CPRs in the Indian context. The focal points in the process are privatization and over-exploitation of CPRs.

According to Figure 1, demographic factors influence CPRs directly by raising demand for land and indirectly by leading to privatization of CPRs. Rapid population change in association with such external factors as public interventions and commercialization induces structural changes in the village communities—changes in occupational patterns, group dynamics, and distribution of economic and political power at the local level (Mandelbaum, 1972). These changes in turn foster disregard of conventions and social sanctions that previously helped to protect CPRs.

With respect to market factors, the increased marketability and profitability of products of CPRs or of CPR-based enterprises, such as livestock rearing,

FIGURE 1 Role of various factors in the depletion of common property resources in the dry regions of India

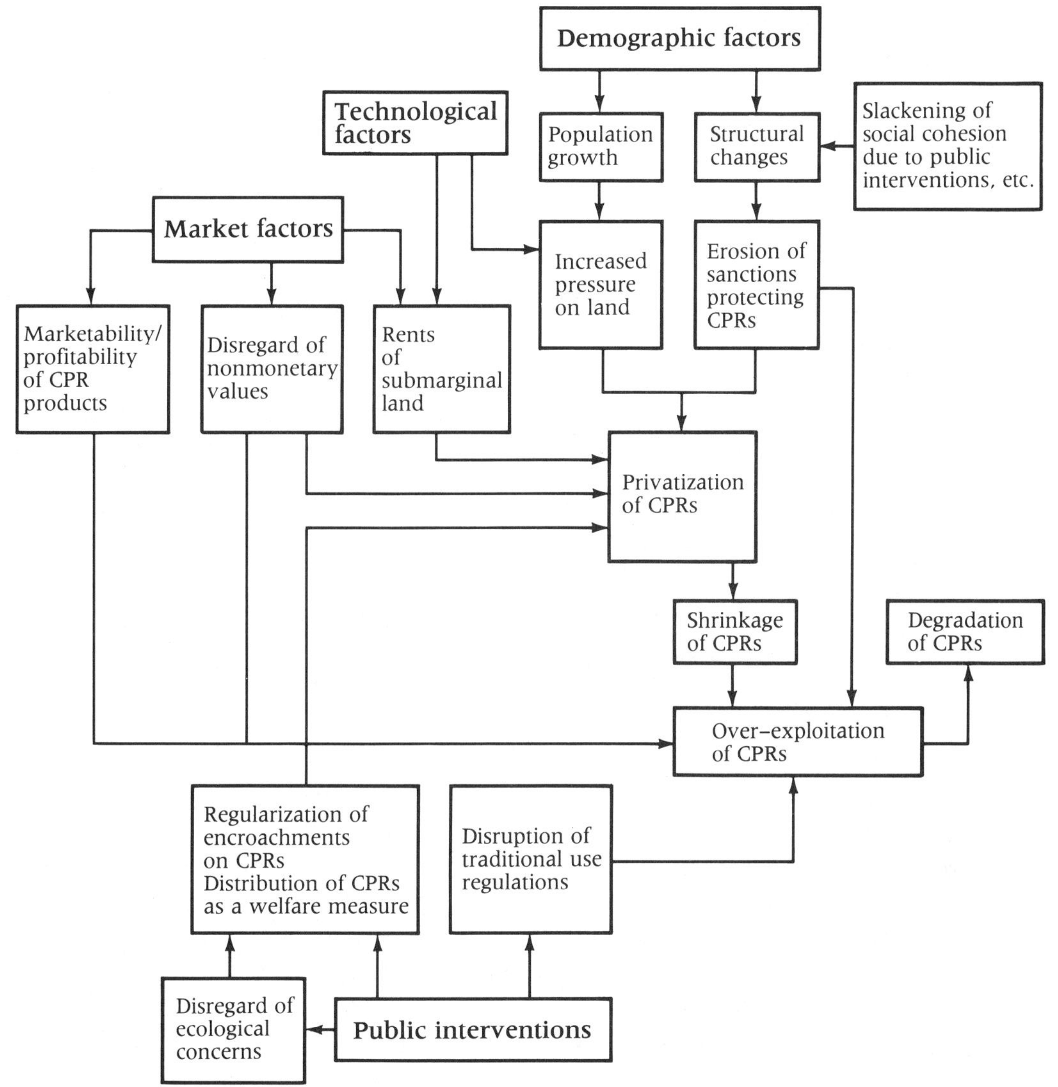

lead directly to over-exploitation of CPRs (Jodha, 1985b). Indirectly, market forces tend to encourage individual initiative and generate disregard for collective sanctions against encroachment on CPRs (Jodha, 1985a). Finally, market forces help improve the property value or rent of submarginal lands historically kept as CPRs. This is turn encourages privatization.

Technological changes such as the introduction of irrigation and tractors also contribute to increased rent on submarginal lands and resulting shrinkage of CPRs (Jodha, 1985b).

Public interventions including institutional reforms, welfare measures, and production-promoting activities can have adverse effects on CPRs. They tend to accentuate the roles of other factors mentioned above in depleting CPRs. For instance, land reforms in India encouraged privatization and shrinkage of CPRs. By disrupting the traditional management systems, they also contributed to over-exploitation, poor protection, and degradation of CPRs. This will be elaborated later.

Data

There is not sufficient empirical evidence on depletion of CPRs in India to fully corroborate the process sketched in Figure 1. The present essay illustrates only some aspects of the process. The evidence presented here is based on data from 82 villages in seven states in dry tropical regions of India. The fieldwork for the study was conducted during 1982–84.[3] Although the villages were widely scattered in the dry regions (see Figure 2), their selection was not random. Village selection was conditioned by the availability of suitable interviewers acquainted with the oral history of the villages and, unlike outside investigators, able to collect information on sensitive issues. Besides collection of village- and farm-level data on current status and use of CPRs, the fieldwork involved physical verification and monitoring of different aspects of CPRs. Reconstruction of the history of CPRs in each area, through oral histories and written records, was also attempted.

CPRs as sources of sustenance

Before we discuss the extent and process of depletion of common property resources in the dry regions of India, it is appropriate to briefly describe the role of CPRs in people's sustenance. CPRs contribute to employment, income, and asset accumulation. They are an important component of people's self-provisioning systems (Blaikie et al., 1985; Gupta, 1985). Because they are taken for granted, however, these contributions are seldom recognized and quantified. Some quantitative details collected as part of the author's study (Jodha, 1986a) are summarized in Table 1. Owing to their degradation and low productivity, the CPRs do not offer high returns to their users. Hence, the rural poor, who have few alternative sources of income, depend more on the low payoff options offered by CPRs. The rural rich, that is, the large landholders, depend very little on CPRs. The intermediate categories of households, not included in Table 1, depend more on CPRs than the rich.

Table 1 suggests that the rural poor receive the bulk of their fuel supplies and fodder from CPRs. CPR product collection is an important source of employment and income, especially during the periods when other opportunities are nonexistent. In certain areas employment on CPRs exceeds employ-

FIGURE 2 Districts and number of villages covered by the study of common property resources in India

ment generated through antipoverty public programs (Jodha, 1986a). Furthermore, CPR income accounts for 14 to 23 percent of household income from all other sources in the study villages—and it is likely this is a significant underestimate. More importantly, the inclusion of CPR incomes in total household

TABLE 1 Extent of people's dependence on common property resources (CPRs) in dry regions of India

State (with no. of districts and villages)	Household category[a]	CPR contribution to household supplies, employment, and income					Gini coefficient[f] for income from	
				Per household				
		Fuel supplies[b] (percent)	Animal grazing[c] (percent)	Employment days[d] (no.)	Annual income[e] (Rs)	CPR income as percent of total income (percent)	All sources	All sources excluding CPRs
Andhra Pradesh (1, 2)	Poor	84	—	139	534	17	0.41	0.50
	Rich	13	—	35	62	1	0.41	0.50
Gujarat (2, 4)	Poor	66	82	196	774	18	0.33	0.45
	Rich	8	14	80	185	1	0.33	0.45
Karnataka (1, 2)	Poor	—	83	185	649	20		
	Rich	—	29	34	170	3		
Madhya Pradesh (2, 4)	Poor	74	79	183	733	22	0.34	0.44
	Rich	32	34	52	386	2	0.34	0.44
Maharashtra (3, 6)	Poor	75	69	128	557	14	0.40	0.48
	Rich	12	27	43	177	1	0.40	0.48
Rajasthan (2, 4)	Poor	71	84	165	770	23		
	Rich	23	38	61	413	2		
Tamil Nadu (1, 2)	Poor	—	—	137	738	22		
	Rich	—	—	31	164	2		

NOTE: This and all other tables in the essay are based on village/household data from study villages reported by Jodha (1986a).
[a]The number of sample households per village varied from 20 to 36 in different districts. The "poor" include agricultural laborers and small farm (<2 ha dryland equivalent) households. The "rich" include large farm households only.
[b]Fuel gathered from CPRs as proportion of total fuel used during three seasons covering the whole year.
[c]Animal-unit grazing days on CPRs as proportion of total animal-unit grazing days.
[d]Total employment through CPR product collection.
[e]Income mainly through CPR product collection. The estimation procedure underestimated the actual income derived from CPRs (see Jodha, 1986a).
[f]Higher value of Gini coefficient indicates higher degree of income inequalities. Calculations are based on income data for 1983–84 from a panel of households covered under ICRISAT's village-level studies. The panel of 40 households from each village included 10 households from each of the categories, namely large, medium, and small farm households and labor households.

incomes from all other sources reduced the extent of rural income inequality, as indicated by lower values of the Gini coefficient. Despite their valuable contributions to the economy of the rural poor, CPRs are one of the most neglected areas in development planning in the country. The consequence is their depletion both in terms of decline in area and in terms of physical degradation.

Depletion of CPRs

Of the two forms of depletion of CPRs, decline in area is more easily captured by written and oral records of village land use. In contrast, the extent of physical degradation of CPRs, though quite visible and keenly felt by villagers, is difficult to quantify because their productivity in the past was not recorded.

The decline in CPR area in the study villages is reported in Table 2. The information relates to 82 villages from 21 districts from dry regions of seven states in India.[4] The benchmark is the period 1950–52, which was a watershed in the agrarian history of independent India. During the early 1950s, comprehensive land reforms were introduced. The status and management pattern of CPRs underwent drastic changes following land reforms and subsequent public interventions. Because they were a major event in the memory of villagers, the land reforms also served as an important reference point to facilitate recording an oral history of CPRs. The benchmark situation was compared with the status of CPRs during 1982–84, when fieldwork for the study was conducted.

According to Table 2, the area of CPRs declined by 31 to 55 percent in various states. The impact of this change is clearly visible in terms of both decline in proportions of CPR lands in total village area and increase in population pressure on CPRs.

In the absence of recorded information with which to assess degradation or decline in productivity of CPRs over time, a benchmark had to be constructed from oral histories and scattered village records. The comparison of this benchmark with the present situation, shown in Table 3, gives some indication of the qualitative decline of CPRs.

The first indicator of physical degradation of CPRs is the almost complete disappearance of a number of products that villagers used to gather from the commons before the 1950s. According to Table 3, not even half the number of items gathered in the past were available in the early 1980s in any of the study villages.

The variety of CPR products declined because a number of plant and tree species disappeared from common lands. Local names of a limited number of CPR plots indicated their coverage by specific vegetation in the past. Few of those species grow there any longer. Similarly, a number of selected CPR plots that were traditionally used for grazing more productive animals like lactating cattle and working bullocks are no longer able to sustain such animals. With their

TABLE 2 Decline of area of CPR land in dry regions of India

State (and no. of districts)	Number of study villages	Area of CPRs[a] 1982–84 (ha)	CPRs as proportion of total village area		Decline in area of CPRs 1950–52 to 1982–84 (percent)	Persons per 10 ha of CPR area	
			1982–84 (percent)	1950–52 (percent)		1951	1981
Andhra Pradesh (3)	10	827	11	18	42	48	134
Gujarat (3)	15	589	11	19	44	82	238
Karnataka (4)	12	1165	12	20	40	46	117
Madhya Pradesh (3)	14	1435	24	41	41	14	47
Maharashtra (3)	13	918	15	22	31	40	88
Rajasthan (3)	11	1849	16	36	55	13	50
Tamil Nadu (2)	7	412	10	21	50	101	286

NOTE: Table adapted from Jodha (1986a), where more disaggregated details are reported.
[a]CPRs include community pasture, village forest, waste land, watershed drainage, river and rivulet banks, and other common lands. Data indicate average area per village.

TABLE 3 Physical degradation of CPRs in dry regions of India

Indicator of changed status	State (with no. of villages)						
	Andhra Pradesh (3)	Gujarat (4)	Karnataka (2)	Madhya Pradesh (4)	Maharashtra (3)	Rajasthan (4)	Tamil Nadu (2)
Number of CPR products collected by villagers[a]							
In the past	32	35	40	46	30	27	29
At present	9	11	19	22	10	13	8
Number of trees and shrubs per hectare in							
Protected CPRs[b]	476	684	662	882	454	517	398
Unprotected CPRs	195	103	202	215	77	96	83
Number of watering points (ponds) in grazing CPRs							
In the past	17	29	20	16	9	48	14
At present	4	13	4	3	4	11	3
Number of CPR plots where rich vegetation, indicated by their nomenclature, is no longer available	—	12	3	6	4	15	—
CPR area used for cattle grazing in the past, currently grazed mainly by sheep/goats (ha)[c]	48	112	95	—	52	175	64

NOTE: Based on observation and physical verification of current status (during 1982–84) and on past details collected from oral and recoded description of CPRs in different villages. The choice of CPRs where plot-based data are reported was guided by availability of past information about them.

[a]Includes types of fruits, flowers, leaves, roots, timber, fuel, and fodder in the villages. The "past" indicates the period preceding the 1950s; the "present" indicates the early 1980s.

[b]Protected CPRs were the areas (called Oran) where for religious reasons live trees and shrubs are not cut. The situation of CPR plots (numbering two to four in different areas) was compared with other bordering plots of CPRs that were not protected by religious or other sanctions.

[c]Relates to area covered by specific plots, traditionally used for grazing high-productivity animals (e.g., cattle in milk, working bullocks, or horses of feudal landlords). Because of the depletion of such plots, such animals are no longer grazed there.

depleted forage potential and changed vegetative composition, they are now grazed by sheep and goats instead of cattle.

The difference in the number of species found on protected and unprotected CPRs is an important indicator of vegetative degradation and associated resource depletion. Certain CPR areas are protected through religious sanctions against removal of live trees and shrubs from them. The number of trees and shrubs recorded per hectare was three to six times higher on protected CPRs than on unprotected ones. The number of watering points, an important component of common grazing lands, has also declined (Table 3).

The process of change

Among factors contributing to the depletion of CPRs, we consider first privatization due to public policies. The inter-village differences in the status of CPRs caused by demographic and other factors are discussed later. In the following discussion, we do not address separately the decline of CPR area and the physical degradation of CPRs. We proceed with the premise that shrinkage and unregulated use of CPRs inevitably lead to their physical degradation.

Public policy and decline of CPR area

Public policies that followed the introduction of land reforms in the early 1950s encouraged privatization of CPRs and led to a rapid decline in their area. Although public policies did not explicitly advocate privatization, in practice virtually all the programs designed to provide land to the landless and to small landholders have reduced the area of CPRs, because little other land for redistribution could be acquired through land ceiling laws (Ladejinky, 1972) or through voluntary donations.[5] Privatization involved distribution of land to the landless and the rural poor under various welfare schemes, and legalization of encroachment on CPR lands.[6] Although privatization was supposed to benefit the rural poor, in practice, as will be shown later, the benefits to relatively better-off households exceeded the benefits to the poor. Similarly, despite the mention of rehabilitation and protection of grazing lands and village forests in the government's planning documents, very little was done in practice (Jodha, 1986a). Moreover, development programs for these resources were highly "technique"-dominated, with little attention given to user perspectives.

Further, such institutional changes in village administration as the abolition of the feudal system, the decline in the authority of the traditional village headman (Mandelbaum, 1972), and the introduction of formally elected village councils (*panchayats*) led to slackening of the traditional arrangements for conservation and regulated use of CPRs (Jodha, 1985b).

Privatization of CPR lands

Table 4 depicts the distribution of privatized CPR lands in the study villages. The proportion of poor households among the total recipients of CPR lands was generally the highest; however, the proportion of land received by the poor was much lower than that received by other social groups. The average land area received by the poor barely exceeded one hectare per household; the area received by "others" ranged from 2 to 3 hectares. The last column of Table 4 reveals that those who already had more land also received more privatized land.

A detailed inquiry into the current status of CPR lands received by the rural poor revealed that the poor could not sustain the gains from land distribution programs. According to Table 5, if Tamil Nadu is excluded, 19 to 42 percent of the land-receiving poor households disposed of part or all of the land received, through sale, mortgage, and so on. This happened despite legal restrictions in several areas on transactions in land belonging to the poor. The area so disposed amounted to 23 to 45 percent of the land received in different states, excluding Tamil Nadu. Nine to 36 percent of lands so disposed were submarginal and unsuitable for cropping. Most of the households that disposed of land did not have the necessary resources—like bullocks and ploughs—to cultivate it. Lack of resources to develop the land, inadequacy of the land to sustain the family, and pressing needs for cash induced the poor households to give up the land.

The evidence suggests that policies and programs to help the rural poor through privatization of CPRs have not worked as intended. It is quite doubtful whether poor people's collective loss from reduced CPR areas has been compensated by private ownership of CPR lands given to them.

Slackening of use regulations

The loss to the users of common property resources has resulted not only from reduced area, but also from degradation or reduced productivity of CPRs. The latter is partly an indirect consequence of shrinkage and overcrowding on CPRs. More importantly, as a byproduct of public interventions, including land reform and introduction of democratic institutions at the village level, the traditional formal or informal arrangements regulating the use of CPRs have weakened or disappeared (Jodha, 1985b). For instance, the abolition of grazing taxes and levies and of compulsory labor to fence and trench CPR lands and to desilt ponds has led to significant declines in the private costs to CPR users. Similarly, the replacement of the informal authority of the village elders by elected village councils and the replacement of conventions by formal laws relating to land have eroded the traditional social and religious sanctions protecting CPRs. This in turn has led to over-exploitation and poor upkeep of CPRs. Table 6 indicates the

TABLE 4 Distribution of privatized CPR lands to different household groups in dry regions of India

State (with no. of villages)	Total land privatized (ha) (1)	Total number of households receiving land (2)	Share of poor[a] in (1) (percent) (3)	Share of poor in (2) (percent) (4)	Land per household received by		Average land size after receiving new land	
					Poor[a] (ha) (5)	Others[b] (ha) (6)	Poor[a] (ha) (7)	Others[b] (ha) (8)
Andhra Pradesh (6)	493	401	50	74	1.0	2.1	1.6	5.0
Gujarat (8)	287	166	20	45	1.0	2.6	1.8	9.4
Karnataka (9)	362	203	43	65	1.3	3.0	2.2	8.0
Madhya Pradesh (10)	358	204	42	62	1.2	3.2	2.5	9.5
Maharashtra (8)	316	227	38	53	1.1	1.9	2.0	6.2
Rajasthan (7)	635	426	22	36	1.2	3.2	1.9	7.2
Tamil Nadu (7)	447	272	49	66	1.0	1.5	1.9	6.7

NOTE: Table adapted from Jodha (1986a). Three districts were covered in each state except Andhra Pradesh and Tamil Nadu, where two districts each were covered.
[a]The "poor" include agricultural laborers and small farm (<2 ha of dry land equivalent) households.
[b]"Others" include both medium and large farm households.

TABLE 5 Dispossession of privatized CPR lands by rural poor in dry regions of India

State[a]	Dispossessed land		Dispossessing households as proportion of total land recipients (percent)	Distribution of households by reasons for land dispossession			
	As proportion of total land received (percent)	Containing area unsuited to cropping (percent)		No bullocks, plough (percent)	Inability to develop land (percent)	Land too small, poor (percent)	Pressing need for cash (percent)
Andhra Pradesh	44	25	30	81	14	15	16
Gujarat	45	15	29	53	23	17	8
Karnataka	32	24	42	72	30	34	12
Madhya Pradesh	23	9	19	47	17	12	2
Maharashtra	40	16	36	53	27	21	9
Rajasthan	33	36	27	52	25	17	2
Tamil Nadu	17	24	18	77	18	9	—

NOTE: Table adapted from Jodha (1986a). For definition of "Poor" see Table 4, note a.
[a]For number of districts and villages covered see note to Table 4.

TABLE 6 Changes in management of CPRs in dry regions of India

State (with no. of villages)	Number of villages pursuing the following measures					
	Formal/informal regulations on CPR use[a]		Formal/informal taxes/levies on CPR use[c]		Users' formal/informal obligation toward upkeep of CPRs	
	In the past[b]	At present[b]	In the past	At present	In the past	At present
Andhra Pradesh (10)	10	—	7	—	8	—
Gujarat (15)	15	2	8	—	11	2
Karnataka (12)	12	2	9	—	12	3
Madhya Pradesh (14)	14	2	10	—	14	3
Maharashtra (13)	11	1	6	—	10	1
Rajasthan (11)	11	1	11	—	11	2
Tamil Nadu (7)	7	—	4	—	7	1

NOTE: Table based on fieldwork during 1982–84 (Jodha, 1986a).
[a]Measures such as regulated/rotational grazing, seasonal restrictions on use of CPRs, provision of CPR watchmen.
[b]The "past" indicates the period prior to the 1950s; the "present" indicates the early 1980s.
[c]Measures such as grazing taxes and penalties for violation of regulations on use of CPRs. See Jodha (1985b) for descriptive account.
[d]Measures such as contribution toward desilting of watering points, fencing, and trenching.
— indicates nil.

extent of change in terms of discontinuation of several practices that once constituted integral parts of formal or informal management systems for CPRs. According to the table, in most cases not even a tenth of the villages that formerly followed various management practices with respect to CPRs do so at present. The consequences of this change hardly need elaboration.

The role of other factors

Public policies and programs encouraging privatization of CPRs affected all villages. But the actual decline in CPRs differed from village to village even within the same state or district. Table 7 shows the range of inter-village differences in the extent and decline of CPRs. These differences call for examination of other factors that might influence CPRs independently or in association with public policies.

These factors can be grouped into three categories. Population size or its density, growth of population, and village size in terms of number of households represent the demographic factors. Second, the ecological factors comprise such variables as area of the village, extent of submarginal lands,[7] and number of livestock. Finally, the role of market forces is reflected in such variables as distance from a marketing center.

Table 8 presents a comparison of groups of villages with the highest and lowest area of CPRs along with the values for factors influencing the status of

TABLE 7 Inter-village differences in CPRs as a proportion of village area and decline in area of CPRs

State (with no. of villages)	CPR area as percent of total village area		Percent decline in CPR area, 1950–52 to 1982–84
	1950–52	1982–84	
Andhra Pradesh (10)	9–30	5–20	26–56
Gujarat (15)	7–31	2–23	21–69
Karnataka (12)	6–36	4–30	16–50
Madhya Pradesh (14)	29–69	19–47	14–51
Maharashtra (13)	8–43	6–34	14–52
Rajasthan (11)	20–49	8–26	17–71
Tamil Nadu (7)	7–39	5–23	21–65

NOTE: Table based on fieldwork and records (Jodha, 1986a).

TABLE 8 Inter-village difference in the extent of CPRs and possible factors responsible for them, 1950–52

State	Village group[a] (1)	Average value of variable for each group of villages						
		CPRs as proportion of col. (3) (2)	Total area (ha) (3)	Size (and density[b]) (4)	Number of households (5)	Submarginal lands as proportion of col. (3)[c] (percent) (6)	Livestock per 10 ha of (3)[d] (no.) (7)	Distance from market center[d] (km) (8)
Andhra Pradesh	Bottom	13	1134	1036 (91)	194	35	12	7
	Top	25	648	537 (83)	103	43	11	13
Gujarat	Bottom	14	588	895 (152)	192	12	12	8
	Top	27	243	369 (149)	78	32	18	10
Karnataka	Bottom	12	1110	1013 (91)	157	22	8	14
	Top	31	624	529 (85)	97	38	10	19
Madhya Pradesh	Bottom	35	593	293 (49)	56	36	7	14
	Top	52	218	183 (84)	35	49	13	18
Maharashtra	Bottom	15	632	545 (86)	82	13	12	9
	Top	38	246	250 (101)	50	31	15	21
Rajasthan	Bottom	22	642	358 (56)	75	55	12	13
	Top	46	1164	551 (48)	98	60	12	15
Tamil Nadu	Bottom	10	734	1242 (169)	252	16	18	9
	Top	37	390	1343 (346)	198	52	40	13
Correlation coefficient		1.0	.33	−.51	−.56	.56	.16	.57

NOTE: Table based on fieldwork and records (Jodha, 1986a). The table covers 56 out of 82 villages.

[a]Villages in each state were arranged in descending order of the proportion of CPR area to total village area during 1950–52. One-third of the villages at the top end and the bottom end of the list for each state were picked to form two groups, described as the "bottom" and "top" groups of villages accordingly. The average situation of each of the two groups with reference to different variables is presented.

[b]Figures in parentheses indicate number of person/km^2.

[c]Lands as defined in note 7.

[d]Information on livestock and distance from market center relates to 1982–84 because the details for 1950–52 could not be collected for all villages. However, this does not significantly influence the results.

CPRs. Table 9 compares villages with the highest and lowest declines in CPRs over time. The important generalizations emerging from these comparisons are summarized in Table 10.

Table 10 suggests a negative association between the extent of CPRs and such factors as size of population and number of households in the village. This is not difficult to understand. Historically, even in the context of a micro unit like a village, higher population pressure implied land scarcity. Once the supply of better lands had been exhausted, any additional demand could be met only through occupation of submarginal lands, including CPRs. This process was intensified with the rapid increase in population over a relatively short period from the early 1950s to the early 1980s. Increased population pressure, as mentioned earlier, induces a shift from land-extensive to land-intensive farming. CPRs bear the maximum burden of land use intensification.

As observed during the fieldwork, smaller villages were relatively isolated from major markets. They were located in ecologically less favorable environments with greater proportions of submarginal lands. They were also found to have greater social cohesion and respect for conventions relating to use of CPRs. These attributes of smaller villages, originating in ecological, social, demographic, and market-related factors, helped protect the CPRs. This explains both the greater initial availability of CPRs and the lower decline in their area over time in smaller villages.

Table 10 indicates a positive association between the initial extent of CPRs and village area, extent of submarginal lands, and livestock density. Submarginal lands and livestock rearing have a long historical association. Viewed in the context of the traditional practice of reserving submarginal lands for common use, this explains the positive association between the importance of livestock and the extent of CPRs. Furthermore, in the villages with higher livestock populations, people had greater stake in CPRs—a factor that also slowed their decline since the early 1950s (Table 10).

A positive association between the decline in CPRs and the extent of submarginal lands as well as the initial area of CPRs (Table 10) can be explained with reference to public policies, which encouraged privatization of CPRs as discussed earlier. Villages with higher initial extent of CPRs lost more of CPR area because it was easier to execute land distribution programs in them. In addition, villagers had a greater awareness of the status of CPRs where the total CPR area was smaller. This prevented encroachments on CPRs by fellow villagers as well as by the state. This sort of safeguard against the decline of CPRs was rarely found in the villages with large and widely scattered CPR lands.

Table 10 indicates a positive association between the extent of CPRs and distance from a market center. It also shows a negative association between the decline in CPR area and distance from a market center. Nearness to a market center implies emergence of individualistic tendencies among the villagers that may reduce reliance on collective sustenance through CPRs.

TABLE 9 Inter-village differences in the extent of decline of CPR area from 1950–52 to 1982–84 and possible factors responsible for them

State	Village group[a]	Average value of variable for each group of villages							
		Decline in CPR area (percent)	Increase in population 1951–81 (percent)	Number of households in 1981	Area of village (ha)	Initial extent of CPR area (1951) (ha)	Area of submarginal land[b] (ha)	Livestock per 10 ha, 1982–84 (no.)	Distance from market center, 1982–84 (km)
Andhra Pradesh	Bottom	28	66	119	448	106 (24)[c]	172 (38)[c]	15	13
	Top	53	73	277	986	143 (16)	320 (36)	12	9
Gujarat	Bottom	27	64	228	266	55 (20)	42 (16)	17	12
	Top	58	77	181	442	77 (18)	96 (22)	13	8
Karnataka	Bottom	27	49	182	583	102 (17)	113 (19)	9	16
	Top	46	52	216	1131	230 (20)	366 (32)	9	17
Madhya Pradesh	Bottom	20	74	50	261	95 (34)	87 (34)	9	19
	Top	48	72	109	706	294 (49)	348 (49)	7	13
Maharashtra	Bottom	20	57	73	289	82 (29)	81 (28)	14	19
	Top	44	65	87	346	88 (25)	64 (18)	11	14
Rajasthan	Bottom	32	67	125	642	140 (22)	355 (55)	12	15
	Top	65	90	164	1336	539 (40)	721 (54)	9	13
Tamil Nadu	Bottom	27	38	349	733	75 (10)	119 (16)	18	10
	Top	62	38	239	551	146 (26)	219 (40)	19	12
Correlation coefficient		1.0	.36	.31	.52	.50	.51	−.2	−.31

NOTE: Table based on fieldwork and records (Jodha, 1986a). The table covers 56 out of 82 villages.
[a]Villages in each state were arranged in descending order of percentage decline in their CPR area from 1950–52 to 1982–84. One-third of villages at the top and bottom ends of the list for each state were picked to form two groups, described as the "top" and "bottom" groups of villages accordingly. The average situation of each of the two groups with reference to different variables is presented.
[b]For definition of submarginal lands see note 7.
[c]Figures in parentheses indicate proportion of CPR lands and submarginal lands to total area of the village.

TABLE 10 General inferences from the tabular analysis of village-level data indicating association between different attributes of villages and relative position of their CPRs

	Relative position of CPRs			
	Extent of CPR area (1950–52)		Decline in CPR area (1950–52 to 1982–84)	
Attribute of villages	Top	Bottom	Top	Bottom
Demographic factors				
Higher population		✓		
Greater number of households		✓	✓	
Higher population increase over time			✓	
Ecological factors				
Larger area of village	✓		✓	
Larger extent of submarginal lands	✓		✓	
Larger initial area of CPRs			✓	
Greater importance of livestock	✓			✓
Market factor				
Greater distance from market center	✓			✓

NOTE: Table based on comparison of groups of villages in each state having highest and lowest extent of CPRs (1950–52) and groups having the highest and lowest decline of CPRs (1950–52 to 1982–84). For details see Tables 8 and 9.

Regression results

The broad trends suggested above were verified with regression analysis. The data for all 82 villages were subjected to linear regression. The extent of CPR area in 1950–52 and in 1982–84 and the decline in area during this period were treated as dependent variables. They were regressed on the independent variables listed in Table 11.

We expected that the size of the CPR area would be negatively influenced by the size of the population and positively influenced by total village area, distance to market, the area of submarginal land, and livestock density. As indicated by the results reported in Table 12, these expectations were largely confirmed. The extent of submarginal area significantly influenced the extent of CPRs in both periods. Size of village population was another significant determinant of the extent of CPRs. As with most of the variables, however, the estimated effect of population on size of CPR area declined in 1982–84 compared with 1950–52. Other factors such as the importance of livestock and the distance from a market center did not emerge as significant determinants of CPR area. Moreover, these variables were less influential in explaining the variation in CPR area in the 1980s than in the 1950s.

Demographic factors declined in importance in the 1980s, while the area of submarginal lands (an ecological variable) remained the only highly signifi-

TABLE 11 Means and standard deviations of CPR area and its determinants in 82 study villages

Variable	Unit	Mean	Standard deviation
Dependent variables			
CPRs (early 1950s)[a]	Hectare	159.79	143.50
CPRs (early 1980s)	Hectare	88.45	62.40
Change in CPRs (early 1950s to early 1980s)	Hectare	−71.34	−88.25
Independent variables			
Population (1951)	Number	571.00	381.00
Population (1981)	Number	916.00	569.00
Household (1951)	Number	110.00	73.00
Household (1981)	Number	177.00	110.00
Total village area (1951)[b]	Hectare	618.11	406.18
Submarginal area (1951)[b]	Hectare	225.78	216.92
Livestock per 10 ha (1981)	Number	13.21	5.90
Market distance (1981)	Km	13.61	6.00
Change in population from 1951 to 1981	Number	345.00	205.00

NOTE: Table based on fieldwork and records (Jodha, 1986a).
[a]This was also treated as an independent variable while regressing change in CPR area on other variables.
[b]Extent of area remained constant over time.

TABLE 12 Determinants of CPR area in 82 study villages during the early 1950s and early 1980s

Explanatory variable	CPR in 1950s	CPR in 1980s
Total area of village	0.11 ** (2.08)	0.05 (1.68)
Population	−0.08 ** (−2.63)	−0.02 * (−1.98)
Submarginal area	0.47 ** (6.32)	0.19 ** (4.65)
Livestock density (livestock/10 ha)	2.00 (1.18)	0.48 (0.53)
Market distance	−0.57 (−0.47)	−0.52 (−0.81)
Intercept	12.89 (0.36)	23.85 (1.30)
R^2	0.82	0.74
F ratio	76.01	45.85
Number of observations	82	82

NOTES: Table based on fieldwork and records (Jodha, 1986a). Figures in parentheses are *t* values; * and ** denote statistical significance at the .05 and .01 levels respectively.

TABLE 13 Determinants of change in CPR area in 82 study villages from early 1950s to early 1980s

Explanatory variable	Regression coefficient
Change in population	−0.05 ** (−2.16)
CPR in 1950	−0.58 ** (−29.47)
Household number in 1950	−0.11 (−1.76)
Intercept	−25.86 ** (−4.93)
R^2	0.94
Number of observations	82

NOTES: Table based on fieldwork and records (Jodha, 1986a). Figures in parentheses are *t* values; ** denotes statistical significance at the .01 level.

cant determinant of CPRs in that decade. With the rapid privatization of CPRs since the early 1950s, the extent of CPRs in many villages had shrunk by the early 1980s to a level that was quite independent of population pressure and of such factors as distance from a market and importance of livestock. The only factor limiting the privatization of CPRs for arable farming was the submarginal quality of land. It is no wonder, therefore, that this variable remained a highly significant determinant of CPR area in the 1980s.

Table 13 presents the regression coefficients of variables influencing the decline in CPR area from the early 1950s to the early 1980s. The initial (i.e., early 1950s) extent of CPR area was the most significant factor influencing the decline in CPR area. The effect of the percentage change in village population from 1951 to 1981 was also statistically significant and in the expected direction. The number of households in 1951 did not emerge as a significant variable in the regression. This could be partly due to interregional differences in the relative size of villages, the impact of which could not be captured in the regression analysis.[8]

Conclusion

This analysis of village- and farm-level data from 82 villages demonstrates the very important role played by common property resources in the sustenance of rural people in dry tropical regions of India. In recent decades, CPRs have declined both in their area and in their productivity. Population growth leading to increased pressure on land has historically contributed to decline in CPRs, which often represented an extensive pattern of land use. This is borne out by the comparative analysis of the initial period (1950–52) in the present case. Following the initial period, however, the effect of rapid population growth must be assessed in conjunction with public policies, which probably would have led to decline of CPRs even in the absence of population pressure. The erosion of traditional effective CPR management systems is clearly a consequence or side effect of public interventions. An important implication of the study is that as long as public interventions are a key factor affecting CPRs, a reorientation of the former should be emphasized to rehabilitate CPRs. A restriction on further privatization of CPRs, introduction of use regulations supported by some element of user cost, and fiscal incentives to village councils should form part of CPR strategies for the future.

Notes

The author thanks R. A. E. Mueller, John H. Foster, R. P. Singh, and T. S. Walker for their valuable comments on an earlier draft. He acknowledges the computational assistance of M. Asokan and Ch. Vijay Kumar.

1 This is partly indicated by the following traditional practices observed in a number of villages. As an indicator of concern for the community's collective needs, the area of permanent pastures (i.e., a grazing CPR) was periodically revised to match the increase in number of cattle. Similarly, to prevent intensive use of highly erodible shallow soils, ponds with vast catchments were dug. Ploughing and cutting of shrubs were prohibited in the catchments of ponds, and statues of deities were erected in such areas as warnings. Vegetation in such places, called Oran, was protected through religious sanctions. For examples of people's concern for CPRs in other countries, see Ostrom (1986) and Livingstone (1984).

2 The land-extensive character of traditional farming systems is reflected in the use of long-duration, indeterminate cultivars, seasonal fallows, crop-bush fallows, and mixed farming (Jodha, 1986b).

3 Jodha (1986a) describes the methodology and procedures used in conducting the study. The present essay at times draws on the more detailed results of the study reported elsewhere (Jodha, 1985a; 1985b; 1986a).

4 The states and districts are as follows: (1) Andhra Pradesh: Anantpur, Mahbubnagar, and Medak; (2) Gujarat: Banaskantha, Mehsana, and Sabarkantha; (3) Karnataka: Bidar, Dharwad, Gulbarga, and Mysore; (4) Madhya Pradesh: Mandsaur, Raisen, and Vidisha; (5) Maharashtra: Akola, Aurangabad, and Sholapur; (6) Rajasthan: Jalore, Jodhpur, and Nagaur; (7) Tamil Nadu: Coimbatore and Dharmapuri.

5 Under a movement called *Bhoodan*, initiated by Vinoba Bhave, the voluntary donations of surplus land by landowners for redistribution to the poor was encouraged.

6 State acquisition of CPR land for construction of public facilities like schools and roads or for reforestation is another way in which CPRs are curtailed. This is not discussed here.

7 The submarginal lands, for our purpose, include those that are rocky, gravelly, or extremely sandy, as well as those having problems like salinity, water-logging, and highly undulating topography. Traditionally such lands were kept as village common.

8 For instance the larger villages in low-population-density districts of Madhya Pradesh, Rajasthan, or Maharashtra, with their relatively lower extent and higher decline of CPR area, proved to be smaller than the smallest villages in the districts in Tamil Nadu or Andhra Pradesh having higher population densities. The impact of village size (i.e., number of households) on CPRs lost its significance when villages from all regions were pooled for regression.

References

Blaikie, P. M., J. C. Harriss, and A. N. Pain. 1985. "Public policy and the utilisation of common property resources in Tamilnadu, India," Overseas Development Group and School of Development Studies, Norwich, England: University of East Anglia.

Boserup, E. 1965. *The Conditions of Agricultural Growth*. Chicago: Aldine.

Bromley, D. W., and P. Chapagain. 1984. "The village against the centre: Resource depletion in South Asia," *American Journal of Agricultural Economics* 6, no. 5: 869–873.

Ciriacy-Wantrup, S. V., and R. C. Bishop. 1975. "Common property resource as a concept in natural resource policy," *Natural Resource Journal* 15: 713–727.

Gupta, A. K. 1985. "Socio-ecology of stress: Why do CPR management projects fail?," presented at the Conference on Management of Common Property Resources, Board on Science and Technology for International Development (BOSTID), National Research Council, Washington, D.C., 21–27 April.

Hardin, G. 1968. "The tragedy of the commons," *Science* 162, no. 3859: 1243–1248.

Jodha, N. S. 1985a. "Market forces and erosion of common property resources," in *Agricultural Markets in Semi-Arid Tropics, Proceedings of the International Workshop, October 24–28, 1983, ICRISAT Center, Patancheru, A.P., India*. Patancheru: ICRISAT, pp. 263–277.

———. 1985b. "Population growth and the decline of common property resources in Rajasthan, India," *Population and Development Review* 11, no. 2: 247–264.

———. 1986a. "Common property resources and rural poor in dry regions of India," *Economic and Political Weekly* 21, no. 27: 1169–1181.

———. 1986b. "Research and technology for dry farming in India: Some issues for the future strategy," *Indian Journal of Agricultural Economics* 41, no. 3: 234–247.

Ladejinky, W. 1972. "Land ceilings and land reform," *Economic and Political Weekly* 7, nos. 5–7: 401–412.

Livingstone, I. 1984. *Pastoralism: An Overview of Practice, Process and Policy*. Rome: FAO.

Mandelbaum, D. G. 1972. *Society in India*. Bombay: Popular Prakasham.

McCown, R. L., G. Haaland, and C. De Haan. 1979. "The interaction between cultivation and livestock production in semi-arid Africa," in *Ecological Studies*, Volume 34: *Agriculture in Semi-Arid Environments*, ed. A. E. Hall, G. H. Cannel, and H. W. Lawton. Berlin, West Germany: Springer-Verlag.

Norgaard, R. B. 1984. "Coevolutionary agricultural development," *Economic Development and Cultural Change* 32, no. 3: 525–546.

Ostrom, E. 1986. "How inexorable is the 'Tragedy of commons?': Institutional arrangements for changing the structure of social dilemmas," Distinguished Faculty Research Lecture, Office of Research and Graduate Development, Indiana University.

Runge, C. F. 1981. "Common property externalities: Isolation, assurance, and resource depletion in a traditional grazing context," *American Journal of Agricultural Economics* 63, no. 4: 596–788.

Sandford, S. 1983. *Management of Pastoral Development in the Third World*. New York: John Wiley.

Singh, Chhatrapati. 1986. *Common Property and Common Poverty: India's Forests, Forest Dwellers and the Law*. Delhi: Oxford University Press.

UNCOD (United Nations Conference on Desertification). 1977. "Ecological change and desertification," Conference Document A/Conf. 74/7, Nairobi, Kenya.

World Bank. 1975. *The Assault on World Poverty*. Baltimore: Johns Hopkins University Press.

Population Growth and Rural Labor Absorption in Eastern and Southern Africa

IAN LIVINGSTONE

RAPID POPULATION GROWTH IN THE COUNTRIES of eastern and southern Africa, together with the attraction of the urban formal sector, might have been expected to stimulate high rates of rural-to-urban migration and greatly worsening problems of urban unemployment. This indeed was the expectation of the authors of the influential report of the 1972 ILO Employment Strategy Mission to Kenya (ILO, 1972). In the event, fears of the 1972 Mission did not become a reality, despite the fact that the population growth rate was not controlled (it actually increased) and despite the failure to shift toward more labor intensive techniques of production in manufacturing (a recommendation of the Mission).

In fact, much of the expanding population was absorbed in the rural areas. A follow-up study of Kenya by the present writer (ILO/JASPA, 1981) found that the rural areas served as a very effective "sponge," absorbing and retaining a great part of the additional population and labor. Moreover, the labor was not even freely available within the rural sector: expanding population numbers coexisted with acute shortages of labor felt by large-scale farms and plantations in a number of areas of Kenya, inducing employers actually to mechanize labor-intensive operations. Also like a sponge, however, the rural areas will at some point become saturated. The labor excess may then begin to appear quite rapidly, and in many places at once. This possibility motivates our interest in the processes that have been in operation within the rural sector affecting income distribution, on-farm and off-farm employment, and access to land.

While the principal focus of this essay is on rural labor absorption in Kenya, parallel processes are going on in other eastern and southern African countries. Given the slow pace of urban industrialization throughout the region, it is not only Kenya that must look to its rural sector for substantial additional labor absorption over the next few decades. The policies in all countries of the region, many with agricultural resource bases far inferior to Kenya's, must be to extend the effectiveness of the rural "sponge" as long as possible.

Mechanisms underlying rural labor absorption in Kenya

With diminishing opportunities for agricultural expansion at the extensive margin, Kenya's rural sector has for decades shown a declining average amount of land held per household. If we measure land in terms of equivalent hectarage of high-potential land, between 1969 and 1979 the number of districts with an average of one hectare or more per person fell from 12 to 7, and the number with less than half a hectare per person increased from 9 to 12. By 1989 at projected population growth rates the first figure would have fallen to 2, and the second risen to 19. Median agricultural area per person in equivalent hectares fell from 0.8 to below 0.5 over this 20-year period.

Farm size typically affects the efficiency of production. Some evidence on scale effects in Kenya (referring to "smallholder" farms only—excluding so-called gap farms and large farms[1]) is presented in Table 1. It illustrates the "divisibility" of farms—the relative absence of economies of scale. The first two columns of Table 1 show the standard pattern of declining gross output (farm operating surplus, to be precise) per hectare with size of holding. (This pattern does not necessarily imply a gain in efficiency at smaller sizes, since it may reflect inefficiently intensive application of one factor, labor, to smallholdings.) The last

TABLE 1 Farm operating surplus per hectare and household income per member in smallholder areas, by size of holding, 1974–75, Kenya

Size of holding (ha)	Average farm operating surplus per hectare (K.Shs. per adult)		Average household income per member (K.Shs. per adult)	
	(a)	(b)	(a)	(b)
Under 0.5	5,265	5,929	560	594
0.5–0.9	2,650	2,852	657	686
1.0–1.9	1,142	1,331	634	687
2.0–2.9	973	1,092	699	756
3.0–3.9	663	728	683	721
4.0–4.9	776	812	943	973
5.0–7.9	434	504	728	790
8.0 and over	249	275	912	960
All farms	873	973	683	727

(a) = as estimated in Integrated Rural Survey-1.
(b) = adjusted on the basis of revised livestock valuation.
NOTES: Data corrected for varying household size in terms of adult-equivalents by calculating average income per member. Data cover "smallholder" farms only (as defined by the Central Bureau of Statistics).
SOURCE: Dorling (1979).

two columns give average household income per member; they show income per member on holdings of 0.5–1.9 hectares as only slightly (about 7 percent) less than that for holdings of 2.0–3.9 hectares, which is in turn about 15 percent below that on holdings of 5.0 hectares and over. While there clearly is significant differentiation by farm size among smallholders in income per household member (income in the lowest category is 60 percent of that in the largest), this effect has been substantially moderated by the extent of divisibility.

There are various ways in which productivity per hectare can be and has been raised on smallholdings, apart from applying more labor per hectare. Some entail the application of more management per hectare. Additional activities can be carried out within the same area by interplanting maize and beans, partly for sale and partly for own consumption; fallow can be reduced, a grade cow or chickens can be added, and trees planted along borders for firewood or wattle (Tidrick, 1979, p. 113). There may also be more careful weeding and harvesting, using supervised labor. Such changes and additions do not represent mere static divisibility within a given production function, but are dynamic changes that have altered the production function. It seems reasonable to ascribe them to a Boserup-type response by farmers to population growth. Not much credit is deserved by government extension services, which in Kenya have always been conservatively based and not very effective in reaching the mass of smallholders (see ILO/JASPA, 1981).

A second vital element in the agricultural sector's capacity to absorb more labor has been a switch to higher valued crops, particularly labor-intensive export crops such as tea and coffee. For much of the colonial period smallholders were not allowed to grow tea or coffee, ostensibly because of alleged dangers of spreading crop diseases to European holdings.[2] Those crops have been rapidly adopted since the late 1950s, principally in Central Province: by 1974, 45 percent of small farm households grew some coffee, 18 percent some tea (Collier and Lal, 1980, appendix 1). They contributed to the rural economy both directly and by raising purchasing power, generating further rural economic activity. A different crop switch to raise value per hectare can be observed in Maragoli, Western Kenya, the area of most rapidly diminishing holding size in Kenya, where vegetables for sale as a cash crop have been substituted for maize to a significant degree, because of the latter's land-intensity of production.

A third contribution to labor absorption has come from increasing yields of existing crops. The most important change here affecting small farms has been the adoption of hybrid maize in high-potential areas during the 1960s. Increases in yields ranged from 47 to 300 percent according to one study (World Bank, 1973). Maize, where it is a staple food crop, occupies a large proportion of the total available land, so yield increases can release land to other crops. Increases in food crop yields have not been equally important in all regions. In Coast, Eastern, and Nyanza Provinces, hybrid maize is grown by only a third or less of smallholders. Other crops, such as sorghum and millet (both important in Nyanza), have received little research and development attention.

Livestock and milk production also plays an important part in smallholder adaptive strategies. The Integrated Rural Survey-1 (IRS-1) showed that 33 percent of the value of annual output in the smallholder areas was accounted for by livestock and milk (in comparison to 59 percent for food crops and only 8.4 percent for export crops). Cattle ownership is widespread: in 1974, according to IRS-1 results cited by Smith (1978), nearly two-thirds of all smallholders had at least one animal, as did one-half of smallholders with holdings below 0.5 ha. Livestock is one of the high-return activities, alongside coffee and other export crops. Collier and Horsnell (1986) argue that it has potential for continued growth in per capita agricultural incomes in Kenya.

Finally, there is the contribution of off-farm incomes, both as wage labor on large farms and smallholdings, and in nonfarm activities. According to calculations made by Smith (1978) from IRS-1, rural households in the lowest per adult-equivalent income group, below K.Shs. 250 per annum, obtained 77 percent of their income in 1974–75 as off-farm income, 18 percent as nonfarm operating surplus, and the rest from employment and remittances. Those in the next income category, K.Shs. 250–499, obtained 47 percent as off-farm income and 12 percent as nonfarm operating surplus. It should be noted, though, that the extent of wage employment on farms and smallholdings is a function of the size distribution of landholding. Changes in this distribution may not generate a net increase in the capacity of the land to absorb labor. Large farms, indeed, tend to use less labor per hectare. What is more important is employment in nonfarm activities, whether as hired workers or in self-employment (see Table 2).

The above summarizes some of the changes underlying the continuing absorption of rural labor in Kenya. Little has been said regarding the limits of these processes. They have not operated evenly throughout the country, and the demand for labor in consequence varies considerably by region. Indeed the three main agricultural zones define the boundaries of a segmented rural labor market. Evidence of approaching "saturation" does exist in some areas in the form of increasing landlessness, unemployed youth, overspill into marginal arable land, encroachment into and deterioration of traditional pastoral areas, and increasing rural poverty, despite the relative buoyancy on average of the rural economy. Not so much in Kenya but in many other African countries, such as Nigeria and Lesotho, an indicator of a downward limit to divisibility exists in the form of fragmentation of holdings into plots that are clearly of diminished economic efficiency.

TABLE 2 Prevalence of nonfarm activities in rural households, Kenya, 1977

Frequency of nonfarm activities per household	0	1	2	3	4	5	6	Total
Proportion of households (percent)	50	26	13	5	3	1	2	100

SOURCE: Central Bureau of Statistics, Rural Nonfarm Activity Survey, 1977.

Land rights and labor absorption

These are the production responses that have supported rural labor absorption in Kenya over the last two decades. What have been the land tenure arrangements that have underlain those responses and how have they been changing? Land rights as they exist or have evolved in Kenya have clearly been permissive of substantial rural labor absorption, not obviously inhibiting the process. They are far from uniform throughout the country. How in detail have land tenure arrangements operated in different regions? (The further question of what has been the effect in the opposite direction—of population growth on land rights and institutional/customary arrangements—would be relevant to a full consideration of rural economic and demographic change, but is not taken up here.)

Especially in comparison with other countries, Kenya has had a longstanding policy of granting freehold land titles within the smallholder sector. This policy goes back to the early 1950s under the Swynnerton Plan, and has been reemphasized since the mid-1970s. It has proved a slow and administratively expensive process, with at most piecemeal implementation.

Contrary to the statement of an Assistant Minister for Finance and Planning that land adjudication "has encouraged the participation of smallholder farmers in the modern agricultural production sector by growing of cash crops like tea, coffee, pyrethrum and dairy production" (FAO, 1985), there is no association in time between the expansion of these crops from the late 1950s and the spread of land titles. The labor-using tendency of cash crop production in this setting does not, therefore, appear to be related to the nature of land rights.

Moreover, there is no geographical coincidence between the degree of labor absorption and the introduction of land titles. For instance, the part of Kenya where average farm size has diminished earliest and to the greatest extent, Maragoli, is not one in which land adjudication has been significantly involved. Formal land titles are not a prerequisite for labor absorption.

It is widely pointed out, among others by Benneh (1985) and Mafeje (FAO, 1985, pp. 11–12), that what is nominally "communal tenure" in Africa in fact involves strongly established usufruct rights centered on the family. Mafeje states:

> It is yet to be proved that African cultivators suffered from insecurity of tenure or eviction from land under cultivation. . . . [I]t would appear that those who held the former view mistook communal responsibility in the political sense with the economic production. . . . [I]t would be more accurate to talk of household economy which was characterized by strong entrenched usufruct rights.

The traditional system may, in fact, have contributed positively to the capacity of rural areas to absorb population increase. Mafeje comments again (p. 12):

> On the positive side it can be argued that by granting access to land to every potential cultivator, the usufruct principle guaranteed full employment of existing labour, while not blocking intensification of production factors in a land-surplus economy. Non-recognition of property rights in land, but insistence on use rights, prevented monopolization of unused land by would-be owners.

Indeed, the criterion underlying the usufruct principle, which is the general one in Africa, is specifically employment rather than efficiency. Thus in Niger, for instance, attention is drawn to the "importance (simply) of village membership in the determination of land relations and access to land" (FAO, 1985, p. 28).

In contrast, the distribution of land titles, by facilitating stratification, may weaken labor retention. This may occur first because "the wealthier farmers usually have better access to information about land law, administrative procedures and farm prices. They may therefore buy out poorer and less knowledgeable smallholders" (Feder and Noronha, 1987, p. 146). A second reason is that poor owners of land may suffer from lack of foresight, not merely in respect of themselves but, and with more reason, in respect of their heirs: the advantage of the usufruct system specifically in terms of stabilizing the labor force is that it is cross-generational, certainly as regards primary heirs. Third, one can expect that land titles will tend to reduce subdivision of land among heirs. While problems associated with farm fragmentation are lessened, it is likely also that the extent to which new households can be absorbed in the farm sector will be reduced.

What the impact of the freehold entitlement process has been in Kenya is uncertain. Feder and Noronha (1987, p. 157), quoting a number of observers of the effects in Mbeere, assert that "it was the influential people (including the chief) and the civil servants who used their knowledge of the law to acquire land at the expense of the poorer and less knowledgeable." Cohen (1980) refers to the increase in land concentration that has resulted. King (1977, p. 353) asserts that the redistributive effects of land reform in Kenya have been minimal, tending largely to consolidate the position of existing landowners. What is generally acknowledged is that there has been considerable speculative purchase of land as an asset in Kenya, associated with a low degree of utilization, while increases in land values as a land market has been created put land out of reach of the landless and marginal smallholders (Benneh, 1985).

It could still be argued that land adjudication might increase labor absorption indirectly, by raising efficiency in the use of land and thus increasing productivity and employment in the long run. The possibility that land productivity increases are achievable by tapping economies of scale may be discounted: the evidence points to their absence both in Kenya and more generally. Differences in *X*-efficiency[3] may be exploited, however, by allowing more enterprising farmers to expand at the expense of those with less energy and initiative. Where land is in surplus, the usufruct system is apparently flexible enough to accommodate this, as stated in the cited passage by Mafeje. Under land scarcity, land productivity is higher on smaller holdings (see the estimates of average farm operating surplus in Table 1, showing productivity of holdings of less than 0.5 hectare to be on average more than double that of the next larger size-class). Productivity differences among farms of the same size are unlikely to be as important. For labor absorption, the equity/access effects of the usufruct system are likely to be more important than *X*-efficiency effects.

This leaves effects that might be attributable to improved access to credit through the use of land titles as collateral. More credit could in principle facilitate the process of agricultural intensification, which has been described as central to the labor absorption process. But it is by no means clear that land titles make for easier credit. The process of land adjudication in Kenya has not been accompanied by an expansion of rural credit. One limitation on the use of land as collateral for loans is the difficulty of enforcing eviction, given the traditional nontransferability of land. As Feder and Noronha (1987, p. 144) state, "land is an attractive collateral, provided that the owner-borrower can assure the lender that the land can be transferred." A study of land tenure in Kenya prepared for the Nairobi Round Table concluded (FAO, 1985, p. 30) that "land ownership titles are no more accepted as security for loans" and that "since the economic argument for generalizing registration is not valid because in practice land titles are not used by producers, Kenya, although it is the African country with the greatest experience of a European type of legal system, has returned to the situation of the 1950s." The agricultural intensification that has occurred in Kenya has not depended upon the expansion of freehold property rights.

Rural household viability

As population backs up in the rural sector, the economic viability of households is threatened. The threat is seen principally in land scarcity and growing landlessness, in the migration of members seeking work elsewhere, and in the periodic "drop-out" of households during occurrences of drought or other crises.

The household generally has multiple sources of income. Farm incomes will often derive from both crops and livestock, and, in coastal or lake regions, from fishing. Nonfarm activities often center upon trading of various kinds, including specialist lines such as charcoal selling. The high prevalence of nonfarm activities among Kenyan households in 1977 was shown in Table 2. Half the households in this sample were engaged in at least one nonfarm activity, a quarter in two or more. Women are very often involved in earning nonfarm supplementary incomes, such as from brewing, pottery, sewing, and, especially, market vending or other selling activities.

Remittances are often an important supplement to household income. These may be sent from family members in the city or even from outside the country. While remittances are important in Kenya, they are relatively more important in dual economies such as that of Zambia and in countries such as

TABLE 3 Rural differentiation in North Kigezi District, Uganda, 1972

Class	Number of acres owned (ha)	Number of households	Percent of total households
Rural semi-proletariat	under 2 (0.81)	60	12.0
Bakopi (poor peasants)	2 and under 8 (.81–3.24)	191	38.2
Peasants	8 and under 50 (3.24–8.1)	158	31.2
Kulaks (rich peasants)	20 and under 50 (8.21–20.25)	85	17.0
Landed proprietors	50 and over (20.25 and over)	6	1.2
Average size of holding	11.5 acres (4.66)	500	100.0

NOTE: References in Dumont (1957, p. 486) relating to Hungary and the USSR refer to "poor peasants" with less than 11¼ acres, "medium peasants" with up to 35 acres, and "rich peasants" with over 35 acres.
SOURCE: Weyel (1973).

Malawi, Botswana, Somalia, and Ghana with substantial numbers working outside the country.

Food provision warrants separate consideration in assessing household viability. The bulk of rural households in African countries have very low marketable surpluses. It may of course pay farmers to grow export crops and buy food. For low-income farmers, however, marketing margins add such a large percentage to farmgate prices that it is generally important for them first to supply their own food requirements and then to look for a crop to market. (It should be noted that prices used by national income statisticians for calculation of subsistence output tend to disguise important variations between countries and between regions within a country in the richness or poverty of own-account production/consumption. There are related variations in the security of household food supply. These variations are not moderated by strong rural-to-rural migration, for which social barriers generally constitute much greater obstacles than in the case of rural-to-urban migration.)

It obviously does not make sense to talk of a particular level of income, similar to Malthus's subsistence level, at which a household ceases to be viable. Households can, however, be distributed into class-like categories that roughly correspond to their degree of vulnerability. Many examples could be given of this kind of stratification. One, by Weyel (1973), for North Kigezi District, Uganda, an area of intense land pressure, is shown in Table 3. The categories here follow the well-known classification of René Dumont. The distribution contrasts strikingly with the general picture drawn of Uganda as an "amorphous"[4] peasant economy. Weyel remarks that "class differentiation was more manifest than just emergent."

A categorization based more directly on types of farm and nonfarm activity is given in a paper on Kenya by Collier and Horsnell (1986, p. 10). They found that

> Even at a highly aggregated treatment of activities—food crops; non-food crops; livestock; business; wages; remittances—some 40 per cent of overall inequality is

> accounted for by which activity mix the household is engaged in. There is a pronounced income hierarchy with households only engaged in food crop production at the bottom and non-agricultural wage earning households without substantial agricultural activities at the top.

Collier, Radwan, and Wangwe (1986) reported considerable stratification also in Tanzania, with the interesting finding that the bulk of variation was accounted for by intra-village inequality and comparatively little by inter-village inequality despite ecological diversity in the stratified sample selected. Differences between households were accounted for largely by nonlabor resources, which were reflected in greater sales of high-valued cash crops, higher livestock ownership and income, and in much higher nonfarm income. This again emphasizes the importance of multiple sources of rural income and differences in access to income-earning activities among rural households.

Movement down to a lower rung of income-earning capacity is likely to be associated with a changing assets–liability position. In South Asia this may involve actual financial debt (see Cornia, Jolly, and Stewart, 1987, p. 124). In Africa changes are more likely to be associated with the sale of land or the loss of productive assets such as livestock. Thus Cutler (1984), analyzing 1982–83 data for Ethiopia, suggests that "one useful indicator of approaching crisis is an increased volume of sales of assets, livestock in particular." The existence of "distress sales" of livestock has been widely acknowledged. One policy implication is that employment creation through public works may, by helping to avoid the necessity for such sales, assist in maintaining the productive capacity of the rural household. Credit programs for assistance with livestock replenishment, such as that showing signs of success in Niger, are another possibility. Livestock economies are particularly prone to stratification, with changes in the ownership of assets, especially when there are trends toward commercialization.

In a number of African countries female-headed households make up a substantial fraction of all households. In some districts of Malawi such households now account for more than 40 percent of the total. They are concentrated at the lower end of the income and wealth distribution. Outmigration of men has left women with inadequate resources, short of family labor, and dependent on insufficient and unstable remittances. In the Luapula District of Zambia, female-headed households, accounting for one-third of the total, had about half the average farm size (Chambers and Singer, 1981). In Botswana such households are particularly handicapped in respect of draught power for arable farming.

The urban informal sector and labor absorption

It is now widely conceded, with varying degrees of reluctance, that the informal sector, rural and urban, will need to be seen in the future as a major absorber of increases in the workforce in Africa. This is a matter of arithmetic: the number of formal sector jobs that can be extrapolated on the basis of past growth rates cannot account for more than a relatively small proportion of the expected increases. This would continue an existing trend. Gozo and Aboagye (1985) estimate the increase in informal sector employment in sub-Saharan Africa to have been from 44 percent in 1974 to 52 percent in 1980 (see also Cornia et al., 1987, p. 122).

The urban informal sector parallels the rural sector, both in absorbing secular increases in the labor force and in cushioning the effects of short-term changes in work opportunities in the urban formal and rural sectors. This has a number of corollaries. First, rural-to-urban migration is likely to be more a function of relative opportunities and earnings in the rural sector and the urban informal sector than of formal sector wages, though the latter will have some effect. Second, the rural–urban balance may be a balance of poverty, with poor opportunities in both rural and urban areas. Third, there is stratification of earnings in the urban sector just as there is in rural areas. McNicoll (1984, p. 208), for instance, observes, not specifically with respect to Africa, that:

> [T]here is a fairly clear productivity ladder down which marginal labor force entrants find themselves pressed as their numbers increase in the absence of either an open land frontier or an otherwise expanding economy. . . . Any economy offers a wide range of open-entry, low-productivity occupations, for the most part entailing self-employment but with a minimal requirement for working capital. Handicraft production, micro-scale trading and arbitrage, and personal services of all sorts are the main areas of this activity, highly visible in most poor countries. Such occupations are not indefinitely extensible, and even before private returns are pushed below subsistence it is likely that social returns have become negligible.

The lower end of this ladder corresponds to the "community of the poor" distinguished by House (1981) for Nairobi from an "intermediate sector." Closer scrutiny of House's data (Livingstone, 1984) reveals a ladder of earnings rather than just these two subsectors. Differences in education, skills, capital, or social contacts may determine which rung of the ladder new migrants reach.

One cannot therefore presume that by dint of being informal a sector has the capacity for dynamism and growth. The urban informal sector generally forms a substantial part of an economy that may or may not be growing. It may absorb increases in its labor force in a dynamic way or alternatively with increasing competition among participants, diminishing marginal productivities, and falling supply prices of labor.

Similarities with other African countries

Most of the discussion thus far has been of Kenya, where rural labor absorption has been very effective—albeit deferring rather than resolving problems of rural-to-urban migration and fertility decline. How have other countries of this region fared in those terms?

Malawi demonstrates in a very similar way the importance of divisibility in farm holdings in a process of rural labor absorption in a situation where the urban/industrial alternative is limited—a consequence of its small domestic market, landlocked situation, and competition from cheap South African manufactures. The intercensal population growth rate over the period 1966–77 was 2.9 percent. The bulk of the increase in population was absorbed into agriculture, particularly the smallholder sector. (By 1977 the proportion urban was still below 9 percent.) Table 4 shows that the size of smallholdings declined progressively over the period, especially in the Southern Region, where land pressure has been most concentrated. Here the proportion of smallholdings below 1 hectare increased from 44.5 percent in 1968–69 to 69 percent in 1980–81.

In Kenya subdivision was accompanied by agricultural intensification, which allowed agricultural incomes per capita not only to be maintained, but actually to expand over time. This process is less evident in Malawi, despite some agricultural progress. In Table 5, relating to three districts in Liwonde Agricultural Development Division, farm plots below 0.5 hectare in size (those shown in Table 4 to have increased in the Southern Region to almost one-third of the total) on average yielded only 33 percent of the mean agricultural income and just 9 percent of that for farms bigger than 2.5 hectares. Those of size 0.5–1 hectare on average yielded only 56 percent of the income of farms in the next larger size group. The low rates of labor utilization for the smallest farm categories shown in Table 5 suggest a substantial degree of disguised unemployment associated with land scarcity. The bottom two size groups, containing two-thirds of the families, supply from farm production only 37 and 75 percent respectively of their food requirements, measured in caloric values.

The rural development policies pursued in Malawi have not performed well by the criterion of labor absorption. Emphasis has been on large-scale estate production of tobacco and sugar, supported by low-wage policies. It has been

TABLE 4 Percent of very small-sized holdings, Malawi, 1968/69 and 1980/81

	Percent of smallholdings below			
	3 ha	2 ha	1 ha	0.5 ha
Total Malawi				
1968/69	85	72	37	—
1980/81	95	84.5	55	—
Southern Region				
1968/69	91	80	44.5	22
1980/81	98	93.5	69	32.5

SOURCE: ILO/JASPA, *Youth Employment and Youth Employment Programmes in Africa: Malawi* (Addis Ababa, 1986), p. 28.

TABLE 5 Agro-economic data for three districts, Liwonde Agricultural Development Division, Malawi, 1982

	Farm size (ha)						
	Under 0.5	0.5–1.0	1.0–1.5	1.5–2.0	2.0–2.5	2.5 and over	Total
Agricultural income[a] relative to average (3-district average = 100)[b]	33	78	138	198	253	370	100
Percent labor utilization[c]	11	24	36	43	53	72	29
Percent of food requirements provided[d]	37	75	116	139	174	254	92
Percent of families	29	38	19	8	3	3	100

[a] Total value of agricultural production less production costs.
[b] Mangochi, Machinga, and Zomba districts.
[c] Based on labor requirements for crop production less family labor availability in terms of man-hours.
[d] Family food requirements compared with food available from production, both measured in caloric values.
SOURCE: Etema (1984) derived from Ministry of Agriculture, Liwonde Agricultural Development Division.

suggested (Livingstone, 1985) that this retarded labor absorption in agriculture by reducing effective use of the available land, restricting access to higher-valued cash crops, and limiting the forward consumption linkages that would have followed from a more general increase in rural purchasing power. Similar criticisms have been made by Kydd and Christiansen (1982).

The problems of rural labor absorption are also severe in Ethiopia. The population is heavily concentrated in the highland areas of high arable potential where land scarcity has led to very small holdings. In a survey of four districts, Rahmato (1984) found that 93 percent of holdings were below 0.5 hectare in one (Bolloso) and 36 percent in another (Manna). The agricultural sector is heavily subsistence-oriented: a survey of the 13 districts carried out by the Institute of Development Research in 1975–76 showed the proportion of households engaged only in food crop cultivation to be 78 percent or more in 9 out of the 13.

Ethiopia's prospects of absorbing labor in an expanding urban economy by developing export-oriented manufacturing are limited by locational factors (Ethiopia is technically not landlocked but is effectively so given the central location of most existing industry and poor communications to the sea). At the same time lack of domestic purchasing power constrains import-substitution. Despite the lack of strong pull factors, however, there is relatively rapid and continuous migration into urban areas, impelled by land scarcity in the highland areas, and periodic drought and low productivity in the low- and medium-potential agricultural areas. The urban population in Ethiopia is projected to increase from just below 15 percent in 1982 to over 22 percent in 1994.

Once again past policies have not been well designed to improve the situation. The broader effects of the weakness of the agricultural sector were commented on by an ILO Mission, which stated (ILO/JASPA, 1981, p. 386) that "incomes and the productivity of labor in the countryside are extremely low and hence both the marketable surplus and the internal rate of savings in agriculture are insufficient to finance either its own expansion or the development of the rest of the economy." The view of the Mission was that Ethiopia "has systematically under-invested over a long period in rural infrastructure—roads, power, irrigation, storage and processing facilities—and in the health, education and training of its rural population." Past policies, which include an agricultural strategy based upon capital-intensive state farms, were not directed toward maximizing effective absorption of labor in the rural economy. Indeed, major emerging problems of erosion suggest that even the existing resource base is in jeopardy.

While land scarcity is a spreading condition in Africa, many economies in the region can still be properly described as "land surplus." Is there a labor absorption problem, and therefore a population problem, in such economies? Tanzania is a case in point. Thomas (1974) has analyzed changes between the 1957 and 1967 Tanzanian population censuses, and it is clear that the same trends have been in evidence since then. Population pressure was high in the densely populated mountain areas and on the shores of Lake Victoria. Despite this, the mountainous areas were not experiencing massive outmigration. Although the regional pattern of population change was "variegated," with resettlement, seasonal and short-term labor migration, and urbanization all contributing to a picture of "apparent flux," much of the growth of population is taking place in situ, at the location of past population clusters. This means that notwithstanding the availability of land in Tanzania as a whole, the need to find further means to absorb labor in such high-density areas as Kilimanjaro is vital.[5] Other countries in Africa also show this coexistence of acute local land pressure and areas of land surplus.

Conclusion

In present-day Africa, with low rates of industrialization and exceptionally rapid rates of population increase, against a background of variable resource bases, a straightforward strategy of maximizing output growth (for example, as put forward in the Berg Report—World Bank, 1981) is inadequate. Nor is it particularly helpful to proffer a choice between growth and equity, with the implication that an optimum tradeoff exists. It is important to emphasize labor absorption—more than formal sector employment creation—because the former can simultaneously promote output and equity. An example of a basic policy that could do both is the unimodal strategy argued for by Johnston (1985), under which agricultural research and extension efforts are directed toward the situation of the average farmer rather than a limited number of "progressive" farmers with atypical endowments of resources. Actual government policies widely pursued throughout Africa have often diverged from this, at considerable cost.

Notes

1 "Gap farms" are those whose size falls between the smallholder and large farm categories as defined by the Central Bureau of Statistics.

2 The introduction of coffee and tea on small farms was therefore not primarily "Boserupian," being the result of withdrawal of negative policies by government.

3 Unlike differences in unit costs associated with the scale of operations, *X*-efficiency refers to unit cost differences (at whatever scale of output) deriving from the efficiency of the entrepreneur in combining and managing the factors of production under his or her control.

4 The term was used by Polly Hill (1968) to make the same criticism of descriptions of peasant agriculture in West Africa.

5 It may be noted that one of the earliest systematic reviews of the economy, the 1961 World Bank report on Tanganyika, did make the overly simplified deduction that population-surplus areas and population-sparse areas could be easily equilibrated through settlement schemes. These were indeed taken up subsequently as a major plank of policy but proved unsuccessful (Cliffe and Cunningham, 1973).

References

Benneh, G. 1985. "Emerging land tenure systems in Africa and their implications for development of peasant agriculture," paper presented at the FAO Regional Workshop on Experiences in Agriculture Sector Planning in Africa, Arusha, Tanzania.

Chambers, R., and H. Singer. 1981. "Poverty, malnutrition and food insecurity in Zambia," in E. Clay et al. (eds.), *Food Policy Issues in Low Income Countries*, World Bank Staff Working Paper No. 473, Washington, D.C.

Cliffe, L., and G. L. Cunningham. 1973. "Ideology, organisation and the settlement experience in Tanzania," in L. Cliffe and J. S. Saul (eds.), *Socialism in Tanzania*, Vol. 2. Nairobi: East African Publishing House.

Cohen, John M. 1980. "Land tenure and rural development in Africa," in Robert H. Bates and Michael F. Lofchie (eds.), *Agricultural Development in Africa: Issues of Public Policy*. New York: Praeger, pp. 249–300.

Collier, P., and P. Horsnell. 1986. "The agrarian response to population growth in Kenya," paper presented at the IUSSP seminar on Economic Consequences of Population Trends in Africa, Nairobi, December.

________, and D. Lal. 1980. *Poverty and Growth in Kenya*. Baltimore: Johns Hopkins.

________, S. Radwan, and S. Wangwe. 1986. *Labour and Poverty in Rural Tanzania*. Oxford: Clarendon Press.

Cornia, A., R. Jolly, and F. Stewart. 1987. *Adjustment with a Human Face: Protecting the Vulnerable and Promoting Growth*. New York: Oxford University Press.

Cutler, P. 1984. "Famine forecasting: Prices and peasant behaviour in Northern Ethiopia," *Disasters*, 8 January.

Dorling, M. J. 1979. "Income distribution in the small-farm sector of Kenya—background to critical choices," paper presented at the seminar on Alternative Patterns of Development and Life Styles in Eastern Africa, Society for International Development, Nairobi.

Dumont, R. 1957. *Types of Rural Economy*. London: Methuen.

Etema, W. 1984. "Food availability in Malawi," paper presented at the conference on Food Self-Sufficiency in Sub-Saharan Africa, Dar es Salaam.

FAO. 1985. *Report of the Round Table on the Dynamics of Land Tenure and Agrarian Systems in Africa, Nairobi*. Rome: FAO.

Feder, G., and R. Noronha. 1987. "Land rights systems and agricultural development in sub-Saharan Africa," *Research Observer* 2, no. 2. Washington, D.C.: The World Bank.

Gozo, K. M., and A. A. Aboagye. 1985. "Impact of the recession in African countries: Effects on the poor," Geneva: ILO, mimeo.

Hill, Polly. 1968. "The myth of the amorphous peasantry: A Northern Nigerian case study," *Nigerian Journal of Economic and Social Studies* 10, no. 2.

House, W. J. 1981. "Nairobi's informal sector: An explanatory study," in A. C. Killick (ed.), *Papers on the Economy of Kenya.* Nairobi: Heinemann Educational Books.

ILO. 1972. *Employment, Incomes and Equality: A Strategy for Increasing Productive Employment in Kenya.* Geneva.

ILO/JASPA. 1981. *Rural Development, Employment and Incomes in Kenya,* International Labour Office/Jobs and Skills Programme for Africa, Addis Ababa.

Johnston, B. F. 1985. "Agricultural development in tropical Africa: The search for viable strategies," paper prepared for the Committee on African Development Strategies and the Council on Foreign Relations and the Overseas Development Council, March.

King, Russell. 1977. *Land Reform: A World Survey.* London: G. Bell and Sons.

Kydd, J., and R. Christiansen. 1982. "Structural change in Malawi since independence: Consequence of a development strategy based on large-scale agriculture," *World Development* 10, no. 5: 355–375.

Livingstone, I. 1984. "The rural and urban dimensions of the informal sector: A discussion of Kenya," School of Development Studies, University of East Anglia, Discussion Paper No. 154.

———. 1985. "Agricultural development strategy and agricultural pricing policy in Malawi," in K. Arhin, P. Hesp, and L. Van der Laan (eds.), *Marketing Boards in Tropical Africa.* London: Kegan Paul International.

Mafeje, A. 1985. "African customary land tenure systems: Obstacles to development? Current land reform policies: Their strength and weaknesses," in FAO, 1985, pp. 11–15.

McNicoll, G. 1984. "Consequences of rapid population growth: An overview and assessment," *Population and Development Review* 10, no. 2: 177–240.

Rahmato, D. 1984. *Agrarian Reform in Ethiopia.* Uppsala: Scandinavian Institute of African Studies.

Smith, L. 1978. *Low Income Smallholder Marketing and Consumption Patterns—Analysis and Improvement Policies and Programmes,* FAO Marketing Development Project, Ministry of Agriculture, Nairobi.

Thomas, I. D. 1974. "Population growth, population retention and migration in mountain areas of Tanzania: Some illustrative data and directions for research," in V. F. Amann (ed.), *Agricultural Employment and Labour Migration in East Africa,* Makerere University.

Tidrick, G. 1979. *Kenya: Issues in Agricultural Development.* Nairobi: Ministry of Economic Development and Planning.

Weyel, V. 1973. "Land ownership in Nyakmengo, Ruzhumbura (North Kigezi), a preliminary survey," in *Research Abstracts and Newsletter: A Journal of Policy Communication* 1, no. 3, Makerere Institute of Social Research.

World Bank. 1973. *Agricultural Sector Survey—Kenya, Report No. 1,* 245a-KE. Washington, D.C.

———. 1981. *Accelerated Development in Sub-Saharan Africa: An Agenda for Action.* Washington, D.C.

CLASS STRUCTURE AND LABOR RELATIONS

Agrarian Change and Class Conflict in Gujarat, India

JAN BREMAN

ONLY A FEW DECADES AGO RURAL INDIA was described in terms of poverty and stagnation, a society imprisoned in tradition. Gunnar Myrdal's well-known study published in 1968 provides a prominent illustration of this school of thought, which dominated Western debates on development around the middle of the twentieth century. While his *Asian Drama* had a wider setting, the author based his argument mainly on the southern part of this populous continent, more precisely on the incapacity for self-transformation that seemed to characterize former British India. Myrdal noted with concern that development efforts were largely nullified by the enormous population growth caused by a fall in the death rate and a continuing high birth rate. Major institutional reforms were urgently needed, in his opinion, to overcome the underutilization of labor, which hindered any increase of agrarian production. What he had in mind was a radical redistribution of agrarian resources, which would place land in the hands of the tillers. The objectives of such a basic reform were twofold: a political one, namely, reduction of the rigid inequalities in the social structure; and an economic one, namely, the optimal use that a much greater mass of owners would make of the means of production allotted to them. A closer and more realistic assessment of policies and politics, however, convinced Myrdal that government lacked the will and the capability needed for transforming the rural system in this way. In his opinion, the failure to introduce any fundamental change into traditional property relations and the ingrained attitudes of agrarian people who were perceived not to be "development-oriented," was due to the "soft" character of the Asiatic state.

That social dynamics were nevertheless not entirely lacking was shown by the concentration of landholdings among a comparatively small village elite, mostly members of upwardly mobile peasant castes who had in recent times gained political power. Myrdal was well aware of the growing weight of this class and the concomitant deterioration in the position of sharecroppers and agricultural laborers. His skepticism regarding the chances that the social balance would be tilted in favor of the great mass of small producers led Myrdal to advocate an agrarian strategy that would enable the small landowning elite to provide leadership to, and form the social basis of, the process of economic growth.

Although not elaborately or even well argued, this policy recommendation became a major turning point in Myrdal's analysis. The veiled bias in favor of capital owners already shown by bureaucratic agencies had to be changed into purposeful and systematic promotion of the interests of dominant peasants in order that their potential to develop into genuine rural entrepreneurs of modernization might be realized. The increasing social differentiation, an inevitable result of this scenario, would have to be compensated by providing protection and security to the labor force working on the land. Using the catch slogan of "welfare capitalism," Myrdal suggested that the authorities should "give a small plot of land—and with it a dignity and a fresh outlook on life as well as a minor independent source of income—to members of the landless lower strata" (Vol. II, 1968: 1382). The Swedish economist thus implied that, since a more radical path to Indian development was unattainable, social inequality could in any event be reduced or at least curbed by capitalist methods. By holding out the prospect that a better life was within reach of all those who owned even a little land, he seems indirectly to argue that these lowest classes, which form the majority of the rural population, would see the need for self-restraint in their demographic behavior and consequently would practice birth control out of their own interest.

The course of events has corroborated Myrdal's opinion. The route he favored, much more actively promoted by Indian policymakers in the 1970s and 1980s than he had foreseen (Chakravarty, 1987: 24), has indeed facilitated a shift toward a capitalist mode of production in agriculture. "Betting on the strong" was the phrase used by Willem Wertheim at a very early stage to describe the strategy of entrusting the rich peasants with implementing the Green Revolution (Wertheim, 1964: 259–278). These true "sons of the soil" have come out from under the shadow of large landlords, moneylenders, and traders to whom they were tied by pre- or early-capitalist dependency relations during the colonial era. As major beneficiaries of a series of land and tenancy reforms that were introduced shortly before and after Independence, they acquitted themselves of the task assigned to them as "progressive farmers," namely, to modernize agriculture and to increase the agrarian surplus. Moreover, and this is an important addition, these peasant entrepreneurs started to invest part of their savings away from agriculture, a development that has stimulated the diversification of the rural economy as well as strengthened rural–urban linkages.

The emerging capitalism is a major trend but still does not everywhere dominate the structure of agrarian production. The transformation process has occurred at varying rates in different parts of the country. Regional disparities have increased in a way that makes one suspect that more is at stake than a mere difference in stages of the same growth pattern. Uneven development is very much a feature of the general trend toward capitalism (Patnaik, 1986: 18–20). Second, it is questionable whether the aggressive, expansion-oriented behavior of the peasant-capitalists lends itself to imitation by the great mass of petty producers who have far fewer resources at their disposal. If we are properly to understand the outcome of the economic restructuring that is going on, it is essential to keep track of mobility trends—frequently more downward than upward—as well as mechanisms of consolidation in the broad middle range of society. Nevertheless, I agree with Myrdal that the force and the speed with which capitalism is gaining ground can best be shown by emphasizing the situation at the polar ends of the agrarian hierarchy.

Surat District in the south of Gujarat is one of the bulwarks of peasant capitalism. I have been able to document the advance of this mode of production over the last 25 years in a number of micro-studies based on recurrent fieldwork. The landowning elite and the landless proletariat, although disproportionate in size, together represent roughly two-thirds of the region's agrarian population. In combination, the capital and labor provided by these two classes produce the greater part of the agricultural yield.

The rich peasants with holdings of 8 acres or more mostly belong to the local dominant caste. For the last two generations they have exercised power at the district level. By setting up their own economic and political associations, they have lately been able to extend their influence far beyond the local domain. These entrepreneurs have been militant in expressing their displeasure at what they consider to be willful neglect by the state of the countryside and agriculture. They propagate a rural populism that is strongly opposed to the idea that urban-industrial interests should be given priority in the guise of a growth-pole strategy. Their leaders, like those of other movements of peasant producers that are coming to the fore in other parts of India, have little faith in politicians and exert pressure on them by occasionally mobilizing their rank-and-file. Their demands, expressed in a typical *kulak* idiom, stress the peasant owner as the focal element. The interests of other rural classes, insofar as these are considered at all, are secondary to those of the sons of the soil. Charan Singh was a leading proponent at the national level of this brand of agrarian populism (Singh, 1986; Byres, 1988). He mapped out a path to development based on close integration between agriculture and rural industry, in a way that would keep the autonomy of the peasant-cultivator intact. From among them a new type of manager would emerge, leading what would primarily be small-scale industry. The employment resulting from this growth model is supposed to contribute to the welfare of all.

Does reality in South Gujarat agree with this scenario? The anticipated rise of agrarian production has indeed taken place and, thus self-propelled, the rural economy has undergone a true metamorphosis. The expansion in employment opportunities, however, has not kept pace with the growth of the landless population. The further broadening of the bottom of the agrarian hierarchy is due not so much to high fertility as to immigration. One consequence of the transition to capitalism is that labor becomes mobile over long distances. Myrdal made no allowance for this in his study and even explicitly excluded the possibility, as becomes clear if one reads his notes on migration. In his opinion, the life of the village population is determined by immobility and isolation, the only exception being departure for urban destinations. Poverty, more than any other factor, prevents people from escaping from their birthplace:

> [T]he poorer the people, the stronger the barriers to migration; poverty squeezes the margin for risk-taking, blunts the incentive to try new things, and rigidifies all restraints upon initiative. (Vol. III, 1968: 2140)

My own findings in South Gujarat are quite contrary to those of Myrdal. The massive circulation of labor, mostly for one season but sometimes for longer periods, shows that the hinterland of the capitalist enclaves has become very strongly tied to this mode of production. Intra-rural circulation, a neglected theme in literature on labor migration, shows how the new economic dynamics have radiated throughout large parts of India. Both the mobility of labor and the value attached to this factor of production under a capitalist regime explain why poverty, instead of disappearing, now governs the life of more people in the countryside than ever before. The findings of my fieldwork-based research in South Gujarat are, in my opinion, also indicative of the transformation process that is taking place in other regions of India (Chakravarty, 1987: 27). The misery that I encountered refutes the rationality that Myrdal, perhaps against his own better judgment, attributed to the logic of capitalism.

A new landscape

Ten years ago I was in the midst of investigation into the effects on labor relations of the changes in the agricultural economy that had occurred in the plain of South Gujarat, on India's west coast. The emergence of a new landscape was due to the completion of an irrigation system that secured for an extensive part of this fertile region a sufficient year-round supply of water. A network of canals now cleaved the district, originating in a reservoir that had been constructed by building a dam upstream in the River Tapti. This intervention in the natural ecology brought the forces of production to a higher level of development. The area of perennially irrigated land was also expanded through the increased availability of oil and later electric pumps, which made it possible to dig new and much deeper wells.

The transition to wet-land agriculture brought with it sweeping changes in the cropping pattern. The cultivation of millet, cotton, and, to a lesser extent, groundnuts, which had prescribed the agrarian cycle for almost a century, made way for the production of paddy, bananas, and, increasingly, sugarcane. In many villages this last crop now takes up half or more of all arable land.

That agriculture became more intensive is only partly attributable to the expansion of irrigated land, although the significance of this factor can hardly be overestimated. Production has also been increased as a result of the Green Revolution, for which the central plain of South Gujarat was one of the earliest concentration areas. The promotion of the new package of inputs, starting with the Intensive Area District Programme in 1962–63, meant that farmers in the area had high-quality seed varieties at their disposal, more fertilizer, agricultural extension services, and generous credit facilities with which to modernize their equipment, methods of cultivation, and farm management.

Another important factor was the expansion of the road network as part of the improved infrastructure, which stimulated the changeover to mechanized means of transport. In touring the region one cannot overlook the many indications of modernity. There is dense traffic on the roads: heavily laden lorries and tractors with trailers bringing agricultural produce to depots for processing or further transport; the crowded but frequent buses that have made the countryside more accessible and, conversely, easier to leave; and the horde of bicycles, mopeds, scooters, motorbikes, and cars, all of which emphasize the mobility made possible by the new pattern of communication. The background to this lively picture is the intensely cultivated fields, on which water, machines, and workers are present almost year-round. The supply of nonagricultural commodities and services has increased considerably. The growing number of village shops now sell more and better consumption goods, thus showing the rise in purchasing power of at least part of the rural population. The landscape that I had come to know during fieldwork in the early 1960s (Breman, 1974: 231–233) has changed almost unrecognizably in the subsequent quarter-century.

Demographic growth has placed increasing pressure on agrarian resources. In the central part of the plain, population density now exceeds 350 persons per square kilometer. On the one hand, agricultural dynamism has created new employment opportunities, but on the other hand agricultural mechanization has reduced the demand for unskilled labor. Due to economic diversification, however, and notwithstanding an annual rate of natural increase of 2.6 percent that has caused the population almost to double during the last 25 years, the total scope for rural subsistence has actually increased.

The development of the forces of production gave a powerful impetus to the building trades. Raw materials largely originate within the region and are processed in small-scale industries into stones, bricks, roof and floor tiles, cement pipes, timber, and so on. The demand for such products, used in the construction of roads and irrigation canals as well as for utility buildings and houses, has increased enormously. The motorization of transport and mechanization of agriculture, requiring greater technical skill, have caused maintenance and repair workshops to be opened up at busy traffic junctions. Extension of the electricity network to rural areas has made it possible to disperse the diamond industry, which until a few decades ago was located in the cities of Surat and Navsari. In the 1970s in particular, workplaces where the raw diamonds were cut and polished opened up in increasing numbers of villages. The net result of all these activities was that at the beginning of the 1980s approximately one-third of the nonurban population of Surat District worked outside agriculture, considerably more than the average for the rural sector in the state of Gujarat as a whole.

Even more significant for the trend toward diversification was the rise of new agro-industries. By far the most important of these are the sugar factories that process all cane produced in the region. One of these factories also produces the chemical solvent acetone. In addition, the sugarcane waste, rice husks, and grass serve as raw materials for the six paper factories that have been set up in the district. Their productive capacity is modest, and they specialize in manufacturing strawboard and packing paper. The new agro-industrial complex also has a stimulating effect on other branches of industry. In Bardoli, the central town of the plain, for example, a local smithy has grown into a large engineering works specializing in the manufacture of machines for the paper factories. After successfully installing a number of complete machine halls in the region, this family-operated business has received orders for similar equipment elsewhere in India and even abroad. Finally, cattle-keeping has become more businesslike in character, modeled after the famous AMUL milk cooperative in Central Gujarat. Milk is delivered to, and processed by, a dairy cooperative in Surat city, which has set up its own cattle-feed factory. And eggs collected from large poultry farms are sold in distant urban markets.

In the countryside of South Gujarat, agriculture and industry have thus become more closely interwoven in terms of capital and management. This interconnection can best be illustrated by the cultivation and processing of sugarcane, which, during the last few years, has become almost a mono-crop in many villages. At the beginning of the 1960s the area planted with sugarcane amounted to barely 1,000 acres (1 acre = 0.40 ha), but two decades later it had multiplied to 142,000 acres. The first factory to make a modest start in 1957 has grown into the largest cooperative enterprise for this commodity in all Asia, with a daily production capacity of 7,500 tons. At present there are nine such factories in the region, and more are under construction. Without exaggeration one may conclude that in South Gujarat, as in various other parts of India, agriculture has taken the shape of industry over the past few decades. Before examining the consequences for labor of such changes, I shall first discuss the landowning class, which spearheaded the transition to the new capitalist mode of production that has become predominant in the plain of South Gujarat.

Peasant entrepreneurs

In 1960 rather more than 50 percent of all farms in the district had less than 25 percent of the cultivable area at their disposal. At the top of the agrarian ladder, about 10 to 15 percent of rich peasants commanded roughly half of the available land. Most of them belong to the dominant Patidar caste. Although the members of this community represent only some 7 percent of the total population, they dominate the social hierarchy.

Their elevation to their present status as a rural elite goes back no more than two to three generations. At the beginning of the present century the Patidars had a reputation of being simple and devoted cultivators of comparatively low origin, whose ancestors arrived in the region within living memory after famine had driven them from their more northerly native habitat. The introduction and rapid expansion of cotton cultivation, dating from the 1860s, marked the commercialization of agriculture in Western India under colonial domination. Linkages with the world market were channeled through traders established in the cities or even abroad, to whom the peasants became tied through the credit that they offered.

The economic and social advance of the Patidars began in the first decades of the twentieth century when they set up their own cooperatives, first for the purchasing and later for the processing of cotton. The growing control that the producers thus exercised over marketing was an important step in the penetration of a more capitalist mode of production in South Gujarat's countryside, further accelerated by tenancy reforms shortly before and after Independence.

With the adoption of irrigation agriculture, the cooperatives, whose advantages had been proven, continued to be the economic model in which the Patidars coordinated their farm management. It helped them to enlarge their scale of activities. A cooperative in Bardoli, whose members specialized in growing bananas, was not content to sell only to the local market; at the end of the 1960s, it leased a freighter with which to export the members' produce to the Gulf States of the Middle East. Patidar landowners, who meanwhile had gained social recognition as a vanguard of progressive farmers, controlled the cooperatives and provided most of their managerial staff.

Cooptation along caste lines has ensured that the new agro-industry with its cooperative basis has grown into a Patidar stronghold. This applies particularly to the factories that process the ever-increasing amounts of sugarcane. The rich peasants have themselves raised the necessary capital for these industries, liberally supplemented with government subsidies and credit. The purchase of a share in the cooperative gave the sugarcane grower the right to deliver his cane to the factory. Cash advances and inputs supplied by the factory, such as seed, fertilizer, and tools, together with agricultural extension services, indicate managerial intervention in farm production, but without eliminating the landowner as private entrepreneur. All decisions regarding cultivation are made by the peasants, but in doing so they can count on the generous facilities that the cooperative makes available to its members. Each factory has an agricultural staff whose aim is to increase production and productivity. This is done by singling out model farmers who are prepared to experiment with new seed varieties and planting techniques; by organizing meetings at which new methods are discussed; by arranging excursions to more advanced production areas in neighboring Maharashtra State; and so on. The result is that the yield has increased considerably, amounting at present to over 40 tons per acre. My research shows that costs of production have increased far less than the prices received by the members of the cooperatives, which, in 1989, amounted to Rs 430 per ton of cane delivered. A share in the region's oldest sugar factory, with a nominal value of Rs 500, was traded early in 1989 for about 50,000 rupees or more. There is no better way to highlight the enormous profits that the peasant capitalists reap from growing sugarcane.

The prosperity of the dominant landowning caste-class has visibly increased. A growing number of wealthy families have replaced the simple houses in which they formerly lived by large bungalows of two or three storeys. Tiled floors, modern sanitation, urban-made furniture, refrigerators, kitchen utensils of stainless steel, television, video-cassette recorders, and, last but not least, a motorbike or scooter at the doorstep, all emphasize the comfortable life of the rural elite. In addition to the quantity and quality of consumer durables, the great variety of food and other daily consumption goods illustrates the achievement of a prosperous lifestyle that was formerly found only among the urban elite. Equally noteworthy is the reduced participation of household members in agricultural work. The gradual withdrawal from the economic process of women and children, the former particularly in regard to work outside the house, has been carried to such an extent that even the male members of the main landowning caste now avoid working on the land and restrict themselves to supervising their hired nonfamily labor.

It would be incorrect to conclude from this that the Patidars are about to give up agriculture as a profession. Such a conclusion would also be contrary to the investment behavior of this entrepreneurial class. The yield of their agrarian resources has largely been used to improve the value of their land and to modernize and augment their capital stock. In addition, they seek nonagrarian outlets for their profits. The need to diversify arises from their anxiety to find sources of wealth to enable the next generation to maintain and preferably to improve their present level of prosperity. In preparation for this they send their sons to college. Equipped with a Bachelor's degree in science or commerce, the male offspring are then sent to learn a trade or enter a profession, usually with the notion that they should become self-employed or, better still, owners of small industrial enterprises or trading establishments. Women belonging to the dominant caste restrict their childbearing—now frequently by sterilization. They do so after having borne two sons, one of whom will continue the farm, and as few daughters as possible in the hope of avoiding the high dowry and marriage expenses that may be demanded by a partner from the same social class. This conservative demographic strategy explains why the Patidars' share in population growth is a modest one.

The increasing rural surplus wealth is partly drained away and converted into urban capital, but this does not necessarily mean that the peasant-entrepreneurs have lost control over these investments. The surplus that reaches the cities is increasingly accompanied by Patidars of the younger generation looking for an opportunity to open a workshop or to find some other way to get a footing in the urban milieu as small-scale entrepreneurs and businessmen. While the drain of rural capital toward the urban economy at first sight seems to corroborate Michael Lipton's thesis of a pattern of uneven development between countryside and city (Lipton, 1977), I would argue that the direction in which money flows is only one dimension of a more complex shift in the balance of social forces, eventually leading to a new class alliance between segments of the rural and urban bourgeoisie.

The economic advancement of the Patidars has also increased their self-esteem. Known as *bhumiputra* (sons of the soil), they make no excuses for their rural roots. With their larger scale of activity and expanded power base, however, villages for them are now no more than platforms for action in a wider sphere. Patidars control the institutional infrastructure right up to the district level and even beyond. One encounters them in all major places of economic, political, and social transactions. This minority is accustomed to making its presence felt. Their dominance over public life in the region is highly visible: industries and agricultural cooperatives bear the name of their caste; they manage the local sections of the various political parties; they serve on the managing boards of schools and institutions. In all respects, the heart of the plain is Patidar-land.

The caste solidarity on which the advance of the cooperative movement in Western India is based also forms the foundation for political action. In the 1920s the Patidars successfully resisted attempts by British colonial authorities to raise land taxes. After Independence, the continuing social progress of this landowning caste made them no less critical of the state and its bureaucracy. Their displeasure is expressed through the *Khedut Samaj*, a nonparty political union of peasant producers, which recruits its rank-and-file in Surat District largely among the Patidars. According to the populist thinking prevalent in this milieu, state power has been too long in the hands of party politicians and bureaucratic policymakers who are biased in favor of industrial-urban interests. The leadership, which owes its allegiance to the rich peasantry, stresses the basic unity of landowners and landless, but this conceals acute contradictions within rural society. Although the current tilt in the balance of power toward the countryside has thus led to a shift in the sectoral division of interests, the social disparity has remained much the same.

The landless existence, past and present

The rural proletariat is greater in number than the various classes to which the landowners of Surat District belong, either separately or in combination. According to the 1981 census, the landless represented 54.5 percent of the agricultural population, and their majority is even greater in the densely populated central part of the plain. Although South Gujarat is distinguished by a particularly high concentration of agricultural workers, the proportion of landowners is also gradually declining in other parts of the state. For Gujarat as a whole in 1961 there were 1.24 landowners to each landless worker, but by 1981 the ratio had dropped to 0.84.

Which mechanisms are responsible for the growing proletarianization of the agrarian population? It is only natural to associate the trend with the advance of the capitalist mode of production. One familiar interpretation is that this restructuring along class lines is the result of the accumulation of agrarian property at the top of the hierarchy, while an increasing number of small peasants lose their holdings and slip downward into a landless existence. Such a dual segmentation I believe does not exist in reality, however, not even in the region where production relations are permeated with capitalist features. In the region of my fieldwork, for example, medium and small holdings (of 4–8 acres and of less than 4 acres respectively) continue to exist. The strategy adopted by the households belonging to the small peasantry to prevent their regression on the social ladder is characterized on the one hand by intensifying production and, on the other hand, by supplementing the family income with off-farm employment, for which the diversification of the rural economy has created many new opportunities. In the case of petty landowners in particular, the effort to preserve their holdings for the next generation involves heavy self-exploitation—long working hours at low pay, and maximum participation of household members in the labor process. But this strategy also entails pushing some members of the younger generation into a landless existence, preferably away from the home village so as to veil their downward mobility. In other words, the process of proletarianization is fuelled by the expulsion of "superfluous" members from resource-poor peasant households, but without this expulsion leading to any drastic shift in landownership patterns.

A large proportion of the landless population in the South Gujarat plain belongs to the tribal caste of Halpatis, whose ancestors were denied access to agrarian property many centuries ago. They played an important part in the opening up of new land, a gradual process that was completed early in the twentieth century, but their role was mostly as laborers in the employ of landlords. Was the doubling of the landless population of Surat District between 1951 and 1981 perhaps due to a disproportionate demographic increase of those castes to which the bulk of agricultural workers traditionally belong? This second explanation of the growing pressure at the base of the agrarian system certainly is in line with prevailing opinion in higher social echelons. It is a familiar cliché that the landless have no inhibitions about the size of their families and that severe, even draconian, sanctions are necessary to force them to keep their demographic behavior in check. However, there is nothing to indicate that the Halpatis, who, more than any other caste in South Gujarat, are identified with landlessness, distinguish themselves from the rest of the rural population by a conspicuously high birth rate. In 1971, in the core of the central plain around Bardoli, the Halpatis made up one-third of the population, somewhat less than had been the case 20 years earlier. The 1971 census also showed that within the agricultural population in the region three out of every four belonged to the landless class. Neither local proletarianization nor a high birth rate among the lowest social categories can explain the speed with which the ratio of landowners to agricultural workers has declined. Rather, the explanation seems to lie in the qualitative changes in labor relations that have attended the erosion of patron–client ties and in the inflow of long-distance seasonal migrant workers toward this rural growth pole of capitalist production.

The changing shape of the agrarian workplace cannot merely be explained in quantitative terms. During the last half-century production relations in the central plain have undergone important qualitative changes that have strongly increased pressure on the agrarian base of existence. In the past, the Halpatis of South Gujarat lived in a state of bondage. *Hali* was the name given to a farm servant who had become tied to a landowner through indebtedness. From an early age he was bound to work for his master for an indefinite period in exchange for a subsistence allowance. The relationship was usually formed when a rich peasant enabled a worker to marry by meeting all the expenses involved. The loan would be made primarily in kind. The master usually also had a right to the services of the hali's wife, who was employed as a maid in the household, and of their adolescent children. Girls helped their mother without pay; sons were taken on to look after the cattle on reaching the age of eight or ten. Together, and with considerable interchangeability, the bonded family did all the work on the land, in the house, and in the master's compound. All day, and if need be also in the evening or at night, they were at the beck and call of their master. In return, the master was obliged to provide his bondsmen with a livelihood. In addition to a daily ration of grain, this included various emoluments such as morning or noon meals, tobacco, tea, now and then cash to be spent on drink, firewood, a set of clothing each year, medicines if a member of the family should fall ill, a hut on the master's land, waste material for thatching the roof, and a tiny plot for the servant to cultivate for his household's own use.

The character of such employment was so comprehensive that neither the wage, the nature of the work, nor the working hours were laid down in any detail. Work was not always available, and the grain ration that the hali received during the off-season was charged as an advance on future wages. In this way, his indebtedness gradually increased over the years, partly because the master charged interest on the amount outstanding. Concepts such as debt, interest, loan, and advance were in any case only of nominal significance. Neither party wanted the relationship of dependency to be terminated by repayment of the debt. Bondage was attractive to the landowners because they were assured of a permanent and cheap labor force. In turn, the Halpatis preferred this mode of attachment because it provided them with security, however minimal. Colonial reports show that casual wage laborers, who formed a small minority, were far worse off from the economic point of view and were less respected socially. Elements of patronage were inherent in this form of voluntary servitude (Breman, 1974). The master not only committed himself to looking after his servant and servant's family, but also accepted the moral obligation to protect them against third parties. In turn, he could count on the unconditional loyalty and total subordination of his clients.

By stressing the personal character of this vertical relationship and the intimacy and warmth of affection among members of the households of successive generations of master and servant, it is easy to draw a picture of this hali system that is far too idyllic. The norm of reciprocity, which balanced the lopsided division of rights and duties, existed merely as an ideal. In practice, the highhandedness and compulsion the master showed toward his landless clients emphasized the sharp inequality between the two parties. The surplus value appropriated by the master from his attached laborer was not only economic in the restricted sense of the word. The docility and deference the hali was obliged to show to his master increased the latter's superiority. In this sense, patronage has to be seen as the sociopolitical dimension of a precapitalist relationship that was essentially exploitative.

This type of servitude has been linked too easily with a subsistence economy and a feudal lifestyle of dominant landowners, which came to an end when the transition was made to production for the market. In fact, there are indications that peasants in South Gujarat used the profit from growing cotton for the world market to replace their own labor power by that of halis, thus showing off their increasing prosperity to the outside world. As the consequences of capitalist penetration of the agricultural economy began to make themselves felt in terms of farm management, that is, in the increasing monetization of wages and contractualization of all economic relations, the former patron–client ties between substantial landowners and landless workers became less significant. From then on, the rich peasants acted as employers and no longer as patrons indulging in acquiring inferiors who, from the viewpoint of economic rationality, were increasingly found to be redundant.

The transition to what was formally "free" labor came when landowners would no longer provide their workers with security because, according to the principles of the new mode of production, the benefits did not offset the costs. Most certainly, the disintegration of the system of attached labor was also, and at least equally, due to rebellion by the landless laborers against the state of complete dependency in which they had been imprisoned. To free themselves of the stigma of social inferiority that, according to the new system of values, was linked to their earlier bondage, the younger generation of agricultural workers showed a distinct preference for a life as daily wage earners, which frequently made it necessary for them to change their employment and employer.

The dark side of this emancipatory trend has been that the worker's value is now determined solely by his labor performance, and the price paid for labor is fixed at an extremely low level as a result of the ample supply. The patron's provision of daily necessities to the halis, even in the off-season, does not apply to casual laborers, who have to use their own resources to survive on days on which no work is available. The change in relations between demand and supply is thus due not only to structural but also to cultural factors, amounting to a reassessment by both parties of the utility or desirability of vertical ties. Emancipated in its noneconomic dimensions, labor is now recognized only as a commodity, but its supply far exceeds the need.

This conclusion seems to contradict my earlier assertion that total employment in the rural economy of South Gujarat has increased. Only a small minority of the local proletariat, however, seems to be eligible for new income opportunities outside agriculture. Those jobs are largely filled by people coming from other parts of Gujarat or even from outside the state. Halpatis are denied access to the new forms of employment, ostensibly because they lack the required skills or industrial mentality. The massive influx of migrants also threatens the Halpatis' traditional position as agricultural workers. This is particularly well-illustrated by sugarcane, which is cut by an army of more than 100,000 workers who come to the plain each year for a period of six to seven months. This annual migration results from the refusal of cooperative factories to rely on local labor in the harvesting campaign. The reasons for this replacement of local by migrant workers will be discussed later on. Its effect is that Halpatis are undergoing a process of economic marginalization—one that victimizes the women even more than the men.

As a result of de-patronization, Halpati agricultural workers also find that they are socially isolated. Without the mediation of their former masters, they find no one to listen to their claims to a fair deal. In the past the Halpatis were paid a wage at a bare subsistence level, but were indentured to landowners who were willing to ensure the livelihood of their inferiors, even if no work was available. Such generosity was of course due to self-interest, which was well-understood in terms of the precapitalist mode of production. It is in that respect in particular that the stage has been set for a new drama. The landless have always lived in a state of poverty, but for many the daily misery has become even greater because they no longer have the basic security on which they could formerly depend. Their newly gained freedom has meant little more than the freedom to starve.

State intervention has brought no change for the better. The government both at the national and regional levels recognizes that the landless have the right to a livelihood, but the protection this should entail is nonexistent in practice. Legal measures have fixed the minimum wage at a level that permits little more than survival and that makes no allowance for nonworking members of the household. Moreover, as I have explained elsewhere, there is no effective control over compliance with this ordinance, with the result that the actual wage paid is generally much lower (Breman, 1985). Neither does government support the Halpatis' claim to preferential employment in their own territory. The official point of view is that the free movement of labor should in no way be hindered and that entry into the production process should depend completely on the spontaneous effects of the demand-and-supply mechanism. The passive stand taken by the state apparatus turns to active opposition, however, whenever efforts are undertaken to improve the situation of the landless proletariat. Organizations that seek to mobilize labor are essential to any emancipatory trend at the bottom of the rural hierarchy. However, the authorities are prone to judge such movements primarily as a threat to law and order, taking forceful counter-action at an early stage.

It would be incorrect to suggest that the landless in the central plain of South Gujarat are a docile mass whose fighting spirit has been broken. Portraying them as living in a state of passivity, amoral individualism, and fatalism (Rudolph and Rudolph, 1987: 387) is both inadequate and unfair. Time and again, the Halpatis have shown that they do not submit meekly to their social marginalization, but they have little chance of translating their rebelliousness into permanent and collective action. Sporadic, spontaneous, and strongly localized are the main characteristics of the persistent strikes and other forms of labor protest. The "weapons of the weak" (Scott, 1985) also include a series of actions and attitudes with which the landless show their dissatisfaction and militancy in more individual fashion. Their increasing willingness to enter into open confrontation with the rural elite, however, shows that the economic, political, and social hegemony of the entrepreneurial class of peasant capitalists does not go unchallenged (Breman, 1985a).

Labor nomads

I now turn to the migrant workers. The pull induced by the economic growth poles that have emerged in various parts of India is not the only reason for large-scale labor circulation to which I have referred. Expulsion as a result of greater pressure on resources in the home regions is a major reason for migration, particularly of those at the bottom of the agrarian hierarchy. This has been due partly to population growth but also to other factors. The migrant workers who come to the plain of South Gujarat every year in large numbers do so because rural ecological conditions make it impossible to find work in their home areas. There are recurrent water shortages in the hinterland, caused not only by monsoon failure but also by deterioration of the environment in consequence of large-scale tree-felling during past decades. The almost barren wasteland that has resulted from deforestation is exposed to uncertain rainfall; as a consequence, the productivity of the land, which was never very fertile, has now declined even further.

It is paradoxical that the new canal irrigation system, which provides so much benefit to the rich peasants on the plain, has led to a veritable flight of labor from the region where the water is derived. The artificial lake created by building a dam in the Tapti River robbed a great many tribal cultivators of their land. They have been resettled in squalid housing colonies. Deprived of their livelihood, these former peasants have no choice but to follow the water to the plain, where they sell their labor power. Their semi-proletarianization results directly from the type of development policy adopted by the state. Local government has attempted to alleviate the lack of employment in the hinterland with the construction of public works, but the number, duration, and scale of such projects is insufficient for the purpose.

Labor mobility is usually associated with rural-to-urban migration, to such a degree that many authors consider migration almost synonymous with urbanization. This is incorrect, because many of those who are pushed out of their village economy continue to live in rural areas. A minority of the land-poor peasants who flowed into Surat District seem to have settled there permanently. Far greater numbers of migrants, however, go to the plain as seasonal workers, sometimes year after year. The background to this pattern of intrarural circulation is formed by growing regional inequality in the level of development; in turn, these "labor nomads" further accentuate the inequality. Again, this is vividly illustrated by the massive use of migrant labor in the harvesting of sugarcane in South Gujarat.

The presence of these migrants is a surprising phenomenon in view of the underutilization of the indigenous proletariat to which the majority of the agrarian population in the central plain belongs. Why do landowners and other rural employers prefer to rely on outsiders? The peasant-entrepreneurs say that the local landless are incompetent, lazy, indifferent, physically weak, impertinent, and untrustworthy. Such terms illustrate the animosity that is felt toward the Halpatis, a social class whose members are penalized with economic sanctions for their refusal to continue to behave as inferiors. Their replacement by imported labor, however, is also and even primarily due to more rational economic considerations. The migrants provide a workforce that is fully and unconditionally available at the lowest possible price for the duration of the annual harvest campaign.

Factory agents in the hinterland enter into agreements with labor brokers stipulating that, at the end of the rainy season, the latter will journey to the plain accompanied by a specified number of cane-cutters whom they have contracted during the preceding months. These labor brokers then also act as gang bosses, accountable for their recruited workers, and take charge of the open-air camps in which the cane-cutters bivouac during their stay. Finally, they commit themselves to leaving the plain again together with their work gangs at the end of the season, in early May. Once cane-cutting has begun, the factories operate continuously day and night. The cane is cut only during daylight hours, but its transport continues during the night. The workers frequently have to return to the fields at night in order to load the fleet of bullock carts, tractors, and lorries that keep production going. A workday averages ten to fifteen hours throughout a seven-month period. Even the day of rest that workers are allowed once a fortnight is not a concession to the gruelling work regime, but is caused by the need to clean the factories.

Living in an alien environment, under continual supervision in the fields and in the camp by a gang boss who has promised meticulous observance of the factory's directives, the labor nomads have no space into which they can withdraw at the end of the working day. In comparison, it would be far more difficult to control the local landless, who could not be denied a social life of their own. To begin with, the agricultural laborers who belong to the plain have household obligations that make demands on their time. Some years ago and with obvious reluctance, one of the sugar cooperatives conceded to the repeated request made by leaders of a local Gandhian movement that a number of local agricultural workers be employed in cane-cutting. This experiment was soon brought to an end when the Halpatis resolutely refused to work into the night hours.

For each ton of cane cut, the members of a team consisting of two adolescent or adult members, usually a man and a woman, are paid a fixed amount. For his mediation, the broker receives 10 percent of the total earnings of his gang from the factory. When converted into a daily wage, this cane-cutting tariff is still less than the minimum wage that can legally be claimed by agricultural workers in the plain.

On arrival the workers in a gang are handed two mats and a few bamboo sticks with which to construct their shelter. "Hut" is too grand a term for these tent-like structures, which are too small to accommodate man, wife, and children together. The adults usually sleep in the open air around the embers of the fire on which they have cooked their evening meal. A group of such shelters gives the field or wayside allocated for their bivouac the appearance of a camp, but there is no drinking water, washplace, or other sanitary facilities. When the canefields in the neighborhood of the camp have been harvested, the migrants move on to another spot and set up a new "camp." Gangs of workers who have shared the same encampment for some time now go in different directions, most probably not to meet again during the campaign. This continual rotation is characteristic of the nomadic life of the cane-cutters.

The women in particular lead an extremely hard life. Although the man normally uses the cutting knife, the partner with whom he forms a team, usually his wife, will sometimes relieve him in order that he may rest. It is also her task to strip the cane of its leaves, to break it into pieces, to bundle the pieces, and to lay them out on the field ready for transport. During the short break, when the men lie on their backs exhausted and silent or smoking a *bidi*, the women have to care for the infants who have come to the field with them and to suckle the very youngest. When the workers return to camp at the end of the day, the women carry the headloads of wood needed for the cooking fires. Finally, when they reach camp, the women remain busy at such tasks as preparing the evening meal and doing the washing.

The privations suffered by the dependent family members of migrant workers who have to remain in the hinterland perhaps exceed those of the cane-cutters who go to South Gujarat partly in the hope of saving enough money to get them through the monsoon months. Those who are forced to remain behind are among the most vulnerable—the young children and the aged, not yet or no longer able to take part in cane-cutting. Their labor power is useless and as consumers they are not allowed to go to the plain. Anyone who pities the army of cane-cutters should realize that those who have to stay at home, the aged and the very young, are probably even worse off.

The pressure to leave the hinterland is primarily economic, although existing dependency relationships expressed in indebtedness can be an important incentive to seasonal migration. Free choice is curtailed, however, the moment the cane-cutter signs a work contract with his broker. On doing so he is given earnest money, and thus commits himself to work for the length of the campaign. The amount of money labor brokers need to recruit the members of their gang is greater than the sum advanced to them by the factory. They complement it with a loan from a moneylender, and naturally charge the workers for the high interest rate that has to be paid. The result is that, for each rupee that a cane-cutter receives at the start of the campaign, he has to repay two at its end. During the season the debt often increases because the migrants borrow small sums from their gang boss. Wage payment is deferred until the end of the campaign when the migrants are returning home. The members of a gang are paid only a living allowance once a fortnight, a grain ration supplemented by some money (not more than one or two rupees a day per capita) for cash expenses. This is almost always inadequate, and the gang boss then provides a loan. As a result, he usually has a great deal of money on loan to the cane-cutters. He does this with the express purpose of disciplining his gang, with the tacit approval of the factory.

Although the binding of labor through indebtedness is legally prohibited, the cooperative enterprises of South Gujarat employ their workforces on this basis. The stakes for the labor broker-cum-gang boss-cum-camp boss make it understandable that he occasionally uses extra-economic force. In the last resort, physical violence is used to force the contracted cane-cutters to journey to the plain and, once there, to prevent them from returning home prematurely. Labor brokers who are excessively harsh, however, are likely to engender violent reaction, lending support for the often-expressed view among factory fieldstaff that cane-cutters are impulsive and irrational. Supposedly, although cutters will not take umbrage at the coarsest abuse, they are likely to explode in fury for some futile reason and injure or even kill their antagonist with the cutting knife.

I do not consider the sort of dependency relations that have been discussed above to be a continuation of the former feudal-type relations between master and servant. The new bondage that has emerged is not necessarily in conflict with the development of a capitalist mode of production. In other words, it is possible for labor to become mobile without being able to exercise free choice over terms and conditions of work. Cooperative managements naturally enough deny any charge that agro-industry is guilty of coercion and illegal binding of labor. They argue that cane-cutters are employed by their gang bosses and that interpersonal relations among the workforce are beyond the control of the factory. Management contracts only with the intermediaries, and maintains that whatever arrangement these gang bosses may make with the cane-cutters is not its concern.

The government has offered no protection to the local proletariat against the massive influx of outside workers. Neither has any official action been taken toward ending the exploitation and oppression to which migrants are subjected. Existing legislation provides the basis for such action, but implementation so far has been haphazard at best. This applies both to the act that places restrictions on the movement of labor between states, and to the regulation intended to keep the operations of labor brokers in check. At their own request, the cooperative sugar factories have been given official exemption from compliance with these laws. No reason for this leniency has been given. The influence exercised by the sugar barons is illustrated by a secret memorandum found in 1987 by a commission of enquiry set up by the Gujarat High Court, which explicitly forbids the rural labor inspectorate in Surat District from prosecuting management of the sugar enterprises for failure to pay the minimum wage.

In contrast to local agricultural laborers, who have at least some political influence, since they are rooted in the same social milieu as the employers, the huge army of seasonal migrants is politically powerless. As a prominent politician of Gujarat told me: cane-cutters do not vote, either in the villages where they come from, or where they work. Nevertheless, the agro-industries are apparently not entirely insensitive to the growing criticism about migrant labor conditions. They fear that the negative publicity might cause the government to tighten its control over abusive labor practices. That there has so far been no sign of any such corrective measures is undoubtedly due to the strong pressure exerted by the lobby of peasant capitalists at both state and central levels.

The cane-cutters, who stay only temporarily in the plain, who are segregated from the local landless in the fields and in their camps, and who, finally, are penned-up in their gangs and disciplined by gang bosses whose principal loyalty is to the factory, have no opportunity to build collective resistance. They do not even share a common background. Unlike the situation ten years ago, many harvesters now come from within Gujarat itself—in my estimation about 30 percent. Nearly all seasonal migrants belong to tribal and other low castes who own little if any land. By tapping different labor reservoirs, the factories are beginning to reduce their dependence on Maharashtra, the hinterland that has traditionally been the main source of recruits.

The age composition of the workforce also militates against emergent solidarity. Understandably enough, the factories prefer to use workers who are at the peak of their physical strength—young adults between 16 and 30 years of age. I have noticed that each year many migrants are taking part in cane-cutting for the first time, and even the older people have less than ten years' experience. Only very few keep up this vagrant lifestyle, with its heavy workload, for a long period. Circulation, in other words, not only is a reflection of the seasonal nature of the industry but is also a major feature of the workforce composition. The obvious explanation is that only the very strongest are prepared repeatedly to subject themselves to the harsh labor regime of the campaign.

But such a conclusion assumes a greater freedom of choice than is available to the cane-cutters. The lack of any alternative means of livelihood forces them, year after year, to make the journey to South Gujarat. The physical exhaustion that they suffer is the reason the sugar factories bring in new blood after a few seasons. Constant rotation of the workforce is also encouraged because cutters with longer experience may show a spirit of resistance that newcomers, not yet accustomed to the work, lack. Just as segregation into camps prevents the gangs of harvesters from coming into contact with the local landless, so the regular replacement of old by young migrants tends to prevent worker mobilization.

The key to capitalist dynamics in rural Gujarat lies in the continual effort by the cooperatives to increase not only production but also productivity. During the last 20 years, the area of arable land in the plain that is planted with sugarcane has expanded enormously and per acre yields have increased by 25 to 30 percent over the same period. The peasant-entrepreneurs are naturally interested in potential productivity gains in the harvesting operation. Under the existing technology it seems inconceivable that labor-intensity can be raised any further. With the best will in the world, two adults cannot cut much more than half to two-thirds of a ton of cane by hand each day and prepare it for transport, over a period of six to seven months. Any further increase of the workload would cause their physical collapse. Mechanization is not yet under serious consideration, but I anticipate that the incentive to adopt it will gradually increase. Initial experiments have already been made. Although mechanical harvesting will certainly not be cheaper—over the years harvest labor has amounted to only 15 to 20 percent of the total cost of production—management could decide on mechanization to avoid having to provide cane-cutters with better treatment and better pay. For the time being, employers in agro-industry merely use the mechanization option as a threat against troublemakers such as those who urge compliance with existing labor legislation. Mechanization of harvesting would undoubtedly make the majority of migrant workers superfluous. Lacking any alternative means of livelihood, they would face a grim future.

Prosperity and pauperization

Introduction of the sugarcane crop has so far benefited mainly, if not only, the class of noncultivating proprietors. A decade ago the cane-cutters were paid a wage that amounted to approximately two-thirds of the rate that agricultural employers were then legally obliged to pay. During the spring of 1988, according to my latest fieldwork, two-thirds of the present minimum wage is what they are still paid. It would be inaccurate to deduce from this, however, that the degree of exploitation has not changed during the intervening years. The price received by the peasant capitalists for their sugarcane has risen more rapidly than the costs of growing and processing. While the producers thus have a greater profit margin, the underpayment of labor has persisted. As members of the dominant caste-class, the producers demand what they consider to be their due: more capital, even if the surplus value derived from labor is at the expense of the workers' livelihood.

"The crushing of cane and of labour" was the title under which I published the results of my first investigation into the employment of seasonal migrants by the cooperative sugar factories of South Gujarat (Breman, 1978). My new findings show that this conjunction is still linked to the success of the agro-industry in the region. The continuing and even increasing social polarity counteracts the idea so cherished by development experts and policymakers that economic growth, although in the first instance primarily benefiting those who own or have access to capital, will gradually trickle down to the have-nots. In the situation that I have described above, which is by no means unique to South Gujarat, I have seen no signs of any trickle-down effect among those who own no commodity other than their labor power (see also Bardhan, 1984: 188–199).

Behind the facade of the social harmony model, with which government legitimizes its development schemes, is hidden an essential unwillingness to introduce any change into the extreme inequalities between rich and poor, powerful and powerless. As a result, those who speak on behalf of the subjugated classes are soon branded as guilty of subversive activities. Even when their intention is no more than to represent the interests of the poverty-stricken people who cannot make their own voices heard, and although they try to do so within the boundaries of legality, the authorities are often inclined to regard such intervention as inadmissible agitation, an unjust invasion of proper law and order. The employers are averse to any reference to class contradictions and claim, paradoxically enough, that efforts to achieve harmony must regulate labor relations. In their perception there is no place for opposing interests; conflicts, should they unexpectedly arise, must be brought to an end in consultation between employers and workers.

In any effort to reach a compromise, however, such outsiders as trade unionists, politicians, and other "activists" can never be permitted to play a role. Those who proclaim themselves to be the sons of the soil and who form the hard core of the New Agrarianism (Rudolph and Rudolph, 1987: 354–364) will tolerate no interference in their domain of hegemony. They fiercely resist any attempt by government agencies, however guarded and reluctant, to move into the social arena as impartial arbitrator.

Indifference to the fate of the landless is too mild a description for the unreserved hostility that the Patidars show toward the subaltern classes. They expect their leaders to follow a collision course leading to further polarization, and all agreed when the main farmers' union in Gujarat demanded some years ago that the minimum wage should be abolished, even though it was already inadequate for anything more than survival. The ideas expressed by members of the dominant caste in rural areas smack of Social Darwinism. They amount to saying that it is not poverty that is the problem but the presence of a large underclass, which exists in conditions of misery only due to its own shortcomings. In their opinion, society would benefit if it were relieved of that millstone. The rapidly forgotten slogan from the early 1970s, to abolish poverty, seems to have been replaced by the sentiment, often expressed by those who have property, to be rid of those who have not.

An ideology that denies the poor the right to an adequate wage easily lends itself to suggesting that the granting of suffrage to such people was a mistake. This type of thinking has evident implications for population policy. The exercise of free choice that Myrdal had in mind when he pleaded for a government-supported program of drastic birth control cannot be the guiding principle on which to build a population policy, according to the prevailing ideology. There is definitely a need to restrain the demographic behavior of the masses that live in poverty, but this underclass is considered to lack the mentality of responsibility which is a precondition for voluntary action. A restraint on their disproportionate growth would comply with the wish voiced by the more prosperous that this supposedly counter-productive and criminally minded surplus should be removed from sight or even from society as a whole. Such attitudes of Social Darwinism, which I often encountered while doing fieldwork, make it understandable why the rural elite profoundly sympathized with the forced sterilization program that was practiced during the Emergency of the mid-1970s. Not the landowners, but agricultural laborers and other vulnerable categories, were the groups on whom local authorities concentrated their efforts to meet the quotas laid down at higher levels within the bureaucracy.

The new mode of production in the South Gujarat countryside has given rise to a hardening of social relations. The rich peasants have no hesitation in using physical force to neutralize the claims for emancipation that are being voiced ever more loudly by those at the bottom of society. Such a reaction provides new fuel for the bitterness that has long been felt among the landless Halpatis. The resultant heightening of tension increasingly gains the character of class conflict, a process that is also noticeable in other parts of rural India. The newly established order lacks social legitimacy, and in consequence may well also lack political stability.

It would be much too simplified to posit that the state is an instrument in the hands of the ruling class, but it cannot be denied that development policies place a premium on the ownership of capital (Bardhan, 1984a: 32–39). To present the state as "a third actor" (Rudolph and Rudolph, 1987: 397–401) is tantamount to overlooking its anti-poor bias. The so-called target group approach—expressed in terms such as growth with equity, basic human needs, employment guarantee schemes, positive discrimination—can also be seen in a more ominous light. The barriers thrown up by government agencies to prevent any mobilization of the landless into their own organizations have prevented the rise of a countervailing force that would be a precondition to reducing the strong inequalities between the classes. Seen from below, the state is not soft but extremely harsh.

The social stagnation of which Myrdal spoke, if it ever existed, is definitely a past state of affairs. Nevertheless, more people now live in poverty in rural India than ever before. This part of the population, declared to be superfluous, could be called a new class of Untouchables: people who are ephemeral in economic transactions, who have become marginalized in rural society or even pushed out of the village economy, and who have no access to the welfare services and government facilities that can be claimed by those considered to be full-fledged citizens. Far from being afforded protection against intimidation and terrorizing, they have to make themselves invisible if they are not to be prosecuted by those who uphold law and order. The quantitative surplus that they represent is a social rather than a demographic problem.

Note

The fieldwork on which this essay is based was sponsored by the Indo-Dutch Programme on Alternatives in Development. The author is grateful to Jean Sanders for assistance in preparing the English translation.

References

Bardhan, Pranab K. 1984. "Poverty of agricultural labourers and 'trickle-down' in rural India," in *Land, Labor and Rural Poverty: Essays in Development Economics*. Delhi: Oxford University Press.

———. 1984a. *The Political Economy of Development in India*. London/Delhi: Basil Blackwell/Oxford University Press.

Breman, Jan. 1974. *Patronage and Exploitation: Changing Agrarian Relations in South Gujarat, India*. Berkeley/Delhi: University of California Press/Manohar.

———. 1978. "Seasonal migration and co-operative capitalism: Crushing of cane and of labour by sugar factories of Bardoli," *Economic and Political Weekly* 13: 1317–1360.

———. 1985. "I am the Government Labour Officer . . . : State protection for rural proletariat of South Gujarat," *Economic and Political Weekly* 20: 1043–1055.

———. 1985a. *Of Peasants, Migrants and Paupers: Rural Labour Circulation and Capitalist Production in West India*. Delhi/Oxford: Oxford University Press/Clarendon.

Byres, Terence J. 1988. "Charan Singh (1902–87): An assessment," *The Journal of Peasant Studies* 15: 139–189.

Chakravarty, Sukhamoy. 1987. *Development Planning: The Indian Experience*. London/Delhi: Oxford University Press.

Lipton, Michael. 1977. *Why Poor People Stay Poor: A Study of Urban Bias in World Development*. London/New Delhi: Temple Smith/Heritage Publishers.

Myrdal, Gunnar. 1968. *Asian Drama: An Enquiry into the Poverty of Nations*. 3 vols. New York: The Twentieth Century Fund.

Patnaik, Utsa. 1986. *The Agrarian Question and the Development of Capitalism in India: First Daniel Thorner Memorial Lecture*. Delhi: Oxford University Press.

Rudolph, Lloyd I., and Susanne Hoeber Rudolph. 1987. *In Pursuit of Lakshmi: The Political Economy of the Indian State*. Chicago/Bombay: University of Chicago Press/Oriental Longman.

Scott, James C. 1985. *Weapons of the Weak: Everyday Forms of Peasant Resistance*. New Haven/London: Yale University Press.

Singh, Charan. 1986. *Land Reforms in UP and the Kulaks*. Delhi: Vikas Publishing House.

Wertheim, Willem F. 1964. "Betting on the strong?," in his *East-West Parallels*. The Hague: Van Hoeve, pp. 259–277.

Agrarian Responses to Outmigration in sub-Saharan Africa

AKIN L. MABOGUNJE

IN THE LITERATURE ON THE EFFECTS of rural population change, the common emphasis has been the economic and institutional consequences of rapid growth in population and the labor force. Interest has focused on such consequences as increases in labor productivity, changes in land tenure systems, and effects on common-property resources (e.g., Boserup, 1965; Binswanger and McIntyre, 1984). The obverse situation in which the effects to be studied are consequences not of population growth but of substantial labor outmigration has attracted less attention—largely because such outmigration has tended to occur in the later phases of agricultural development, by which time the focus of research interest has shifted decisively to the urban economy.

Yet, in Africa various policies and activities during both the colonial and post-colonial eras have induced considerable rural labor exodus at early stages in the development process. At the same time, fertility levels have remained fairly high while mortality rates have declined sharply. It is thus a matter of considerable interest to investigate the consequences for rural living conditions of this combination of demographic circumstances. What impact have these conjunctural processes had on rural institutions, particularly those affecting labor relations and class formation? How have these processes been modified by the adverse economic circumstances of recent years and the forced reduction in the "urban bias" of policy, with presumably a resulting slowdown in the rate of rural outmigration?

It has been suggested that the contemporary phenomenon of massive outmigration from the rural areas of sub-Saharan Africa may represent a continuation in a different guise of earlier migratory movements provoked by domestic slavery and slave trade on the continent. Nonetheless, in order to understand the types of agrarian response evoked by these relatively "free" flows of people, it is important to position the modern rural exodus correctly within the relevant historical context in which it originated. Beginning particularly after the Berlin Conference of 1884 (when the European powers partitioned Africa into "spheres of influence"), both the pattern of rural outmigration and the evolving character of agrarian response to it can be viewed as the consequences of the integration of the local economies of sub-Saharan Africa into the expanding world capitalist system. This process entailed the restructuring of the prevailing precapitalist modes of production and their subordination to international capitalism through the operations of a world market for the surplus produce extracted from these earlier modes (Stichter, 1982).

In sub-Saharan Africa, restructuring involved the establishment of one of four modes of agricultural production within what came to be identified as the colonial territories of various European nations. These were in addition to mining activities, which were equally significant in some areas. The four modes were plantation agriculture, settler farming, outgrower schemes, and peasant production (Datoo, 1977). Plantation agriculture involved the alienation of land, usually to a large-scale, monocultural, foreign-financed enterprise, and the initiation of a labor movement with resultant entry into wage earning. Settler farming, while it shared with plantation agriculture the characteristics of being foreign financed and sometimes large scale, usually comprised a mixed farm with integrated arable and stock-farming rather than monoculture and was run by a yeoman farmer rather than a salaried manager. Furthermore, while the settler farm required land alienation this was within an area specifically set aside for colonists. And though it also relied on migrant laborers, this labor was supplemented with a system of villenage or squatters. Outgrower schemes involved the cultivation of plantation crops by peasant farmers in an interdependent, contiguous set-up to a plantation, where processing could not readily be separated spatially or organizationally and yet it was necessary to increase the plantation's output without further alienation of land. Finally, there was peasant production of exportable commodities on their own land, usually on small farms and with hardly any change in production organization or technology.

These four colonial modes of production, beyond their more obvious effects on such matters as race relations, the land tenure system, and land availability, generally exerted traumatic and momentous impact throughout the rural social structure. For in sub-Saharan Africa, labor rather than land has always been the factor of production in short supply. The mechanisms of procuring labor for these various modes of colonial production had far-reaching consequences for traditional institutional arrangements and rural class formation and provoked diverse responses to ensure the continued viability of agrarian economies in most parts of the region.

This essay explores these consequences. The first section describes the changing nature of the factors encouraging massive outmigration from rural areas in sub-Saharan Africa and examines the changing destinations of the migrants. The second section considers the agrarian responses to these movements, especially as they relate to the restructuring of the labor supply system for meeting the needs of the local economy. The third examines labor demand adjustments, especially as they have been conducive to rural class formation. The fourth considers other consequential changes in rural labor relations, with special reference to the role and status of women. The fifth section examines the role of the state in mediating these diverse responses and in guiding them in directions compatible with the overriding capitalist objectives of colonial and post-colonial economies of these countries. A concluding section evaluates the implications of these various changes for population processes and rural development in sub-Saharan Africa.

Changing patterns and causes of rural outmigration

Outmigration from the rural areas of most countries of sub-Saharan Africa began as a response to demand for wage-earning labor of an emergent proto-capitalist economy. Starting in the last quarter of the nineteenth century, the mechanisms displacing migrants from their rural homes were many and varied. In those areas of western and central Africa where plantation agriculture was the means of drawing the local economy into the capitalist world economy, outmigration resulted both from the extensive expropriation of land formerly occupied, even if scantily, by African settlements and from the establishment of a system of wage payments for labor. That this twin process of alienation and imprestation was resisted by the Africans is self-evident. A high degree of coercion was needed to initiate it. For instance, the situation in Zaire, the former Belgian Congo, was described in the following terms:

> [O]n April 14, 1911, an agreement was signed between the colony and the Lever Brothers company of Port Sunlight in England, creating the oilworks company of the Belgian Congo (SHCB). For ten years, this company was to be allowed to select and lease up to 750,000 hectares of the "national" lands bearing *elaeis* oil palms in five "circles" with a 60-kilometer radius. . . . The area involved was about one-fourth the size of Belgium.
>
> The recruitment of the necessary labour force for the palm groves and the company's oil-pressing brought about the decomposition and destruction of the pre-colonial societies, due to the prolonged displacement of the population, the mortality caused by long walks, heavy work, hunger and insufficient housing, the lack of respect for local traditions and the decline of the home regions. In Kwilu, the exploitation of the rich southern regions in the lightly-populated zones of Kikwit led the firm to hire fruit-cutters who were Mbunds, Kwese, Pende, and even some Angolan workers. Confronted with the population's resistance to working for the firm, the colonial administration assisted the firm both in increasing food production by imposed cultivation, and in forced recruitment of the peasants. (Kabunda, 1975: 310–311)

Similar processes of labor recruitment were adopted by the French colonial administration in West Africa. Indeed, until 1946, every male in French West Africa between the ages of 18 and 60 was subject by law to an annual corvée, which required him to contribute a certain number of days' labor to whatever enterprise the administration assigned him (Thompson and Adloff, 1958: 491–492). Gradually, throughout the continent, forced labor as a means of mobilizing workers for agricultural production came to be regarded as unsatisfactory and was replaced by the levying of taxes on the native population. Officially, this system was justified on the grounds that the colonized people should contribute to the costs of their own administration. But in reality, the tax compelled the local populations to find ways of raising money and, hence, to participate in the emerging monetized, capitalist-oriented economy.

The tax was usually assessed on the basis of the number of adult members in each tribe, kin group, or village (a head tax) or on the basis of the number of dwellings (a hut tax) (Stuckey and Fay, 1981: 5). There were two ways in which the indigenous populations could raise money. One was to work as wage laborers in the mines, plantations, settler farms, or even farms of other tribe members already engaged in producing for the world market; the other was to raise cash crops for market sale on the plots of land that had remained in their own possession.

All over sub-Saharan Africa, the need to raise money for taxation and other purposes soon led to major geographical differentiation between areas that could produce export commodities for the world market and those that could not. The former became more developed and the destination of large waves of migrants from all parts of their countries and beyond. Such regions included the settlers' "White Highlands" in Kenya; the regions reserved for white colonists in the Mengo and Busoga and the Toro districts of Uganda; the oil-palm, rubber, and cocoa-producing coastal belts of Nigeria, Ghana, and the Ivory Coast; and the cotton and groundnut belts of northern Nigeria, Senegal, and Gambia. By contrast, most other areas where ecological conditions permitted the production mainly of subsistence crops became the reserves of cheap labor. These areas became characterized by massive outmigration, either permanent or seasonal (the *cin rani* phenomenon) (see, e.g., Berg, 1965). Areas of heavy outmigration included the Mossi regions of Burkina Faso, Mali, the northern regions of Ivory Coast, Nigeria, and Ghana, northern Uganda, Rwanda, Burundi, and northern Tanzania, as well as virtually all areas outside the white-occupied or reserved enclaves in southern Africa.

Following the colonial era, new paths to development were pursued by the recently independent countries. Emphasis was placed on industrialization, on the improvement of educational and health facilities for the population at large, and on the expansion of transportation and communication networks. This new emphasis was strongly biased in favor of urban centers; as a result since the early 1960s urbanization has been very rapid in sub-Saharan Africa. Indeed, while the weighted average annual growth rate of the total population rose from 2.4 percent in 1960–70 to 2.8 percent in 1970–82, that for the urban population rose from 5.5 percent to 6.1 percent over the same interval (World Bank, 1984: 82–85). In all of sub-Saharan Africa, the number of children enrolled in primary school as a percentage of their total age group jumped from 36 percent in 1960 to 78 percent in 1981.

The appropriateness of the type of education being imparted to the children under these expanded programs is often questioned. Curriculums may have poorly prepared the majority of children for employment in rural areas. Certainly the educational program has promoted a style of life more easily achievable in urban than in rural areas. The attractiveness of the city was easily perceived not only by the children but also by their parents. Indeed, in a study of the migration of school-leavers to Benin City, capital of Bendel State, Nigeria (Makinwa, 1981), 57 percent of respondents claimed that their movement was financed by their parents or grandparents.

This massive movement from rural to urban areas added a new dimension to the earlier migration pattern, which consisted largely of rural-to-rural labor migration. (That movement continues, though its relative significance has been greatly reduced.) For one thing, the rural-to-urban migration is more widespread and affects both the areas of former export agricultural production and the areas of subsistence agriculture. How much the impact differs between these two kinds of area has not received adequate research attention. It is clear that the situation would also vary from country to country. In some countries, for example Nigeria, Gabon, and Zambia, recent windfall revenues from such export minerals as petroleum and copper gave this townward pull of rural migrants a much greater fillip, with significant consequences for the national economy, especially in terms of the decline in agricultural production.

One final factor that has critically influenced outmigration is the regime of prices for agricultural produce both on the world market and within individual countries. Commodity prices on the world market have tended to decline, especially with the current stagnation in world trade. For subsistence crops, most national governments encourage lids on the prices of food crops to benefit urban consumers. These depressed prices for agricultural commodities create an economic climate very unfavorable to rural productive activities and hence encourage increased outmigration.

We now turn to discussing the agrarian response to the various streams of outmigration from rural areas. Three general remarks are made by way of introduction. First, given the prevailing movement toward the closer integration of national economies into a world capitalist economy, much of the agrarian response can be expected to reinforce tendencies making for commoditized labor and land, with the support of the state. Second, given that labor rather than land is the already scarce factor in the rural areas of most sub-Saharan African countries, it is likely that the massive additional losses of labor will be of great significance both for rural labor relations and for emergent rural class formation. Third, everywhere in sub-Saharan Africa, the urban-industrial emphasis in development planning that characterized the first two decades after independence has left countries saddled with relatively high international debt burdens. The debt overhang has forced many countries into an agonizing reappraisal of the role of agriculture and has induced new attention to rural growth and welfare in their overall national development strategies. This reappraisal has had a reverberating impact on agrarian systems, on rural demographic patterns, and on the interactions of the two.

The restructuring of traditional labor supply institutions

The most obvious and pressing consequence of rural outmigration is on labor relations and the rural labor market, which must adapt to the massive loss of the able-bodied young who constitute the bulk of rural labor. In the preindependence era, the main external labor demand was for young males. Especially in the colonial territories of East and Central Africa, where emerging capitalist economies were based on mining, plantation agriculture, and even settler farming, this preference went hand-in-hand with an active discouragement of female migration as a means of preventing permanent residence of migrants at the destinations. In such circumstances, seasonal or circulatory migration of no more than two to three years' duration became the norm to which rural institutions had to respond appropriately.

In considering the nature of agrarian responses, it is important to bear in mind the varieties of institutions that were affected on both the supply and the demand sides. In most parts of sub-Saharan Africa, labor supply is principally governed by the extended family, the age-grade, and the village community. The extended family—what Polly Hill (1972: 17) calls "farming units"—would appear to be more relevant in this context than the nuclear family, since it is within the former that the social relations of production and reproduction were traditionally maintained. Although individual households or nuclear families were established and recognized within the extended family, household decisionmaking was invariably structured around the extended family system.

Where a young, married male was migrating alone, it was assumed that the extended family in his absence would keep an eye on the needs and behavior of his wife. It was also expected that the wife and her children would contribute the labor for food production. Among the Bemba of Zambia, for instance, the male migrant often cleared the farm and felled the trees before he departed (Richards, 1939). Where the migration was seasonal, as in much of northern West Africa, he might also prepare the ground and sow the crops before leaving (Rouch, 1956; Skinner, 1960). The wife and the children then had the responsibility of weeding, caring for, and harvesting the crops. This pattern led to women gradually being empowered to assume the status of head of household and lessened their fear of marital dissolution. It placed a great obligation on the migrant to maintain strong ties with his home base and rural kin. Among the Basotho of Lesotho, most of whose young males worked in South Africa most of the time, migrants were required to fulfill their *bohali* or bridewealth obligations as a means of guaranteeing the conjugal stability of their home and validating in jural terms the relations of mutual dependence of their affines (Murray, 1981: 147). On balance, in areas where there was no strong cultural resistance to women taking up farming responsibilities in the absence of their husbands, male labor migration has been accompanied by high rates of marital dissolution and illegitimate childbearing.

Age-grades and similar cooperative working arrangements provide another institutional framework for the supply of agricultural labor. In regions where these institutions are prominent, outmigration has produced a different range of responses. Among the Mambwe of Zambia, for instance, the labor migrant's wife could take his place in the cooperative work-parties, helping others in the cultivation of their fields and thereby ensuring assistance in return in the cultivation of her absent husband's fields (Watson, 1958: 34–35). By contrast, Hill (1972: 101–102) has reported that the institution of *gandu*, a cooperative work-party in Hausaland of Nigeria involving a man, his brothers, and their sons, would seem to have no mechanism for reabsorbing a migrant (except perhaps where he had gone away for Koranic studies or on pilgrimage to Mecca), and the institution was prone to break down completely when the head of the *gandu* died.

Among some ethnic groups, the village community operates as an institution mediating labor supply relations, especially in balancing external demand against internal needs. Thus, among the Tonga of Malawi, corporate groups of kinsmen exercised control over migration to ensure that the minimum number of men required for local agricultural work stayed behind (Van Velsen, 1964; see also Boeder, 1973). Such men were usually fairly evenly spread over the hamlets. The policy behind this was explicitly stated by the Tonga, namely that they were reluctant to leave their hamlets devoid of men and hence they attempted to organize labor outmigration in such a way that there would always be at least one man left per hamlet.

Of course, not every rural community in sub-Saharan Africa has confronted the phenomenon of outmigration in such a deliberate manner and made positive responses to accommodate it within the community's socio-political and cultural framework. Where no such responses were evoked, serious adverse consequences became evident, sapping the very integrity of the community. Indeed, one of the earliest studies of the economic effect of outmigration, in Northern Rhodesia (Zambia), stressed the general impoverishment of social life and the serious nutritional deficiencies that accompanied it (Richards, 1939; see also Wilson, 1941 and Miracle and Berry, 1970: 91). Similarly, a study among the Mossi of Burkina Faso found that the growing of cotton and other late crops was increasingly neglected either in the anxiety to migrate from home as soon as possible or for fear that corvée assignment would preclude any crops planted from being harvested (Skinner, 1960).

In the early postindependence era, the greater freedom of movement within countries and the strong orientation toward urban centers gave a new twist to problems of labor supply in the rural areas. In the first place, rural outmigration now extended to much younger age cohorts, particularly primary school-leavers who were moving to urban centers either to seek employment or to continue their education. Many of these migrants constituted a permanent loss to the rural areas since they had not been initiated into the labor system of rural production. In the second place, this form of outmigration affected rural areas in all regions of the country, both those that had experienced significant outmigration in the colonial period and those that had been the destinations of such migrations. One's intuitive impression is that the former areas were better able to cope with this development than the latter. In her study of Bendel State, which, as an area of agricultural export in the colonial period, used to be a destination for rural migrants in southern Nigeria, Makinwa (1981: 129–130) noted the almost whining protest by the older villagers against what outmigration had done to productive activities. It was claimed that communal labor was no longer available since there were no young workers to do the difficult tasks. Most village studies recorded the disproportionate presence of the aged in the available rural labor force. Rural farm hands, often hired with remittances sent back by migrants, hardly made up for the family labor power lost to the rural areas by outmigration.

On the whole, it would appear that agrarian communities, especially in the last two decades, have been unable to restructure rural labor supply institutions appropriately or to bridge the gap between labor supply and demand created by rural outmigration. The result has been stagnation and decline in agricultural production. Other factors have contributed to this decline, notably the lack of adequate price incentives, the deterioration of infrastructure such as roads and water supply, and the weak research and extension services. These, however, have all been part of a mutually reinforcing trend pushing young labor out of rural areas faster than rural institutions can adjust to the loss. In the event, in many sub-Saharan African countries not only agricultural production for export but also subsistence food production has suffered negative growth rates or rates far below those of population growth. As a result the period since 1970 has been one of food crisis in the region, albeit exacerbated by such natural hazards as the droughts of 1973–75, 1979, and 1983. With the growth rate of per capita food output negative in 27 out of 39 sub-Saharan African countries between 1969 and 1979, the stage was set for massive food importation through trade and aid. This situation brought the region up sharply against the glaring distortions in its development strategy and has been instrumental in provoking a reassessment of labor markets and employment opportunities in rural areas throughout sub-Saharan Africa.

Labor demand relations and emergent class formation

Wage labor in agriculture imposes a certain relation between the two individuals involved in the transaction and the land on which the hired labor has to work. The worker sells his labor for a wage while his employer treats the purchased labor power as a factor of production to be combined with other factors, notably his farmland. For the result to be economical, efficient, and beneficial over the long term, it must give rise to an increasing output from the land. For this to happen, the hirer of labor must have security of tenure to his land and a guarantee of proprietary rights to any increase in land productivity that he brings about. It is this security and guarantee that underpin farmers' ability to continue to hire labor, and thus the expansion of the rural wage-labor market as a whole. The need for unrestricted access to secure land (in other words, for commoditizing land) gradually stratifies the rural population into a landowning and labor-hiring class and a wage-earning and sometimes landless class. Hence the close relation that exists between labor demand forces, land tenure, and emergent class formation in many rural areas of sub-Saharan Africa.

The traditional land tenure system in these countries is, of course, based on communal rather than individual rights to land. It is the community, whether defined as the tribe, the clan, or the lineage group, that owns land; the individual has only the right of use or the usufruct of the land. In detail, however, the situation is more complicated and differs by region. Gluckman (1943), in his review of the land tenure system among ethnic groups in southern Africa, noted that the chief was regarded as owning the land in trust for his people. The land was allocated by him to sub-chiefs, who in turn allotted shares to village headmen and through them to heads of families, who then distributed land for use to their dependents. The number of steps in this hierarchy of land rights depended on the depth of the political hierarchy. Indeed, Gluckman argued that in order to bring out the strong connection between land rights and social status these rights should be called "estates of holding" and distinguished as primary, secondary, and so on to show that they occur in a hierarchy. While the cultivator was protected in his use of the land and could transmit rights in it to his heirs, he could not transfer these rights to strangers without consulting all those who held residuary rights in the land if vacated by the cultivator. Similar systems of land tenure have been reported for the Bemba of Zambia (Richards, 1939) and the Yao of Malawi (Mitchell, 1954). Ownership by lineage groups is more characteristic of ethnic groups in West Africa and a number of others in eastern and central Africa such as the Kikuyu of Kenya, the Lungu of Tanzania, and the Luvale of Zambia.

One of the most striking consequences of rural-to-rural migration in Africa has been the tendency in the destination areas for the traditional land tenure system to change in the direction of individual ownership of land, commoditizing it and allowing its alienation through pledging, mortgaging, leasing, or outright sale. In these areas, therefore, together with wealth accumulation from export crop production using migrant wage labor, considerable class stratification has become evident. Successful households can purchase additional agricultural land, giving rise to sharp differences in household economic circumstances and making for striking contrasts in class orientations. The process of class formation remains somewhat fluid, however, with many social transactions continuing to be mediated through kinship and other precapitalist structures.

By contrast, in regions of major outmigration, especially in eastern and central Africa, institutional responses initially were such as to inhibit the commoditization of land, tending instead to conserve traditional norms. During the colonial period, because of the uncertainty surrounding employment in the white-dominated urban areas, the lack of social security, and the inability to acquire land and houses in the towns, outmigrants apparently looked for their ultimate security to their communal land. Where there was simultaneously increasing pressure of population on the land, especially in those areas of southern Africa operating the "estates of holding" type of communal tenure, the chief was obliged to ensure that land was available in reserve for the outmigrants when they returned. In Lesotho, for instance, one way of doing this was to forbid unmarried men to have land and to restrict married men to three fields of two acres each (Allan, 1965).

In West Africa, where the lineage system of communal tenure did not impose similar obligations on the headman, traditional norms in the source-regions of outmigration were preserved, largely because there was no great demand for land. These were areas where most agricultural production involved subsistence crops for which cash returns had not reached a level that would give real commercial value to the land. Where the household mechanism for outmigration did not entail that the land was prepared for cropping before the migrant departed, land simply was left fallow and underutilized. No major rights were infringed if a neighbor or a stranger were to use such land for cropping since the crops grown were invariably annual. The only condition likely to be stipulated was that permission to do this be secured from the resident head of the lineage or the village head. Rural outmigrants did not always retain a secure use right to their original land. In northern Nigeria, for instance, following the British conquest of the Fulani emirates in 1902–03, the ultimate rights to the land of the territory were regarded as having been transferred to the British Crown in accordance with Islamic law. The concept of lineage land disappeared and migrants had no "natural rights" of use on their return to their homeland unless they had departed temporarily (paying the tax in their absence) or had gone to Mecca. Hill reports information provided by Alkalin Sokoto in 1929 that rights of returning migrants had diminished since the British Occupation (Hill, 1972, citing a Memorandum on Rural Land Policy in the North).

Traditional land tenure institutions could not of course be insulated from change. Returning wage laborers brought with them the new attitudes to land that they had encountered at their destinations. They were aware of the economic possibilities of growing newly introduced crops such as rice, tobacco, benniseed, and soyabeans for cash. The need to acquire the best land for these activities began the process of commoditization. A market in land emerged. Sometimes, the process of purchase would last over many outmigrations, with the migrants remitting money home when possible to complete the purchase price. Notables and traditional chiefs who had kept or acquired some real power in the colonial administration were able through their position to appropriate and alienate communal lands to themselves or to increase the amount of land cultivated for their personal benefit by members of the community operating under customary relations (Kohler, 1971). In northern Nigeria, until independence, villagers who migrated to farm elsewhere were obliged to turn over their farms to the village head, who could dispose of them as he wished (Hill, 1972: 85). Equally important was the role of rural moneylenders or traders, who were able to acquire large estates and become successful farmers through usurious loans that led to indebtedness and eventual forfeiture of land by farming families. This process was often the cause rather than the effect of outmigration (see also Gervais, 1984: 137).

In recent years as economic crisis in most of sub-Saharan Africa has made food crop production more profitable, the tendency for land to be bought and for land prices to rise in regions of outmigration has grown. The process has been stimulated in most countries of the region by the growing demand for animal protein in Europe and other industrialized regions, which has made large-scale mechanized cultivation of maize for livestock feed a very profitable venture. The existence of vast, underutilized, foodcrop-producing land in areas of major outmigration has prompted an increasing number of relatively wealthy, urban-based entrepreneurs to acquire extensive farmland. Such a development creates new sources of local labor demand and provides the headman and other important members of lineages with ready cash—making the alienation of land to such entrepreneurs irresistible.

The gradual erosion of traditional land tenure systems could thus be one of the unintended byproducts of current economic circumstances in which increasing self-sufficiency in food production has become an imperative for most countries of sub-Saharan Africa. Its effect in stimulating nascent rural class formation in the traditional regions of major outmigration is a factor of increasingly widespread social significance. In Burkina Faso, for instance, Gervais (1984) reported the growing importance of the class of functionaries and members of the educated elite who use their family plots or new plots given them by the traditional headman to cultivate cash crops on an area that is large compared with that of other peasants. Their high salaries give them the resources to buy not only more land but also the seeds, fertilizer, and machinery (e.g., diesel pumps) with which to increase crop yields. This situation has become common in many sub-Saharan African countries, facilitated by the ease with which the new class of farmers can negotiate loans from commercial banks and gain access to agricultural inputs provided by the state at greatly subsidized prices. In short, outmigration, by weakening the integrity of communal control over land, has made it easy to promote a new regime of individualized landholding and the emergence of a new class of landowning, labor-hiring individuals whose style of operation and living standards set them far apart from the common run of subsistence farmers.

The changing role and status of women

One of the most critical agrarian responses to outmigration in sub-Saharan Africa has occurred within the rural household—in the adjustment in the role and status of women. In a demographic sense, this has entailed the attempt to reconcile the woman's childbearing function with her equally critical economic role. Although the evidence is not conclusive, there are indications that marriage tends to be delayed in areas of heavy outmigration and the birth rate to be somewhat reduced. More significantly, because many households come to be headed by women as a result of the outmigration of their husbands, agricultural production has had to adjust toward tasks that are less labor intensive and involve fewer male-specific obligations.

Riddell (1970) noted that in four Liberian villages he studied, following the outmigration of males, agricultural planning functions were carried out by women. In his research among the Sabo people, also in Liberia, McEvoy (1971: 300–301) argued that "it is clear that the long-established practice of labour migration among the Sabo has been and is a major factor contributing not only to the 'formation' of particular female-headed households in Sabo villages, but also to the persistence of this household type as a socio-cultural phenomenon of considerable significance in modern day Sabo village social life and organization" (see also MacGaffey, 1983). He noted that 20 percent of households in Saboke villages were headed by females who had stayed behind to manage the households while their husbands were away. He further observed that in consequence there has been a marked shift in cropping pattern from subsistence crops such as rice to less labor-intensive and less seasonally peaked crops such as cassava. A similar shift from plantain cultivation to cassava growing has been observed among women cultivators in Eastern Zaire—for much the same reason (Newbury, 1984). In Mauritania, Dussauze-Ingrand (1974) reported a fall in cropped acreage and a shift to cattle grazing among the Sarakolle as a result of extensive international migration. He observed, however, that this did not imply an occupational shift among the women but reflected new patterns of capital accumulation as remittances allowed labor to be hired for these tasks.

Women's increasing decisionmaking role in critical areas of production as a consequence of outmigration raises the issues of the relationship between the nuclear and the extended family and of the extent to which the latter can continue to retard the emergence of a rural class society. In societies where women are not allowed to assume a vital decisionmaking role in agricultural production, there is an attempt either to shorten the duration of migration (or abandon it) or to acquiesce in the break-up of the extended family structure. Goddard (1973a) observed that the first option was the more acceptable to migrating families around the Sokoto region of Nigeria, hence the emphasis on seasonal migration (see also Connell et al., 1976). In some other societies, the nuclear family was responsible for economic organization and decisionmaking while the extended family retained social significance. Carter (1970) noted in one Liberian village that extended households were becoming much more important as a factor permitting greater flexibility in migration patterns.

One way in which outmigration tended to weaken the extended family as against the nuclear was through the impact of remittances sent by migrants. Apart from those going to headmen or aged relations, remittances were usually sent to the nuclear family for investment either in human resource development, notably in education for the children, or in asset improvement or acquisition such as the construction or improvement of huts, increasing the number of cattle, expansion of storage facilities, and so on. Particularly through the education of children, migrant remittances may generate serious inter- and intra-generational differentiation, which has tremendous significance for the manner in which individual nuclear families respond to the local agrarian situation.

A shift in emphasis among economic activities is another way in which the extended family, and women in particular, have had to adjust to the fact of male outmigration. Gulliver (1967), for instance, reported that among the Ndendenli of Tanzania craftwork such as the making of cooking tools and utensils almost disappeared as migration took away those who had the skills, and the items were replaced with cheap, manufactured substitutes. Among the Mossi of Burkina Faso, the growing use of plaited grass for both hut and compound walls and the serious shortage of huts are also attributed to the absence of men, who traditionally prepared sun-dried bricks for these purposes (Skinner, 1960: 385–386). Similarly, the decline of cotton-cloth production among the Mossi has been associated with outmigration, although the relationship between the two is complex.

In social and economic terms, therefore, outmigration has set in motion agrarian responses whose long-term effect is to strengthen the integration of the nuclear family, consisting of a man and his wife or wives, into the capitalist system. Outmigration has tended to commoditize the labor of the men and to make their lives revolve more around the wages they are paid at their destination. Their long absences from home have weakened the fabric of their social support system and forced their wives to become more involved in the rural production system and therefore more vulnerable to new capitalist challenges. The extended family system, though still providing surveillance support for absent husbands, is less and less able to perform its traditional economic role (Goddard, 1973b). The diversity of skills that such a system once sustained, especially of handicrafts and cottage industries, gradually fell into disuse as the young sought only wage-earning, unskilled labor (see Arrighi, 1970: 212). On the other hand, the nuclear family, in spite of its greater social vulnerability to the vagaries of economic change (i.e., its lack of the safety net provided by the extended family), became the institution more adaptable to capitalist manipulation. Consequently, as the impact of outmigration deepens, the nuclear family increasingly emerges as the primary social organization (Ali and O'Brien, 1984; Goddard, 1969). Whether in terms of labor supply or access to land, this institution becomes the focus of rural change, gradually replacing kinship, gender, age, or guild as the unit of organization. Its prevalence in a given rural situation usually sets the stage for more radical capitalist transformation of a type helped along by significant intervention of the state.

The role of the state

The point has already been made that the phenomenon of outmigration and the pattern of agrarian response to it can only be understood within the framework of the penetration of a capitalist into an erstwhile precapitalist social formation. Within such a framework, the state, especially the colonial "state," is not a neutral agency. Indeed, as Halfani and Barker (1984: 42) have noted, the role of the colonial state was to reconcile the indigenous people of sub-Saharan Africa to the interests of metropolitan companies and, where settlers were present, to sort out differences that might arise between companies and settlers. With regard to the indigenous people themselves, the state served as the handmaiden for capitalist penetration of the social formation by facilitating the development of a differentiated peasantry, from which a class of capitalist farmers and another of agricultural wage laborers could emerge.

The pattern of agrarian response, where it tended in the general direction of capitalist social relations, was thus reinforced by state action. A central mechanism of the state for fostering agrarian responses compatible with capitalist objectives was the colonial legal and judicial system. In virtually all sub-Saharan African countries, the state made clear that the nuclear family was the preferred unit of social organization. Statutory marriage law conferred state recognition and formal respectability on this institution. In the French colonial state, statutory marriage was made a prerequisite for attaining the status of assimilated "citoyen" with its varied privileges as against the status of the majority of disenfranchised "sujets." Although this distinction with its privileges was meaningful only in respect of urban dwellers, it did cast its shadow on the situation in the rural areas.

A major rural presence of the legal and judicial system was its power, either through statutory provisions or case law, to buttress the individualization and commoditization of landholding, with consequent potential for alienation. It has already been emphasized that this trend was resisted for a long time, even in regions of massive outmigration. However, especially in the postindependence era and with the renewed emphasis on food crop production, the state has tried to provide uniform legislation to make land more easily available to the emergent wealthy elite class desirous of moving vigorously into agricultural production. In Nigeria, for instance, the Land Use Decree of 1978 attempted to transfer all rights to land to the state as a means of facilitating the conveyance of large tracts to individuals through statutory occupancy certificates of determinate duration. This decree made it possible for powerful and wealthy individuals to move into areas where outmigration had reduced the pressure of population and to secure rights to vast acreages of land. The incursion of large-scale capitalist producers into erstwhile underdeveloped agricultural regions tended to accelerate the process of commoditization of land in the affected rural areas, transforming the local peasants into (initially) part-time wage laborers as well as landowners (Shenton, 1986: 137–139).

More directly, state intervention in areas of rural outmigration took the form of capital development. This often involved programs for the establishment or improvement of infrastructure: roads, water supply, irrigation channels, health and educational facilities, and markets. Sometimes these programs were accompanied by credit schemes that offered farmers loans to expand their production, as well as extension services, a subsidized supply of inputs, and introduction of a mechanized system of cultivation. In many countries, the catalyst for such development was international aid agencies. Yet, because the undeclared agenda of the state was more efficient capitalist agricultural production, the effect on the peasant was far from uniform. Those peasants who could exploit some initial advantage attracted to themselves a disproportionate share of the various amenities (Weigel, 1982). Those less fortunate often found themselves with diminishing access to resources and income and on a downward spiral of insolvency, land expropriation, and reduction to the status of landless peasants.

The role of the state in modulating agrarian responses to outmigration is thus very often to increase class differentiation among farm households. Sometimes this outcome was explicitly pursued, especially where the area concerned was the site of some state- or internationally sponsored project. In such a situation, the emerging rich peasants were characterized as "progressive farmers," and extension service attention was lavished on them. Yet their enhanced success may have been at the cost of diminished attention to the rest of the population.

Thus, in many countries of sub-Saharan Africa, state intervention in regions of outmigration is prompting the emergence of a class system of capitalist farmers and semi-proletarianized peasants. Reyna describes such an emergent system in Burkina Faso (formerly Upper Volta) as follows:

> Land ownership differentials are established in two ways: when officials take advantage of their positions, or where individuals gain control over larger amounts of more fertile land due to the history of their kin group's land occupation. Once land differentials emerge, the demand for and availability of financial capital influence the rate of private class formation. In areas of highest agricultural commodity demand (e.g., the southwest and the forest), farms based on wage labour rapidly emerged, while in those of lower demand (e.g., the Mossi Plateau, the Eastern ORD and Tenkodogo) such farms have occurred more slowly. These two factors are themselves conditioned by Upper Volta's position in the world capitalist economy, because the country lacks, as it did in colonial times, a comparative advantage in a primary commodity that core industries substantially demand and hence the financial capital which frequently materializes to satisfy such demand. (1983: 223)

The development of state-promoted agricultural ventures in traditional regions of major outmigration has far-reaching agrarian consequences. Indeed, unlike the "separate identity" of colonial plantations and settler farming, the recent changes in production strategies in most African countries are laying the foundation for intense class antagonisms in rural areas. Some countries are already sensitive to the catastrophic potential of such rural transformations and are promoting vigorous programs to improve the economic and social conditions of small producers. In Nigeria, the administration has established and amply funded a Directorate of Food, Roads, and Rural Infrastructure, based for maximum impact in the Office of the President. Its mandate is to mobilize small peasant farmers at the community level and to make them the centerpiece of a strategy of rural development. The programs of the Directorate are seen as very different from those of the World Bank–assisted Agricultural Development Projects, which have not been able to escape the class distortions that tend to accompany externally promoted projects (see Loxley, 1984).

Conclusion

This essay has explored various ways rural institutional forms in sub-Saharan Africa have influenced societies' adaptation to outmigration and the possible effects of outmigration on those institutions themselves. The scope of the presentation has been limited to tendencies observed in selected studies of specific communities from colonial times to the present. Its major contribution may be to illuminate the inadequacy of our knowledge and information about what is happening in the region, especially as these relate to population processes and the patterns of socioeconomic development in the last three decades. The best way to conclude this essay therefore is to highlight four issues that deserve greater attention if we are to be better informed as to how rural societies in sub-

Saharan Africa respond to and accommodate the strains and challenges of outmigration.

First is the issue of how far the support system provided by the extended family is being weakened and how far the nuclear family is emerging as the central institutional unit for labor relations and for critical economic decision-making in rural areas.

Second is the issue of the role and status of women within this new institutional setting. Given the improvement in transportation and communication facilities, which should lessen the duration and totality of the husband's absence in comparison to earlier years, what changes have been taking place both in the obligations and the privileges of women in the household? How do these affect their reproductive behavior and their attitudes toward family size?

Third is the issue of class formation, especially as an outcome of the state's attempts to stimulate increased rural productivity. What are these efforts doing to land distribution in rural areas? How much of the land is being appropriated by state functionaries, local notables, village headmen, traders, moneylenders, and other members of the rural elite? Are these people emerging as a rural capitalist class? Are they actively engaged in primitive accumulation, promoting the commoditization of rural land and its private acquisition while encouraging the emergence of a class of landless, rural wage labor?

Finally, what are the effects of these processes on the welfare of the masses of peasant producers and their families? In particular, what are their effects on nutritional status—by social class, by sex, and by age group? Is the promise of increased food productivity through capitalist development of the rural areas likely to trap the laboring classes in a state of poverty in the midst of plenty?

These issues are not new in terms of research specifications, but they have received too little attention, especially in the context of the emerging, wide-ranging impact of nationally directed rural development programs that are meant to minimize the deleterious consequences of a too-early massive rural outmigration in sub-Saharan Africa. It is to be hoped that these issues, because they are at the forefront in the complex mix of problems confronting countries of the region, will be accorded renewed research priority and the findings allowed to inform the newly evolving strategies of self-reliant rural development being promoted by governments everywhere in sub-Saharan Africa.

References

Ali, Taisier, and Jay O'Brien. 1984. "Labor, community and protest in Sudanese agriculture," in *The Politics of Agriculture in Tropical Africa*, ed. Jonathan Barker. Beverly Hills: Sage Publications, pp. 205–238.

Allan, William. 1965. *The African Husbandman*. London: Oliver & Boyd.

Arrighi, G. 1970. "Labour supplies in historical perspective: A study of the proletarianization of the peasantry in Rhodesia," *Journal of Development Studies* 6, no. 3: 197–234.

Berg, Elliot J. 1965. "The economics of the migrant labour system," in *Urbanization and Migration in West Africa*, ed. Hilda Kuper. Berkeley: University of California Press, pp. 160–181.

Binswanger, Hans P., and John McIntyre. 1984. "Behavioral and material determinants of production relations in land-abundant tropical agriculture," Agricultural and Rural Development Research Unit, Discussion Paper No. ARU 17. Washington, D.C.: The World Bank.

Boeder, M. L. 1973. "The effects of labour emigration on rural life in Malawi," in *Rural Africana: Current Research in the Social Sciences*, No. 20. African Studies Center, Michigan State University.

Boserup, Ester. 1965. *The Conditions of Agricultural Growth*. Chicago: University of Chicago Press.

Carter, J. E. 1970. *Household Organisation and the Money Economy in Loma Community, Liberia*, unpublished Ph.D. thesis, University of Oregon.

Connell, John, B. Dasgupta, R. Laishley, and M. Lipton. 1976. *Migration from Rural Areas: The Evidence from Village Studies*. Delhi: Oxford University Press.

Datoo, Bashir A. 1977. "Peasant agricultural production in East Africa: The nature and consequences of dependence," *Antipode: A Radical Journal of Geography* 9, no. 1: 70–78.

Dussauze-Ingrand, E. 1974. "L'Émigration Sarakollaise du Guidemaka vers la France," in *Modern Migrations in Western Africa*, ed. S. Amin. London: pp. 239–257.

Gervais, Myriam. 1984. "Peasants and capital in Upper Volta," in *The Politics of Agriculture in Tropical Africa*, ed. Jonathan Barker. Beverly Hills: Sage Publications, pp. 127–141.

Gluckman, M. 1943. *Essays on LOZI Land and Royal Property*, Rhodes-Livingstone Papers, No. 10. Rhodes-Livingstone Institute, Livingstone, N. Rhodesia (Zambia).

Goddard, A. D. 1973a. *Population Movements and Land Shortage in the Sokoto Close-Settled Zone, Nigeria*. Department of Geography, University of Liverpool.

———. 1973b. "Changing family structure among the rural Hausa," *Africa* 43, no. 3: 207–218.

———. 1969. "Are Hausa-Fulani family structures breaking up?," *Institute for Agricultural Research Newsletter*, No. 11, 3 June, Ahmadu Bello University, Zaria, Nigeria.

Gulliver, P. H. 1967. "The case of the Ndendenli: Shifting cultivation in southern Tanzania," paper presented at the Conference on Competing Demands for the Time of Labor in Traditional African Society, Holly Knoll, Virginia.

Halfani, Mohammed S., and Jonathan Barker. 1984. "Agribusiness and agrarian change," in *The Politics of Agriculture in Tropical Africa*, ed. Jonathan Barker. Beverly Hills: Sage Publications, pp. 35–63.

Hill, Polly. 1972. *Rural Hausa: A Village and a Setting*. Cambridge: Cambridge University Press.

Kabunda, M. K. Kabala. 1975. "Multinational corporations and the installation of externally-oriented economic structures in contemporary Africa: The example of the Unilever–Zaire group," in *Multinational Firms in Africa*, ed. Carl Widstrand. Uppsala: Scandinavian Institute of African Studies.

Kohler, J. M. 1971. *Activitiés agricoles et changements sociaux dans l'Quest-Mossi*. Paris: ORSTOM, Memoire 46.

Loxley, John. "The World Bank and the model of accumulation," in *The Politics of Agriculture in Tropical Africa*, ed. Jonathan Barker. Beverly Hills: Sage Publications, pp. 65–76.

MacGaffey, Janet. 1983. "The effect of rural–urban ties, kinship and marriage on household structure in a Kongo village," *Canadian Journal of African Studies* 17, no. 1: 69–84.

Makinwa, P. K. 1981. *International Migration and Rural Development in Nigeria: Lessons from Bendel State*. Ibadan: Ibadan University Press.

McEvoy, F. D. 1971. "History, tradition and kinship as factors in modern Sabo labour migration", unpublished Ph.D. thesis, University of Oregon.

Miracle, M. P., and S. S. Berry. 1970. "Migrant labour and economic development," *Oxford Economic Papers* 22, no. 1: 86–108.

Mitchell, J. C. 1954. "Preliminary notes on land tenure and agriculture among the Machinga Yao," *Journal of the Rhodes-Livingstone Institute*.

Murray, Colin. 1981. *Families Divided: The Impact of Migrant Labour in Lesotho*. Cambridge: Cambridge University Press.

Newbury, M. Catharine. 1984. "Ebutumwa Bw'Emiogo: The tyranny of Cassava: A women's tax revolt in eastern Zaire," *Canadian Journal of African Studies* 18, no. 1: 35–54.

Reyna, S. P. 1983. "Dual class formation and agrarian underdevelopment: An analysis of the articulation of production relations in Upper Volta," *Canadian Journal of African Studies* 17, no. 2: 211–233.

Richards, Audrey I. 1939. *Land, Labour and Diet in Northern Rhodesia: An Economic Study of thc Bemba Tribe*. London: Oxford University Press.

Riddell, J. C. 1970. "Labour migration and rural agriculture among the Gbannah Mano of Liberia", unpublished Ph.D. thesis, University of Oregon.

Rouch, Jean. 1956. "Migrations au Ghana (Gold Coast): Enquête 1953–1955," *Journal de la Societé des Africanistes* 26: pp. 33–196.

Shenton, Robert. 1986. *The Development of Capitalism in Northern Nigeria*. Toronto: University of Toronto Press.

Skinner, Elliott P. 1960. "Labour migration and its relationship to socio-cultural change in Mossi society," *Africa* 30, no. 4: 375–401.

Stichter, Sharon. 1982. *Migrant Labour in Kenya: Capitalism and African Response 1895–1975*. Longman.

Stuckey, Barbara, and Margaret A. Fay. 1981. "Rural subsistence, migration and urbanization: The production, destruction and reproduction of cheap labour in the world market economy," *Antipode: A Radical Journal of Geography* 13, no. 2: 1–14.

Thompson, Virginia, and Richard Adloff. 1958. *French West Africa*. London: George Allen & Unwin.

Van Velsen, J. 1964. *The Politics of Kinship: A Study in Social Manipulation among the Lakeside Tonga of Nyasaland* Manchester: Manchester University Press.

Watson, William. 1958. *Tribal Cohesion in a Money Economy: A Study of the Mambwe People of Northern Rhodesia*. Manchester: Manchester University Press.

Weigel, Jean-Yves. 1982. *Migration et production domestique des Soninke du Sénégal*, Paris: ORSTOM.

Wilson, Godfrey B. 1941. *An Essay on the Economics of Detribalization in Northern Rhodesia*, Part I. Rhodes-Livingstone Papers, No. 5. Rhodes-Livingstone Institute, Livingstone, N. Rhodesia (Zambia).

World Bank. 1984. *Toward Sustained Development in Sub-Saharan Africa: A Joint Program of Action*. Washington, D.C.

THE INTERNATIONAL SYSTEM

Rural Development, Population Growth, and the International System

PAUL DEMENY

IN HIS 1930 ESSAY, "Economic possibilities for our grandchildren," J. M. Keynes speculated, uncharacteristically for him, about the very long run. Let us suppose, he wrote, that "a hundred years hence we are all of us, on the average, eight times better off in the economic sense than we are to-day. Assuredly, there need be nothing here to surprise us." Granted such growth, he went on, "the economic problem may be solved, or at least within sight of solution, within a hundred years. This means that the economic problem is not—if we look into the future—the permanent problem of the human race" (Keynes, 1963: 365–366).

Although in a hundred years we will assuredly all be dead, taking an occasional equally heroic glimpse into the distant future of the world economy and, in particular, of the economies of the developing world, is an illuminating exercise. It provides a perspective beyond preoccupation with the pressing needs of the day, and of the next decade.

While the future is always uncertain, a simple extrapolation of past development trends suggests an outlook not inconsistent with Keynes's optimism. His surprise-free assumption implied a sustained rate of per capita income growth of slightly below 2.1 percent per annum. During the nearly four decades between 1950 and 1987, the gross world product, measured in constant prices, grew at the annual rate of somewhat over 4.2 percent (Maddison, 1989). The size of the world's population during the same period has doubled—it grew from 2.515 billion to 5.025 billion (United Nations, 1989). Thus the average annual rate of population growth was slightly below 1.9 percent. Combining the two estimates yields a per capita rate of growth of 2.4 percent per annum. Sustaining such an average rate of growth for a century would generate an eleven-fold increase in income per capita.

A rate of annual income growth per capita that exceeds 2 percent is, of course, exceptionally high by historical standards. Angus Maddison estimates that during the first 50 years of the twentieth century—a period which included two world wars and the deepest economic recession of the modern era—per capita income growth was much slower: only 1 percent per year. (His calculation is based on a 32-country sample that comprises only three-quarters of the world's population but is "probably reasonably representative for the world as a whole.") This pulls down the average per capita growth rate for the entire 1900–87 period to 1.7 percent per annum. But that is still a respectable performance. Sustaining growth at such a tempo for another hundred years would more than quintuple average income levels.

For thinking about rural development, the relevance of such seemingly frivolous extrapolations is evident. Positing the continuation of past trends serves as a reminder that the great structural differences that now characterize the international system in terms of disparities between countries with respect to rural-urban composition and with respect to the relative share of agriculture within the economy at large may become greatly attenuated in the not too remote future. Long-term development trends point toward, and may eventually culminate in, an economically more homogenous world, one in which, to quote Keynes again, "there will be ever larger and larger classes and groups of people from whom problems of economic necessity have been practically removed" (p. 372).

Some salient aspects of these trends, past and, at least by fiat, prospective, will be briefly touched upon in this note. Their cumulative effects are among the major forces transforming important structural characteristics of the international system. These changes may be seen, in first approximation, as originating from the internal efforts of the multifarious actors of the separate economies that make up the global economy. Most of the papers in the present volume describe the processes and institutions of rural development that constrain or amplify the exertions of individuals who compose the rural economy of the still predominantly agrarian countries of the world. But these processes and institutions, and the individual actors within them, are in turn affected by the international system that surrounds them, partly through the gravitational pull of the historically evolving structure of that system, and partly through deliberate human action. The second part of this note outlines some of the conditions that ought to be satisfied if sustained economic progress is to be achieved in the less developed world, and briefly discusses some of the influences emanating from the international system that bear on success or failure in satisfying those conditions of progress. In doing so, special attention will be given to outside influences affecting domestic demographic change.

Toward a nonrural world

Successful development, especially in countries still relatively poor, means achievement of higher income per capita. On the supply side, as can be appreciated most clearly in a closed economy, the key precondition for achieving higher incomes is rising productivity of labor in agriculture. Rising productivity frees labor for nonagricultural pursuits and generates a surplus of food that permits a larger and larger proportion of the labor force to be engaged in the production of nonagricultural goods and services, exchangeable, among other things, for food. As to consumers' demand, its behavior is governed by Engel's Law. As the income elasticity of demand for food is less than unity, and as that elasticity tends to decrease with rising incomes, increasing material affluence brings about a decreasing share of food within total consumption. Furthermore, within the total demand for food, the component of value added that represents the direct contribution of agriculture tends to decrease. Thus, successful development is bound to create economic structures that are largely nonrural, at least if ruralness is understood, as it should be in the present context, as existence closely tied to an occupation within agriculture, in particular to the production of food. Success would also relegate the struggle for subsistence, the primary preoccupation of mankind throughout its past history, to a distinctly modest place in peoples' daily life.

The proposition just made is likely to be termed naive, or even offensive, by biologists, as if it were a denial of man's fundamental biological dependence on nature. It is not. The proposition has to do with economic value, determined by the effort necessary to satisfy specific human needs. Water is more important for life than diamonds, yet pound-for-pound diamonds are dear and water is cheap. In a technologically advanced affluent society food, too, is cheap, and its production claims only a comparatively modest place within the overall scheme of economic activities.

The point may be illustrated by the example of the US economy. As of the late 1980s, the average American household has been devoting 13 percent of its total private consumption expenditures to food and beverages. That is no small item in a household budget, but much of the value purchased represented product generated outside agriculture. Of the total US civilian labor force 21.9 percent were employed in the food and fiber sector, but only 2.1 percent worked on farms. The value added that originated on farms amounted to only 1.5 percent of the domestic economy (US Bureau of the Census, 1987). Corresponding figures for other high-income countries (countries that also possess an agricultural sector that is the main source of domestically consumed food) are less extreme but show essentially the same picture: a rural/agricultural component increasingly dwarfed by other sectors of the economy.

The developing world as a whole should be headed in the same direction. Some of its development experience, like the earlier experience of the now high-income countries, has already demonstrated that the dynamics of wealth creation made possible by modern technology and economic organization can increase levels of living and change economic structure within a historically short period of time. The overall average figures characterizing twentieth century economic growth cited above are not dominated by the economically advanced countries. This is true especially for the most recent decades. Population-weighted estimates of economic growth, disaggregated by broad income categories, show that the rate of progress has been inversely related to level of income. The finding, albeit in a weaker form, remains true also if it is cast in terms of income per capita, an index that reflects welfare improvements more directly. This, despite much more rapid population growth in the lower-income countries. Annual per capita rates of growth of the gross national product (GNP) between 1965 and 1987 for the three broad groups of economies distinguished by level of income in 1987 (a classification that excludes a number of countries, notably the Soviet Union, for which comparable data are not available) indicate that, on the average, the poorest countries advanced most rapidly. Middle-income countries grew more slowly, but still faster than the most affluent countries (World Bank, 1989):

			GNP per capita	
	Population 1987 (millions)	Avg. annual rate of population growth 1965–87 (percent)	($US) 1987	Avg. annual growth rate 1965–87 (percent)
Low-income economies	2,823	2.2	290	3.1
Middle-income economies	1,039	2.3	1,810	2.5
High-income economies	777	0.8	14,430	2.3

Such summary estimates, of course, gloss over a great deal of diversity within each of these broad categories. Within each group, some countries have fallen considerably short of average performance. Because of the flaws of the income measure as an expression of true relative levels of living, the average income estimates exaggerate the differences between the three groups of countries. More importantly in the present context, they also conceal changes in performance over time. In particular, in the developing world—that is, among the low- and middle-income countries—some countries have experienced special difficulties in the 1980s. But the average rates of growth do depict the main trend during the period covered, and it is not without interest to ask where the trends lead.

Extrapolating the 1965–87 average growth performance for a hundred years for the two groups of developing countries shown above would lift the present middle-income economies into comfortable affluence, well beyond the levels now prevailing in the high-income economies. The same exercise would imply an equally radical transformation of the economic landscape in the present low-income countries: it would bring their GNP per capita to something close to one-half of that currently characterizing the high-income economies. In the former group, the share of income allocated to the raw products of agriculture, and the relative weight of the rural/agricultural sector within the economy as a whole, would come to resemble the current American pattern. The countries now categorized as low-income would be no doubt still so described, since people, and statisticians, think about income levels in comparative terms. But on an absolute scale, these countries, too, would enjoy quite high average levels of material welfare. By the same token, food in these countries would claim only a moderate share of household budgets, and the share of agriculture in their GNP, and the share of those employed in agriculture within their total labor force, would be correspondingly modest. Projection of recent trends even on a much less ambitious time scale would still imply substantial reductions in the proportions now classified as poor—a state whose single most important attribute is lack of access to food in adequate quantities. Elimination of poverty would become largely a matter of political choice.

Agricultural transitions

Marx once suggested that the development paths of economically backward countries would retrace the paths pioneered by the leaders. Trends in expectation of life, one of the best indicators of economic welfare, provide perhaps the most compelling statistically documented example of such a replicative pattern. During the last 200 years or so, most countries of the world had registered monotonic improvement in that index, albeit with varying steepness, separated from each other by time lags of varying duration. The ultimate convergence of the lagged trendlines of individual countries into a relatively narrow band that corresponds to a level approximating the lowest achievable human mortality is taken for granted in virtually all long-term population forecasts.

Graphs showing the recorded evolution of such indexes of development as the proportion of the rural population within national totals, or the proportion of the labor force employed in agriculture, have a family resemblance to graphs of mortality trends. Trendlines of the index last mentioned, for example, show monotonic decreases that can be traced with relative accuracy for well over a hundred years in the now-developed countries. The trendlines run almost parallel for long periods, separated by near-constant time lags. Thus, for long, France had been behind the Netherlands by some 40 years; Germany had lagged behind Britain roughly by 80 years. Eventually, the pioneers run out of space to register further absolute decreases: once the agricultural labor force is less than 10 percent of the total, another 10 percentage-point drop is not possible. Thus, in the most recent period there has been rapid convergence among the countries representing the economically most advanced segment of the international system. The record also suggests that latecomers, once structural transformation gathers momentum, tend to register faster progress than was the case in the pioneer countries. The share of the labor force in agriculture fell below 50 percent in Germany by 1870; in Japan, this was a post–World War II achievement. The proportion in the two countries is now nearly equal—below 10 percent. While both of these countries are heavy importers of food, in each, shrinking importance of the relative size of the agricultural sector went hand-in-hand with significant and steady growth of agricultural output, at a rate more rapid than that of population growth.

Many of the middle-income economies appear to be following growth paths that, in terms of the type of indicators referred to above, resemble historical agricultural transitions; in most of them, with varying time lags, the trendlines lie between those traced by the relatively slow but steady progress of nineteenth and early twentieth century Western European countries, on the one hand, and by rapidly industrializing pre-1960 Japan, on the other. In the middle-income group as a whole, for example, the proportion of agricultural labor force dropped below 50 percent by the early 1970s, and is decreasing apace.

In the low-income countries, which comprise well over half of the world's population (a group demographically dominated by China and India), it is much more difficult to discern a clear pattern of an impending absolute, or even a steady relative decrease in the rural component of the economy—developments that would indicate rising productivity of agriculture. Notwithstanding the impressively high average rates of economic growth that improved the welfare of the poorest three-fifths of the world's population during the last few decades, the prospects of many low-income countries for firmly embarking upon a developmental path leading to achievement of material welfare and economic structures resembling those of the more advanced countries remain tenuous. In a number of low-income countries that did make appreciable progress in that direction, the prospects for staying the course appear to be rather less than assured.

Conditions for progress

In making his starry-eyed projection, Keynes had in mind the developed world, primarily Western Europe. Although extending economic trends registered there far into the unknown future was certainly a safer exercise than doing the same for the developing world today, he nevertheless was careful enough to enter two conditions that had to be satisfied if progress toward ultimate economic bliss was to be achieved. His two caveats were that, first, there be "no important wars" and, second, there be "no important increase in population" (1963: 365–366).

From today's perspective, stipulation of a third condition, one concerning public policies—"governments must do no major harm"—seems to be a necessary addition to Keynes's short list. Keynes was certainly not unaware of the mischief erroneous public policies can play in economic life. He must have assumed, however, that, in the countries with which he was concerned, social processes are governed primarily by the spontaneous interaction between large numbers of individual economic agents, none of which can dominate the rest, and that these agents operate within a polity that checks the excesses of self-seeking economic behavior, yet keeps government power adequately circumscribed. The soundness of such an assumption should be held suspect even under the conditions of the Britain of the late 1920s. Making the assumption for the contemporary world, however, would be patently indefensible.

A fourth condition for progress that an amended Keynesian short list might also include should perhaps be phrased in non-negative terms: "economically backward countries must make the most of the advantages of their backwardness." The notion of backwardness expresses a relation—of one's state in comparison to someone else's—rather than merely dissatisfaction with how things are. Thus, for the situation Keynes addressed—that of countries in the vanguard of economic progress—such exhortation is irrelevant. There, evolution is a near-blind discovery process: society finding out through a myriad of trials and errors what works and what does not. Progress generated through such a process, no doubt, appears to be wasteful. But the record of the alternatives, of deliberately guiding the direction of change beyond the outer edges of the existing social and economic frontier, has been not only less successful, often it has also led to devastating consequences. Designs to do better in the future on this score are less than promising. In economically backward countries, however, a significant chunk of the economic and social agenda can be legitimately phrased as an injunction for catching up with the vanguard and, as a corollary, an injunction to learn from the successes of others and to avoid the mistakes others have made before hitting upon the right solutions. In such situations, decentralized trial-and-error is still likely to prove the best adaptive process, but its efficiency can be improved by concerted sifting through the lessons of the experience of other countries. Once a country's ambition can turn not just to "overtaking" but to "surpassing" its economic rivals, such shortcuts in progress will no longer be open to it, and the invisible hand will have to be given its full sway. But until then, developing countries can follow a simpler and straighter evolutionary path than did the pioneers. Exploiting this advantage can be an important source of economic gain.

North-to-South influences

The four conditions listed above for economic progress—which in countries that are predominantly agrarian means first and foremost rural progress—are admittedly rather general; how they apply in different circumstances is open to conflicting interpretations. The list could also be supplemented, almost endlessly, by stipulations of additional desiderata. Yet a reasonably coherent and comprehensive account of the economic, political, and social history of the developing world could be organized, without stretching the legitimate boundaries of the four topics, around the question of how the four conditions had been or are being fulfilled or violated in each country or region.

But to delve into matters of home-grown development processes and policies, and into the related controversies, would lead this note far afield. We shall proceed, instead, by commenting briefly on channels of external influence through which domestic development has been, and is being, affected in the countries of the contemporary developing world, whether by intent or through its sheer presence—by simply "being there." The notion of outside influence will be narrowed here to denote influence emanating largely from developed countries only, rather than from the international environment at large. To use an awkward but convenient label, we will focus on "North-to-South" effects. Little reflection is needed to appreciate that the topic so narrowed is still vast—indeed, its boundaries are nearly coterminous with those of the study of international relations: economic, political, and social. Here we can do no more than offer a few observations, seeking to select their objects so as to intimate the scope of this inadequately explored topic and to suggest the potential interest in its more intensive cultivation. We turn first and last to the two conditions for sustained economic progress specified by Keynes—those having to do with war and population. Between these two topics, we shall briefly discuss effects of the international system on the two conditions we have added to the Keynesian short list: those having to do with the role of government and the exploitation of the advantages of economic backwardness.

Avoiding war

The first of the conditions for economic progress cited above is absence of a "major war." The relevance of that proviso for securing favorable prospects for Third World development is evident, even if the link between war and development is seldom considered in discussing issues of economic change. Its consideration seems necessary in the present context. Whatever its origins, war's effect on development, by definition, involves the outside world; peace, or its absence, is an attribute of the international system.

If the first half of the twentieth century is chosen as the proper reference point for measuring more recent fulfillment of the stipulation "no major war," the post-1945 period is commonly given high marks. But, arguably, such a judgment reflects a distinctly "Northern" point of view. The two major wars that took place in the second and fifth decades of the present century were largely, if not exclusively, fought between the economically advanced countries. The characterization of these conflicts as "world wars" betrays a bit of a Northern hyperbole. Much of sub-Saharan Africa, large parts of Asia, and all of Latin America were not directly affected by these bloody and, in their immediate effects, growth-retarding wars. The economic impact of the two world wars in the regions that remained uninvolved in those military conflicts was on balance often favorable, particularly in South America, where the disruption of Northern supplies gave a strong impetus to industrialization. (Unhappily, some of the major developing countries that benefited from this war-generated windfall appear to have drawn the wrong lessons from this experience for shaping their post–World War II economic policies. Import-substituting industrialization and protectionism in international trade under conditions of peace and rapid economic growth in the industrially advanced countries proved to be inferior choices in overall development strategy.) Furthermore, and most importantly, the two world wars greatly hastened the collapse of the colonial system. The gaining of independent statehood by the former dependencies of the imperial powers in the postwar decades opened new and brighter prospects not only in the cultural and political domains, but also for more rapid economic progress.

That the fulfillment of these prospects fell short of expectations is in part a consequence of the high frequency of localized armed conflict (and of the common anticipation of the likelihood of such conflict, resulting in heavy military expenditures) that characterized the second half of the twentieth century in the developing world. Some of these wars (and preparations for war) were the result of, or were amplified by, the postwar competition between the two Northern "superpowers"; others reflected historical enmities and newly arisen conflicts of interests not related to Northern rivalries. The directly growth-retarding incidence of armed conflict in the South has had no parallel in post-1945 warfare within the North itself. There, it became widely realized that any intra-North armed conflict could quickly escalate to a truly "major war"—meaning war fought with nuclear weapons. Realistically, in 1930, qualifications would have had to be attached to Keynes's blithe identification of a "major war" as an event disastrous for economic welfare—for all their horrors, modern wars have invariably provided a number of important long-term stimuli for economic development and for growth-favoring socio-political change even among their active participants. But qualifications after 1945 were no longer called for. This contributed to the preservation of peace in the North. Lasting peace in the North conferred a significant developmental advantage over the South as a whole during the last 45 years.

This advantage may well continue in the future, and quite possibly become more pronounced. Apart from the recognized futility of winning a nuclear war, recent decades have also brought convincing demonstrations of the high economic cost of maintaining empires and of the superior efficiency of the trading state over the warring state in serving the interests of industrially advanced countries. Northern inclinations to become involved in Southern conflicts are therefore likely to further diminish, along with any possible temptations for using the developing world as an arena for playing out intra-North military, economic, and ideological conflicts. On the other hand, Northern interest and investment in the production of global peace—in effect Northern willingness to provide a valuable but rather costly international public good for the consumption of the international system at large—are also likely to flag. These changes in the international posture of the countries of the North would come at a time when in many parts of the developing world unsettled issues of nationhood and interstate conflicts may induce continuing heavy investment of resources in armaments, along with recurrent attempts to translate such investments into tangible military gains. Economic growth could be significantly retarded by such tendencies in a number of developing countries. If so, almost certainly rural development efforts would suffer disproportionately severe setbacks.

Using government wisely

The most recent decades have witnessed a major expansion of the role of government in the economic and social life of most developing countries. This happened most conspicuously in the case of the newly independent states, but the phenomenon was pervasive. Indeed, overall expansion of the role of the state in the Third World paralleled, and to a significant degree was modeled after, changes in the same direction in the economically more advanced countries. This similarity signals the source of an important influence on Southern development trends emanating from the North: a transfer of ideas, expectations, and models concerning the proper management of economic and social change.

To a first approximation, such transfer may be seen as one carried out by the developing countries: the result of deliberately importing particular models for home use, or at least of receptivity to ideas diffusing from the center and a disposition to selectively adopt them. According to this interpretation, the role of the developed world in this process is essentially passive. Given the long and varied experience of Northern countries, accumulated during their march toward a high level of material development, the "supermarket" of Northern ideas and models offers a great variety of wares that deserve the attention of potential shoppers. Discriminating shoppers can scrutinize what is on display and choose what they think best suits their particular purposes.

It would be logical to expect that among the models available, developing countries would be particularly interested in those institutional constructs that performed well under conditions similar in significant respects to their own current economic and social conditions. Especially important among these was a model of development that assigned government a highly circumscribed role in the economy: one limited to the provision of goods and services in which it had a clear comparative advantage. The core areas of government activity were identified by that rule as a narrow range of important public goods: national defense, maintenance of domestic order, the administration of justice—including, notably, the protection of a legal framework that is hospitable to private economic activity and entrepreneurship by safeguarding property rights, enforcing private contracts, checking monopoly power, securing a stable currency—and the provision of certain services, such as primary education and public health, and of elements of economic infrastructure, such as harbors, highways, and lighthouses. Western variants on this model of the "night-watchman state" showed substantial differences in drawing the precise boundaries of the role of government, but had the dominant common element of extensive reliance on private markets as the chief method of coordinating economic activity. Such reliance on markets left broad latitude for decentralized economic initiative and experimentation, secured efficient utilization of economic information available in a widely dispersed form only, and entailed broad acceptance of outcomes resulting from spontaneous interaction of economic agents, including the distribution of income.

Despite the conspicuous successes this model of economic coordination brought to the countries that adopted and followed it, its appeal for developing countries after World War II was limited. As was noted above, economic organization and management followed, instead, a far more statist orientation. Two influences, overlapping yet distinct, were particularly important in steering development in that direction, both of "Northern" origin. One was socialism, a doctrine that in its practical manifestations entailed centralized management of the economic process according to a more or less detailed blueprint. Socialist planning had wide appeal, not so much because of its ideological dressing but because it seemed to promise a shortcut in moving from conditions of economic backwardness to national economic power. The other was the welfare-state: it promised a more direct route to national well-being with equity through redistribution of incomes and through government-engineered production and allocation of certain goods and services. Considered too important to be left to the vagaries of economic markets, such services as education or health had to be brought under direct political control.

Even in terms of the theoretical formulations of their original advocates, adoption of an ambitious socialist and/or welfarist agenda by the developing countries would have had to be characterized as premature. Socialism, to be workable, was supposed to rely on highly advanced industrialism created by prior successful capitalist development. In particular, it also assumed the existence of cadres with extensive administrative-managerial experience, that is, availability of a resource in especially short supply in developing countries. Similarly, the launching of a credible welfare state assumed a relatively high degree of achieved economic development—in particular, average income levels that made extensive redistributive measures both materially feasible and politically tolerable.

Not surprisingly, results from the socialist and welfare-state experiments in all economically less advanced countries fell far short of what was promised by the planners. The goals of extensive central planning proved to be beyond the reach of scarce administrative capacity. The persistent thinness of welfare services received from the government by the poor in the developing countries, and the shallowness of income redistribution, if any, favoring the poor, reflected the narrow limits of welfare-statism in low-income countries.

But the economic consequences of attempting to accelerate development through extensive reliance on highly centralized economic and political power were, nevertheless, far-reaching. This was especially so for rural development.

Almost by definition, and regardless of the sources of political support that assured its initial ascendancy, centralized power proved to be urban-based power. Pressures on the political leadership seeking self-preservation, as well as the numerous opportunities open to those in power to engage in ordinary rent-seeking activities, tended to bias the use of economic command mechanisms in favor of urban producers and consumers. Taxation and price policies, and direct intervention in factor and product markets, created institutional structures that tended to weaken or even destroy economic incentives for rural producers, resulting in inadequate investment in agriculture, slow growth of agricultural output, and rates of rural-to-urban migration higher than would have been warranted in the absence of the politically engineered, if professedly unintended, urban bias. Services produced and distributed by the government under conditions of severe limits on the overall fiscal capacity of the state followed a logic that similarly contradicted declared intent. They tended to disproportionately favor the politically powerful and economically better-off: for example, many developed country governments provided free university-level training and urban hospital services more generously than they provided elementary education and primary health care in rural areas. Other ramifications of central planning, most notably policies pursued with respect to international trade, credit, and investment, typically reinforced such deleterious effects on rural development.

To characterize such policies as resulting exclusively or even primarily from deliberate if misguided choices by developing country political leaders and by other influential Third World interests (with the North contributing merely its passively displayed smorgasbord of policy models and its accumulated record of development experience) would be, however, incorrect. Development strategies calling for extensive central direction of the economy, and highly ambitious variants of the welfare state, were actively promoted for export to the developing world by influential Northern advocates; for long, such promotion was far more effective than that given to policy designs relying on the market as the primary means of economic coordination. Dispensation of foreign assistance by the North to the South powerfully reinforced the influence of opinion favoring planning. North-to-South economic assistance, as a long-term feature of international economic relations, was essentially a post–World War II innovation. Such aid to sovereign countries—whether rendered through bilateral arrangements or through international organizations—tended naturally to assume the form of grants or of low-interest loans made to governments, rather than taking the form of assistance channeled to the private sector of the economy. Whether directly used by the recipient government itself, or passed on by it to other agents of the domestic economy through a necessarily politicized process, foreign aid came to constitute an important stimulus toward creating institutions in the Third World for central economic planning and for the welfare state. The desire of donor countries and international organizations to assure purposeful and responsible use of grants and credits by the recipient government reinforced this process. Foreign assistance was typically destined to specific economic sectors: accountability necessitated the building up of a correspondingly articulated administrative mechanism.

The practice of central planning and the premature adoption of features of a modern welfare state resulted not only in poor performance of functions newly and unwisely assumed by the government. The effects also tended to spill over into the crucial core areas of government service. Because of the urban bias, government failure to create a stable and effective legal-institutional frame optimally conducive for economic progress tended to especially hinder rural development.

Profiting from being late

Economic history is a tale of unequal development. Differing speeds of economic progress in the past, and especially since the Industrial Revolution, have led to huge differences among countries in levels of material well-being in the contemporary world. Looking at these differences from the perspective of a poorer country—looking from Tanzania toward Sweden, so to speak—is likely to generate varied sentiments and reactions. One of the most constructive among these should be an appreciation of the power of compound interest growth for narrowing and eventually eliminating such differences. Short of benefiting from an unexpected windfall, such as discovery of an exceptionally valuable and easily exploited natural resource, countries must advance by making and sustaining incremental progress in the right direction. The numerous historical examples of the efficacy of such progress demonstrate that there are advantages in being late in the race. That those advantages do not compensate for the pain of being behind, or that the disadvantages which hinder economic progress of latecomers are equally weighty, may be true, but the point is moot. History cannot be undone, and the record of rewriting the rules that govern international economic intercourse is not promising. But, with the encouragement of many past achievements, and with high expectation of success, countries can make the most of the advantages of economic backwardness.

The potential profits from being late are most tangible in the domain of technology, as embodied in physical capital and in knowledge applicable in producing goods and services. Latecomers need not reinvent the wheel or, perhaps more to the point, need not use tractors designed in the 1920s, plant corn, wheat, or rice as if the green revolution had not happened, or make weather forecasts by the methods of yesteryear. The high costs of developing technologies and knowledge now available off-the-shelf need not be incurred again; cul-de-sacs that were traveled by the pioneers need not be reexplored.

Indeed, the remarkable economic progress of the last four decades in the developing world—a progress that permitted accommodation of a more than doubled population size at increased average levels of living—would have been inconceivable without the prior accumulation of a large variety of useful technological advances in the developed world. Through numerous channels and mechanisms, new knowledge applicable in the production process has been diffused among the populations of the developing countries, as were capital goods of an immense variety, embodying up-to-date technological know-how. Many consumption goods of Northern manufacture or origin also became available in the developing world, at prices that permitted direct satisfaction of some wants at remarkably attractive rates of exchange. Pharmaceuticals, transport over long distances, and telecommunications provide only some of the most telling examples from a very long list.

Changes over time in the composition of goods and services flowing across international boundaries with respect to kind, variety, and quality have been so pronounced as to present almost insuperable index-number problems, rendering extended time-series of indexes of production and consumption and of terms of trade between nontraditional goods and services, on the one hand, and agricultural commodities on the other, meaningless or at least highly questionable. Neither of these adjectives could remotely characterize the gains that countries can derive from international trade. That some gains are realized by both parties in any voluntary exchange is true by definition: otherwise countries could improve their welfare by simply abstaining from trade. That the gains are significant is indicated by the markedly faster long-term progress of developing countries that have followed policies hospitable to the promotion of intensive trade relations and other forms of economic interaction with the rest of the world, as compared with countries that interfered extensively with trade, discouraged other forms of external contacts, or sought outright autarky and even endeavored to completely de-link themselves from the international system. Gaining access to the superior goods and services available in international markets is the major means by which less developed countries profit from their relative economic backwardness. The wider the technological gap that is bridged by international trade, the larger are the potential gains from trade and the faster is progress toward narrowing that gap.

It is sometimes suggested that in developing countries external economic contacts have the harmful consequence of exposing the population to the temptations of modern consumer goods. Such exposure then might dissipate savings, increase material expectations, and generate psychological frustrations as aspirations fail to be satisfied. Such "demonstration effects" emanating from the outside would then constitute a significant brake on the speed of economic progress that otherwise could have been achieved. Development, according to other views in line with such criticism, is promoted by enforced parsimony. If they were free from the blandishments of modern consumer goods, developing country populations, and particularly the populations of rural areas where external contacts are most easily kept to a minimum by the authorities, may follow the type of advice epitomized by Mao's exhortation: "oppose extravagant eating and drinking and pay attention to thrift and economy"—thus accelerating progress toward eventual material well-being.

But there is little evidence that demonstration effects hinder economic progress, and even less evidence to support the notion that imposed isolation and austerity promotes economic growth. To the contrary, economic history shows that the desire to satisfy wants is a prime motor of development: greater exposure to consumer goods and the awareness that such goods are available to those who can pay for them tend to increase voluntary saving and investment, prompt innovation and entrepreneurship, and stimulate productive effort of economic agents at large. Thus, the vast range of unsatisfied consumer aspirations in poor countries, and the fine gradations through which they can be met as incomes rise—progressing from simple utensils to refrigerators, from cheap watches to videocassette recorders, from bicycles to motor vehicles—can be an advantage in speeding economic progress: an advantage conferred by relative backwardness and activated through trade with the outside world.

Comparative advantage dictates that for most developing countries initial success in trade has to rely on exports of agricultural products. The precondition in such success then is rural development: increasing agricultural productivity, achievement of an agricultural surplus, and access to industrial products desired by the agricultural population. Rural development generated by the linkage between these factors also creates the best foundation for domestic industrialization. Drawing on the strength of demand for industrial goods emanating from the rural economy, domestic industry can grow naturally and rapidly, provide employment to an increasing share of the labor force as agricultural productivity rises, and, in most countries, eventually capture a dominant share of exports. The balance of the benefits and costs of a development strategy that gives primacy to rural development and builds on its success is shown by much historical experience to be superior to the alternative strategic choice: that of squeezing the rural economy through heavy taxation and of substituting administrative pressures for material incentives to generate an increasing agricultural surplus.

For latecomers to industrialization, the opportunity to examine the experience of the more advanced countries and thereby to draw lessons for shaping development policy should be a most valuable economic asset. Indeed, acquisition of the best foreign technology and its adaptation to domestic circumstances can be said to be a key to successful modernization only if the notion of technology is interpreted in a broad sense—to include institutional structures, legal-political arrangements, and rules and expectations for social behavior most conducive to economic growth. In many developing countries, improving the vigor with which foreign experience concerning the application of such social technologies is scanned and evaluated, and the astuteness with which the most appropriate social technologies are adapted for home use, is a far from fully exploited means for accelerating economic progress and, especially, rural development.

Efforts of this kind would be likely to prove most useful if initiated and carried out country by country, filtered through and approved by a domestic political process that assures wide participation of the affected citizenry. Not many countries are fully capable of living up to this ideal. International institutions concerned with development issues have thus come to serve as an important complement to domestic efforts in shaping development policy. They assemble, evaluate, and transmit information on international experience for the discretionary use of member countries—a service of potentially great value in domestic policymaking. They often also suggest appropriate policy designs and provide advice and related technical assistance in executing recommended programs and policies. Analogous functions of resource transfer are performed by bilateral programs of international assistance. Assuming their wise and selective utilization, including avoidance of the centralizing bias of international assistance noted above, the existence of international development institutions and the availability of concessionary external resources represent an important advantage in development, conferred by the temporary economic backwardness of the developing countries.

Controlling population growth

The "hundred years" specified in Keynes's speculation cited at the beginning of this note will end in 2030. The remaining 40 years until that date are a short enough span of time to safely conjecture that the "economic problem" will not have been solved by then, whatever reasonable interpretation is given to the notion of a "solution." This would not have surprised Keynes. Neither of the two conditions he specified as necessary for a favorable outcome has been fulfilled. There was a world war, and there was an "important increase in population." An even-handed apportionment of the blame between these two factors for delaying the coming of an age of universal affluence would be, however, plainly unrealistic. Whatever effect the war had in that regard was probably trivial compared to the impact of population growth. The magnitude of the latter was far greater than anyone in 1930 would have dared to predict. "Important increase," in fact, is a feeble characterization of what demographic trends have wrought. Since 1930, global population size has increased by roughly two-and-a-half-fold. Almost certainly, the increase by 2030 will be well above fourfold: from 2 billion to about 9 billion.

That in the short span of a single century the world economy could not absorb on top of an initial 2 billion people an additional 7 billion without substantial sacrifice in terms of what would have been potentially attainable income per capita, is plausible enough. It is also likely that population growth—growth already realized by the end of the penultimate decade of the twentieth century, and growth virtually certain to occur during the coming few decades—has either permanently foreclosed certain attractive developmental paths that might have been followed under conditions of much less explosive demographic expansion, or at least has greatly delayed the time of their attainment. But such counterfactual speculations represent not only water over the dam; they also refuse to yield hypothetical historical scenarios that could be fairly set against the path that world development has actually followed. Had world population size reached a stationary state at a level of 3 billion by 1990 (instead of the actual 5.3 billion as a momentary way-station en route to a stationary size of perhaps 10 to 14 billion toward the end of the twenty-first century), would that have heralded the imminent arrival of an economic golden age? Would such a population have brought the prospect of a unified world civilization—with no barriers to the movement of ideas, goods, capital, and also of people—much closer to realization than it is today? It is impossible to tell. Social institutions and political systems at mid-century were simply not geared to deliver the flexible and rapid realignment of fertility to falling mortality levels that would have been necessary for keeping demographic growth moderate, and for achieving an early stabilization of global population size. Had such social institutions and political systems existed and functioned, world development trends would have been very different also with respect to many non-demographic matters. The world that would have resulted would have differed from the one we know in countless unfathomable ways.

Impressed by the explosive character of global population growth in recent decades, some observers discern the dynamics of a biological system out of control. "Swarming" is an image often invoked, along with the corresponding biological checks that uncontrolled growth sooner or later will trigger. Among these checks, running out of food (relative to the numbers of people) still occupies pride of place. But this view of the matter ignores negative feedbacks that operate at the level of individual consciousness and also through social institutions. Neither of these mechanisms could prevent the population explosion from occurring. But man, although often a victim of his passions, is also a calculating animal, whether acting individually or in concert with others. Observation, experience, and reflection concerning population growth and its consequences have been changing the mix, the effectiveness, and the modes of the feedbacks that shape fertility behavior. The results of that change to date fall far short of what might be judged optimal in light of widely varying welfare criteria. It is evident, however, that humankind breeding itself into a condition of generalized misery and mass famine is on the far fringe of the many possible outcomes of the demographic trends that economic and social developments are likely to generate.

That the world has plenty of food to feed its present population given "equitable" distribution is a commonplace proposition. That world agriculture can in time—say, within the next 30 years—produce enough food to support 10 billion people, even with recourse solely to already existing technologies, is widely supported by expert opinion. Given more time, and given the reasonable expectation that technological progress in agriculture will continue (indeed, advances in genetic engineering and in other aspects of biotechnology suggest the likelihood of accelerated future gains), a generous extension of "carrying capacity" beyond the 10 billion figure seems also warranted. Thus, given the commonly accepted assumption about the likely stabilization of global population size by the end of the next century, in the classic race between food supply and population size, food supply wins, much as it did, with remarkable consistency, during the last 200 years.

But such reassurances come hand-in-hand with significant qualifications, warnings about costs, and reminders of opportunities forgone. The average quality of the diet at which the world population is assumed to be adequately "fed" is modest—it permits no guarantee of, say, the present nutritional standard prevailing in Western Europe. The necessary transformation of traditional to modern agriculture stipulated in the calculations will require major investments in physical and human capital and assume major increases in agricultural inputs, especially energy. Intensification of agriculture may involve deleterious environmental consequences and sacrifices of aesthetic values.

The most important qualification in the present context, however, concerns the appropriate unit of analysis for which these and similar issues of the food-population-development equation should be posed. The notion of "world population" for most purposes is a heroic abstraction: the problems typically come packaged in national and subnational units. Above the national level, material exchanges between populations, notably transactions involving food, are effected primarily through international trade. Transactions between countries mediated by political markets are not unimportant, but they seldom flow from entitlements backed by long-term political commitments and stable moral support. They tend to reflect, instead, either cold political calculations that are subject to rapid and unilateral revision if circumstances change, or manifestations of charity and compassion. But compassion in the international arena is a singularly fragile commodity, constrained, even in its institutionalized forms, by an element of whim, short attention span, and recurrent and often well-founded worry among donors that, because of unintended side effects, unilateral aid in the long run may do more harm than good to the recipients.

Countries not self-sufficient in food, therefore must consider international trade—markets governed by commercial considerations—as the only reasonably secure source of the food they need to supplement domestic production. If a country possesses important and lasting strengths in international markets—for example, if it is a choice destination for international tourism, a rich repository of valuable natural resources, or a technological leader in an expanding sector of world industry—there is no reason why it should not rely extensively on food imports, in much the same way urban areas within a country rely on food produced elsewhere. But as the examples of market strength suggest, few countries can feel fully secure in following such a policy. Fashions in tourism may change, oil deposits can be exhausted while copper may lose its value, and staying consistently on or near the top in competing for markets in manufacturing products that incorporate cutting-edge technology is very difficult. For countries clearly less well placed in international trade, developing a significant and persistent dependence on imported food, especially a dependence on imports of cereals which provide the bulk of calorie and protein intake in diets, is far more risky. Such countries, moreover, are likely to be potentially most competitive in exporting agricultural products, at least during the early stages of their industrialization. If so, they must develop their agricultural sector not only to satisfy, on balance, their domestic food requirements, but also to achieve a significant surplus in their external agricultural trade to help finance their industrial development.

For developing countries, gaining entry into international markets for agricultural products and maintaining there a solid foothold is not easy under the best circumstances. (In the contemporary world, protectionist policies that shelter Northern agriculture represent, for example, a major stumbling block for developing country agricultural exports.) Under conditions of rapid population growth, achieving these feats can be especially difficult. Demographic expansion combined with even modest improvement in per capita domestic food consumption can absorb all of the annual gains in agricultural output even when such gains are large by conventional standards. Among countries that, on balance, are not self-sufficient in food, rapid population growth can represent an increasingly heavy claim on a country's total export earnings, and lead to an increasing degree of dependence on food imports. There exist also a number of other mechanisms through which rapid demographic growth is likely to slow economic progress measured in per capita terms. Examples include the greatly magnified difficulties of providing high-quality formal education to youth cohorts of rapidly increasing size, and the difficulties of satisfying the heavy investment requirements of productively employing ever larger numbers of young persons entering the labor force. Nevertheless, in most contemporary discussions of the issue of rapid population growth, attention tends to be centered on the strategic question of the change in countries' population size in relation to the state and development prospects of their agriculture.

Not surprisingly, and in part simply reflecting the great variety of situations from country to country, judgments of economists concerning the significance of that relationship show the same lack of consensus that characterizes debates about most enduring economic issues. But to a large extent, the impression is deceptive. In considering the longer-term economic future of the developing world, it is invariably assumed, if often only implicitly, that population growth in every country will be controlled: indeed, that a process leading to demographic stability—in effect, to zero population growth—is already well advanced, and that for all practical purposes that process will be completed within the near future, at least if time is reckoned by historical standards. This virtual uniformity of outlook concerning the world's demographic prospects is lent prestige and tangible expression by the detailed numerical population projections prepared by the United Nations and the World Bank. Frequently updated and reissued, these projections depict the demographic implications of a headlong rush to replacement-level fertility in every country of the world, with variations of timing and speed that, again by historical measures, can be fairly characterized as relatively minor.

Unanimity in the expectation that key facets of demographic behavior will rapidly converge among all members of the international system so as to generate stationary populations everywhere cannot be attributed to a shared understanding of the nature of the forces that will bring about such convergence. In fact, such understanding does not exist. The explanation for unanimity must lie, instead, in the commonly shared Keynes-like conviction that envisaging any other future would be inconsistent with achievement of minimum desirable levels of material well-being in each and all component parts of the international system. Indeed, the most reasonable, and by no means fanciful interpretation of the United Nations/World Bank population projections is that through these detailed numerical demographic scenarios the international system transmits to its member states not a comforting prediction but a normative message: in effect it signals the permissible bounds of their future demographic expansion.

Meaningful as such a message may be, it leaves largely open the question of the mechanisms through which any given population trajectory can be realized. The projections do convey the important assumption that international migration will carry only a minuscule burden of first slowing and then stopping population growth in the developing world. They outline a future evolution of mortality improvements that appear to err on the side of caution. This leaves fertility as the key variable through which control of population growth must be effected. In the experience of the North, fertility transition was accomplished through the seemingly free and uncoordinated decisions of individuals and individual couples. Closer examination reveals an intricate web of cultural, social, and economic inducements that edged microlevel decisionmakers to act in ways consistent with the collective interest, which was to control aggregate population growth. An analogous adjustment process operates also in most developing countries, but typically in an institutional setting that makes the transition process less responsive to extrafamilial pressures.

Issues of population growth have become an object of intensive scrutiny by international forums since the mid-1960s, at first largely upon Northern initiative. The policy design that emerged from these discussions still represents the central approach through which international institutions and bilateral foreign assistance programs seek to help developing countries to reduce their rates of population growth—an objective that by the 1980s had become a declared goal of governments in much of the developing world. That policy reflects a deep pessimism about the efficacy of the classic fertility transition process to bring timely results among contemporary high-fertility populations. Direct government intervention, the policy holds, is necessary to lower fertility. At the same time, the policy betrays an extreme reluctance to recommend tampering with the sensitive deeper layer of institutional structures that underpin high fertility. It focuses, instead, on a technological fix: on government-sponsored provision of better contraceptive devices to couples through specialized services. Availability of such services helps prevent the birth of children who otherwise would have been born unwanted: a layer of fertility held to account for a significant share of total fertility in most contemporary developing countries.

The thrust of this approach exhibits both the strengths and the weaknesses inherent in dealing through international cooperation and assistance with a problem whose causes and manifestations are largely located on the national and subnational levels. The approach draws on strong suits of Northern advantage: capacity to supply devices of a sophisticated technology and related technical know-how; experience with organizing government service programs, albeit in a quite different socioeconomic milieu; and ability and willingness to back up program initiatives with financial assistance, albeit assistance clearly offered as temporary. In has the weakness of a necessarily shallow understanding of the local socio-cultural setting, compensated by deliberate avoidance of delving into policy designs that seek to achieve their objectives through legal-institutional rearrangements rather than through programs requiring novel goal-directed organizations, hardware, and finance.

If such program-embodied assistance were simply an addition to local efforts dealing with the problem at hand, whatever such assistance would accomplish could be taken as a net gain for the recipient country. There is a danger, however, that when outside programmatic assistance is significant, a displacement effect also operates: the program in effect expropriates the problem and becomes its sole embodiment, interpreter, and analyst. It is a poorly understood paradox that Northern assistance persists in offering solutions to development problems packaged around specific goal-directed programs, yet at the stage of their own development that corresponded to that commonly found in the contemporary Third World, the most successful Northern approaches and solutions to problems operated on the constitutional, rather than on the programmatic level. Arguably, by eliciting interest in an approach that served it so well in the past, and by encouraging application of that approach in dealing with problems of both rural development and population growth, the North could make a uniquely valuable contribution to advancing human welfare in the contemporary Third World.

References

Keynes, John Maynard. 1963. *Essays in Persuasion*. New York: W. W. Norton and Co.

Maddison, Angus. 1989. *The World Economy in the 20th Century*. Paris: OECD.

United Nations. 1989. *World Population Prospects 1988*. New York (Population Studies, no. 106).

US Bureau of the Census. 1987. *Statistical Abstract of the United States: 1988* (108th edition). Washington, D.C.: US Government Printing Office.

World Bank. 1989. *World Development Report 1989*. New York: Oxford University Press.

AUTHORS

ESTER BOSERUP is Consultant Economist, Brissago/Nevedone, Switzerland.

JAN BREMAN is Director, Centre for Asian Studies, University of Amsterdam.

MEAD CAIN is Senior Associate, Research Division, The Population Council, New York.

PAUL DEMENY is Distinguished Scholar, The Population Council, New York.

JACK GOODY is Fellow, St. John's College, Cambridge, England.

SUSAN GREENHALGH is Senior Associate, Research Division, The Population Council, New York.

ALLAN G. HILL is Senior Lecturer in Population Studies, Centre for Population Studies, London School of Hygiene and Tropical Medicine.

GORAN HYDEN is Professor, Department of Political Science, University of Florida, Gainesville.

N. S. JODHA is Head, Farming Systems Division, International Centre for Integrated Mountain Development, Kathmandu.

MICHAEL LIPTON is Professor, Institute of Development Studies, University of Sussex, Brighton, England.

IAN LIVINGSTONE is Professor, School of Development Studies, University of East Anglia, Norwich, England.

AKIN L. MABOGUNJE is former Professor of Geography, University of Ibadan, Nigeria.

GEOFFREY MCNICOLL is Professorial Fellow, Research School of Social Sciences, Australian National University, Canberra.

PRABHU L. PINGALI is Agricultural Economist, Agricultural Economics Department, International Rice Research Institute, Manila.

AMARTYA SEN is Lamont University Professor of Economics, Harvard University, Cambridge, Massachusetts.